Advanced Sciences and Technologies for Security Applications

Indexed by SCOPUS

The series Advanced Sciences and Technologies for Security Applications comprises interdisciplinary research covering the theory, foundations and domain-specific topics pertaining to security. Publications within the series are peer-reviewed monographs and edited works in the areas of:

- biological and chemical threat recognition and detection (e.g., biosensors, aerosols, forensics)
- crisis and disaster management
- terrorism
- cyber security and secure information systems (e.g., encryption, optical and photonic systems)
- traditional and non-traditional security
- energy, food and resource security
- economic security and securitization (including associated infrastructures)
- transnational crime
- human security and health security
- social, political and psychological aspects of security
- recognition and identification (e.g., optical imaging, biometrics, authentication and verification)
- smart surveillance systems
- applications of theoretical frameworks and methodologies (e.g., grounded theory, complexity, network sciences, modelling and simulation)

Together, the high-quality contributions to this series provide a cross-disciplinary overview of forefront research endeavours aiming to make the world a safer place.

The editors encourage prospective authors to correspond with them in advance of submitting a manuscript. Submission of manuscripts should be made to the Editor-in-Chief or one of the Editors.

Zoran Keković · Savo Kentera · Marina Dabetić ·
Sanja Grbović
Editors

Critical Infrastructure: Risk and Crisis Management in the New Reality

Springer

Editors
Zoran Keković
Faculty of Security Studies
University of Belgrade
Belgrade, Serbia

Marina Dabetić
Vinča Institute of Nuclear Sciences—National Institute of the Republic of Serbia
University of Belgrade
Belgrade, Serbia

Savo Kentera
Atlantic Council of Montenegro
Podgorica, Montenegro

Sanja Grbović
Faculty of Law
University of Montenegro
Podgorica, Montenegro

ISSN 1613-5113 ISSN 2363-9466 (electronic)
Advanced Sciences and Technologies for Security Applications
ISBN 978-3-032-20947-4 ISBN 978-3-032-20948-1 (eBook)
https://doi.org/10.1007/978-3-032-20948-1

This Springer imprint is published by the registered company Springer Nature Switzerland AG
The registered company address is: Gewerbestrasse 11, 6330 Cham, Switzerland

Contents

Geopolitics of Critical Infrastructure—A New Challenge for Societal Security and Resilience

Zoran Keković

Abstract This paper examines how the concept of societal security and resilience shapes the emerging geopoliticization of critical infrastructure (CI). It analyzes CI across socio-political, socio-economic, and socio-technical dimensions to understand how modern risks—such as hybrid threats, cyber aggression, climate impacts, and supply-chain disruptions—transform infrastructure into a strategic geopolitical asset. The study highlights the increasing vulnerability arising from global interdependence, the use of CI as geopolitical leverage, and the implications for national and regional stability, with emphasis on the Western Balkans. The paper also considers how infrastructure sovereignty interacts with societal resilience in the contemporary security environment.

Keywords Critical infrastructure · Geopolitics · Societal security and resilience · Infrastructure sovereignty

1 Introduction

Critical infrastructure (CI) refers to all systems and resources that are essential for the functioning of society, the economy, and state security. Despite the existence of numerous definitions, a universally accepted definition of CI still does not exist. Nevertheless, all definitions emphasize that the criticality of a given infrastructure lies in its fundamental importance for ensuring the vital functions of society and the state [1].

According to the NATO Committee Reports from 2007 [2], CI generally includes facilities and services of vital importance for the basic functioning of a given society—those without which the operation of the society would be significantly impaired. In the definition offered in 2008 by the Organisation for Economic Co-operation and Development (OECD), CI is described as a set of infrastructures and

Z. Keković (✉)
Faculty of Security Studies, University of Belgrade, Belgrade, Serbia
e-mail: zkekovic@fb.bg.ac.rs

Z. Keković et al. (eds.), *Critical Infrastructure: Risk and Crisis Management in the New Reality*, Advanced Sciences and Technologies for Security Applications,
https://doi.org/10.1007/978-3-032-20948-1_1

services that "form the foundation of economic and social well-being, public safety, and the functioning of government institutions" [3]. The U.S. Department of Homeland Security defines CI as "assets, systems, and networks, physical or virtual, so vital to the United States that their incapacitation or destruction would have a debilitating impact on national security, national economic security, public health or safety, or any combination thereof" [4]. At the national level, the definition that encompasses the largest number of mentioned variables is the one offered by the Government of the United Kingdom: "National CI comprises facilities, services, and systems that support the economic, political, and social life of the United Kingdom, and whose importance is such that their loss could: cause large-scale loss of life; have a serious impact on the national economy; have other serious social consequences for the community; or be of the utmost concern to the Government" [5].

In parallel with the growing risks posed by warfare, hybrid and cyber threats, natural disasters, and technological accidents, as well as the new challenges emerging from ICT development, CI has become a major national security priority, recognized as such by contemporary security strategies. National security now increasingly focuses on the protection of all physical and logical facilities, systems, and networks whose disruption, damage, or destruction would produce significant social and economic consequences for the state and its citizens. As a result, the "CI protection and resilience" approach has expanded over the past two decades and has become globally dominant.

In corporate practice and within security management of organizational systems considered critical, the approach based on protecting infrastructure from various threats in the physical and cyber domains remains prevalent. The concept of organizational resilience to potential threats, aimed at addressing vulnerabilities to enable systems to adapt more quickly to changes relevant to economic and social stability, has been adopted to a much lesser extent. Moreover, due to global interdependence, CI shapes not only domestic but also international relations, alliances, and communities in qualitatively new ways. Therefore, in geopolitical and security terms, the resilience of infrastructure is crucial and may have far-reaching consequences for the stability and security of a state.

Post–Cold War politics demonstrated, and major security developments have confirmed, that the survival of nations and citizens increasingly depends on the security of key societal functions rather than on the traditional focus on territorial defense. The ability to ensure the functioning of government and civil-society institutions, to maintain CI, and to uphold democratic principles of governance, is under immense pressure in crisis conditions [6].

The development of the concept of societal security stems from the realization that the traditional concept of "national security" has not resolved all tensions existing within society as the most complex human collective. Military threats were the centerpiece of national security strategies, requiring hierarchical and highly organized security structures designed to ensure defense or prevention [7]. Hellmann identifies three problems with the "old" concept of national security [8]: (a) The strict separation between internal and external security; (b) The centrality of state borders and territory in determining the concept of security and security management; (c) The linkage of threats primarily to risks against sovereignty and territorial integrity.

As Keković and Kešetović argue [6], the most significant attempt to shift the referent object of security from the state to society emerged from the Copenhagen School, which first proposed that social groups be accepted as referent security objects and introduced "societal security" into theoretical debates. For this school, identity holds the same meaning for society as sovereignty does for the state. Thus, "for society, survival is a matter of identity," and discussions of identity become ways in which society articulates existential threats: "If this happens, we will no longer be able to live as ourselves." Consequently, the field of security studies expands to include threats that endanger identity. Traditional spatial metaphors—internal vs. external security—become increasingly inadequate for conceptualizing risks such as transnational organized crime, migration, environmental challenges, and other cross-border phenomena. However, due to the wars in Ukraine and Gaza, the emergence of multipolarity, and challenges to the liberal international order, the risk of inter-state conflict has again moved to the forefront of political and security agendas across Europe.

According to Keković et al., with the globalization of trade and various other domains such as health, security, and the environment, societal security has increasingly oriented itself toward broader aspects of regional and global security. Globalization has heightened sensitivity to distant events, influenced by reduced communication and transportation costs. In recent years, a discernible trend toward deglobalization and the decoupling of global supply chains has emerged, reshaping assumptions about economic interdependence. As a broader reflection, one could also ask whether this trend represents a structural shift in the global economy or an unintended consequence of the preceding decades of deep globalization.) In analytical terms, the relations between modern international actors increasingly resemble Joseph Nye's notion of *interdependence*, describing a situation in which actors or events across different parts of a system exert mutual influence. This causes the interests of societal security to become intertwined with global security interests and to converge over time [6].[1]

Given the nature of contemporary security threats that primarily affect the functioning of civil structures, the term "societal security" more accurately reflects global changes in the security environment, whereby responsibility for managing security shifts from strictly hierarchical state structures toward horizontally organized civil actors [6]. This very shift highlights that modern threats, directed largely at civilian systems, position CI at the center of societal security. As security governance moves increasingly away from rigid, hierarchical state structures toward horizontally networked civil actors, societal resilience becomes directly conditioned by the resilience and functional stability of CI.

[1] These trends are also reflected in the recent National Security Strategy of the United States (2025), which explicitly raises concerns about foreign ownership of critical infrastructure, signalling a departure from earlier assumptions associated with free trade and economic globalization.

This approach is fundamentally oriented toward threats and potential risks, specifically, their effects on CI. Kauffman observes that "as opposed to the risk paradigm, resilience policies assume that disruption will occur" [9]. In other words, risk is accepted as a given, something that cannot be fully prevented but can be mitigated through strengthening systems so that they are better prepared to cope with disturbances [10]. Ridley notes that unlike the risk paradigm, which can be measured through probability and expected loss, the resilience paradigm focuses on maintaining essential system functionality—absorbing, adapting, and recovering from disturbances, which makes it difficult to measure [11]. Fjäder goes so far as to contrast resilience with security: "security is preventive and proactive, *whereas* resilience combines proactive and reactive measures aimed not at preventing threats, but at reducing their impact" [12]. Holling [13] similarly argues that existing knowledge is insufficient for predicting future unexpected events and risks. He concludes that "the resilience framework can accommodate this change in perspective because it does not require precise predictions, but rather the qualitative ability to design systems capable of absorbing and adapting to future shocks in whatever form they may take". Resilience enables complex systems to withstand and survive disturbances and emphasizes their capacity for renewal [14].

In the field of national security, resilience has been incorporated into the security strategies of economically stable states and focuses on the adaptive capacity of individuals, communities, and systems to maintain acceptable levels of functioning, structure, and identity [15]. Definitions that emphasize the robustness and adaptive capacity of social networks have become among the most promising developments in disaster risk reduction. A disaster-resilient community is an invaluable asset for national security. For example, the 2007 U.S. National Strategy for Homeland Security integrates the structural resilience of CI with the operational resilience of emergency response organizations, government institutions, and private enterprises [14].

Keković and Ninković describe CI resilience as the interaction of national (objects of national importance), community (providers of public benefits and welfare), and organizational (socio-technical systems) resilience [1]. Literature on CI resilience expanded significantly after events such as Hurricane Katrina and the Fukushima Daiichi Nuclear Disaster. A special subset of CI resilience research focuses on critical information infrastructure, given growing societal dependence on digital systems and the rising frequency of cyber terrorism and cyber warfare [1]. Thus, CI resilience becomes part of overall societal resilience, wherein the structural reactions of interconnected national, community, and organizational subsystems must enable society to "resist and adapt" to the challenges of emerging risks. Moreover, in an interconnected world shaped by emerging and systemic risks, the functional value of CI is increasingly determined by its geopolitical dimension.

In the continuation of this paper, we aim to address the following questions:

1. How do emerging risks influence the use of CI as a geopolitical weapon and an instrument of soft power in contemporary international relations?

2. How can states use control over infrastructure as a strategic instrument in geopolitics?
3. What are the indicators of CI resilience in the new geopolitical context, and how do they affect societal security and resilience in the Western Balkans?
4. In what way is the concept of "CI sovereignty" connected to societal security and societal resilience?

2 Emerging Risks and Its Influence on Global and Regional CI Vulnerability and Resilience

Numerous crises confirm the rule that, both within states and between them, security threats are of such nature that one can speak of mutual vulnerability of endangered infrastructure. This results in a shift toward governance models based on "complex networks of functionally distinct actors", involving public and private stakeholders at multiple levels of responsibility. The opportunities for applying this governance concept are particularly wide in large infrastructure projects, especially those requiring population displacement or limiting access to essential resources such as water and food [1].

Vulnerability and dependency, within states and across borders, on the control of critical resources and infrastructure continues to grow, especially in the context of geopolitical tensions, climate change, cyber risks, social instability, and other emerging threats. New, previously unknown or insufficiently analyzed "emerging risks" can pose significant challenges to resilience, safety, and operational and business continuity [16]. Such risks interact with processes such as globalization, digitalization, innovation, and cross-border operations. Because they may directly or indirectly influence each other, emerging risks tend to be interconnected, systemic, and interdependent. In this context, the world can increasingly be viewed as a single complex system—*a* de facto*global "infrastructure-of-infrastructures"* [16].

As a concept rooted in enabling societies to sustain essential functions during crises and wars, CI can today also be understood as a "means of inducing or managing crises". The greatest threats that increase CI vulnerability are "multilayered and interrelated": geopolitical conflicts, cyber threats, and climate disruptions, along with secondary risks such as supply-chain fragility, misinformation/disinformation, and societal polarization. Regionally and locally, these vulnerabilities are exacerbated by dependency, weak institutional capacity, corruption, lack of resilience, and deficits in managerial knowledge.

According to the World Economic Forum's Global Risks Report 2025 [17], global risks refer to events that, if realized, would significantly affect global GDP, population, or natural systems. Key risks include: state-based armed conflict, extreme weather events, geoeconomic confrontation, misinformation and disinformation and societal polarization. More than isolated events, today's CI resilience is shaped by clusters of global risks: cyber threats, climate pressure, geopolitical tensions, digital/energy transitions, interdependence, aging infrastructure, and supply-chain vulnerabilities. From this perspective, the main *drivers* affecting CI resilience, including the secondary risks they generate include: geopolitical tensions and hybrid conflict, warfare risk, climate change and extreme weather, cyber insecurity, supply-chain fragility.

2.1 Geopolitical Tensions and Hybrid Conflict

Geopolitical tensions and hybrid warfare among major rivals, such as the US and China, and regional actors including the EU, Russia, India, and Turkey, have a decisive impact on global and regional supply-chain disruptions, forced alignments with more powerful actors, and the loss of strategic autonomy in controlling critical resources.

Countries increasingly use non-military tools: cyberattacks, disinformation, and disruptions to critical supply chains, as part of their strategy. Great-power competition, particularly between the US and China and between Russia and the EU, manifests notably in efforts toward *decoupling* in trade, technology, and supply chains (e.g., semiconductors, rare-earth minerals, AI technologies). The Russia–EU relationship further illustrates the "militarization of infrastructure", where sanctions and countermeasures restrict access to technology, energy, raw materials, and platforms impacts that disproportionately burden smaller states.

2.2 CI as an Instrument of Warfare

During the 1999 NATO bombing of the Federal Republic of Yugoslavia, key CI sectors–energy, bridges, telecommunications, were targeted. The full-scale war against Ukraine has profoundly reshaped approaches to CI security and resilience. Since 2022, Ukraine's energy sector has been subjected to sustained aerial attacks (cruise missiles, ballistic systems, UAVs), exposing vulnerabilities linked to outdated infrastructure, centralization, and limited redundancy. Anticipating emerging threats, from naval drones to advanced munitions, is essential. Integrating defense with sustainability under a "Build Back Better" paradigm represents a forward-looking model [18].

2.3 Climate Change and Extreme Weather

Climate change acts as a threat multiplier, driving more frequent extreme weather events that strain infrastructure and provoke social instability. Critical systems like energy and water are vulnerable to heatwaves, droughts, and storms, which in turn can spark humanitarian emergencies and unrest. Nations increasingly recognize climate change as a security issue, not just an environmental one. As one World Economic Forum report notes, climate and nature degradation are now being "accepted as security issues", requiring coordinated action to break the negative feedback loop between climate disasters and societal instability [19]. Strengthening infrastructure resilience to climate risks – by hardening grids, diversifying water sources, etc.—has become essential to maintaining societal security [19]. Consequences include **physical destruction of infrastructure**, reduced resilience, humanitarian crises, and migration flows.

2.4 *Cyber Insecurity*

Global cyber insecurity is another persistent driver. Ransomware attacks, zero-day vulnerabilities, and state-sponsored cyber operations (e.g., in NATO–Russia/China tensions) jeopardize energy grids, hospitals, financial systems, and SCADA/ICS control systems. Meanwhile, experts warn that quantum computing could soon undermine current cryptographic protections that shield critical systems. As quantum hardware advances, a sufficiently powerful quantum computer may break widely used encryption, potentially exposing energy grids, finance networks, and other CIs to breaches [20]. AI-driven threats and quantum vulnerabilities thus represent a new frontier for CI resilience, demanding continual innovation in security measures. Potential impacts include: data loss, loss of control over physical systems, psychological shock and societal panic.

2.5 *Supply-Chain Fragility*

Global and local supply-chain fragility remains a permanent systemic risk. The COVID-19 pandemic exposed rapid supply-chain collapse due to dependency on foreign suppliers of chips, energy, medical equipment, and the vulnerabilities of maritime and land routes. Such events dramatically increased the need for diversification and domestic capacity building. Beyond physical sabotage, control of supply chains and chokepoints in high-tech infrastructure has become a potent geopolitical weapon. Nowhere is this more apparent than in the contest over semiconductors, the "brains" of modern infrastructure [21].

In sum, emerging risks from cyberspace to climate change are forcing both advanced and developing nations to innovate in how they identify, prepare for, and bounce back from CI disruptions.

3 Control Over Infrastructure as a Strategic Weapon in Geopolitics

CI has emerged not just as a passive target of conflict, but as an active instrument in geopolitical competition. States and non-state actors alike are leveraging control, disruption, or coercion of infrastructure systems as a strategic weapon. This trend is evident in recent real-world events (2023–2025) where energy grids, communications cables, and high-tech supply chains became battlegrounds for power projection.

On the other side, developing countries are the most vulnerable to this challenges. African nations, for example, face fast-growing digital infrastructure coupled with capacity gaps, making cyber threats an acute concern. Africa is often described as "an open frontier for cybercrime" due to generally weaker ICT infrastructure and limited cybersecurity capabilities [22]. Critical sectors like telecom, banking, power, and transport in many African states are more exposed to hackers and ransomware, which could disrupt services and economic activity.

3.1 State Actors

Due to the geopolitical influence they exert on international political, economic, security, and diplomatic flows, the principal actors capable of using CI as an instrument of geopolitical power include states that itself possess geostrategic resources, such as: China, the United States, the Russian Federation, Turkey, Japan, the UAE, Saudi Arabia, and others.

State actors have increasingly employed sabotage and coercive control of CI to achieve strategic ends. A stark example is the spate of attacks on undersea pipelines and data cables in northern Europe in late 2022, when underwater explosions ruptured the Nord Stream gas pipelines—a critical energy artery from Russia to Germany—in what analysts widely deemed an act of *state-linked sabotage.* One year later, in October 2023, the undersea Baltic connector gas pipeline between Finland and Estonia was damaged, alongside nearby telecom cables, disrupting energy supply and communications [23]. These incursions underscore how undersea infrastructure has become a theater of hybrid warfare, where plausible deniability cloaks state-sponsored attacks. NATO and EU officials have openly warned that such underwater sabotage could be a new form of conflict, prompting NATO in 2023–24 to launch dedicated efforts to police and protect subsea infrastructure [23]. The geopolitical logic is clear: damaging a rival's critical pipelines or cables can weaken their economy and sow uncertainty, all without a direct military confrontation. Control over infrastructure routes—including the threat of cutting them—thus serves as strategic leverage.

China uses CI as an instrument of geopolitical influence through several strategic mechanisms, combining economic expansion, technological dominance, and political leverage. This multidimensional approach has led many scholars to describe China's strategy as a model of "geo-infrastructural power".

Through Belt and Road Initiative (BRI,) China finances and builds major *transportation and logistics infrastructure* (ports, railways, highways) in more than 60 countries, creating long-term structural dependency. Debt-repayment difficulties often lead to concessions of control over infrastructure (e.g., the Hambantota Port in Sri Lanka), enabling China to project influence along key trade routes from Asia to Europe and Africa. Through *digital infrastructure,* Chinese companies (Huawei, ZTE) build 5G networks worldwide. Despite competitive pricing, Western countries perceive them as security risks due to possible state-enabled data access, espionage, or digital sovereignty erosion. Through the Digital Silk Road, China exports telecom infrastructure, creating digital dependencies and expanding its informational presence abroad. Also, deployment of AI-based surveillance systems (CCTV + biometrics) to authoritarian regimes raises concerns about the spread of digital authoritarianism and the erosion of democratic governance through infrastructure. Within *energy investments,* state-owned Chinese companies invest in foreign energy systems, purchasing stakes in electricity distribution networks and gas pipelines, particularly in developing and Eastern European states, gaining access to strategic resources and influencing energy stability patterns. Notably, China produces *over* 90% of the world's gallium and germanium, metals vital for semiconductors and fiber-optics and possesses a virtual monopoly on rare earth minerals. In July 2023, Beijing imposed export licensing on these minerals, and by 2025 had progressively tightened restrictions up to an outright ban for certain countries [24].

Acquisition of Strategic Assets is reflected by buying ports, logistics centers, and railway lines in indebted or developing countries (e.g., Port of Piraeus). Such assets could limit access in crises, impose political concessions, or be used for dual-use (civilian + military) purposes. *Standard-setting power* is a tool to promote Chineese technological standards (AI, 5G, smart cities), creating long-term technological dependency.

To conclude, China poses a major strategic challenge in global CI because it combines: long-term power projection, technological and financial leverage, structural dependency mechanisms and informational penetration. Psychological and strategic influence of CI investments also serve to project China as a stable, reliable development partner, applying long-term "strategic patience."

The United States dominates global digital infrastructure through corporations such as Google, Amazon, Microsoft, Meta, and Apple. Its *financial infrastructure,* centered on the US dollar, acts as a source of soft power through sanctions, SWIFT access, and capital-flow control. The US also maintains a global network of *military infrastructure* across more than 80 countries, reinforcing its position as a security guarantor. Before the transition of 2025, the US (through USAID and other agencies) extensively invested in energy, healthcare, and education infrastructure in developing regions.

The United States and its allies have exploited their dominance in semiconductor design and manufacturing equipment as leverage to constrain rivals. Since 2019, coordinated export controls by the US, Japan, and the Netherlands have cut off China's access to the most advanced chip technology—for example, blocking the sale of extreme ultraviolet (EUV) lithography machines (made solely by Dutch firm ASML) needed for cutting-edge chip fabrication. These measures are explicitly aimed at sustaining China's technological dependence and preventing its military from acquiring superior chips [21]. In 2022–2023, the US further tightened these controls (including on cloud computing and AI chips), and pressured allies to follow suit. The result is a form of *techno-economic containment*, using infrastructure chokepoints, like semiconductor manufacturing equipment, as strategic barricades. China has retaliated by weaponizing its own grip on critical materials that global infrastructure relies on.

Russia's influence is most pronounced in *energy infrastructure*, leveraging gas and oil pipelines (Nord Stream, TurkStream, Druzhba) as tools of political pressure, reward, or coercion, full or shared ownership in European oil and gas assets, which has, however, significantly decreased following the outbreak of the war in Ukraine as a result of sanctions, asset seizures, and forced sales. In the digital domain, Russia invests in alternative internet protocols (RuNet) and uses cyber operations and information control to exert influence in post-Soviet regions.

Post-Brexit, the United Kingdom seeks new geopolitical and investment partnerships. Through UK Export Finance (UKEF), major infrastructure investments in the Western Balkans (e.g., transport corridors, green energy, Morava Corridor) can be interpreted not only as economic ventures but as attempts to secure geopolitical influence and strategic control over regional infrastructure. Large-scale packages (≈£15.75 billion) create long-term dependencies through technology, financing, and operational standards [25].

Turkey positions itself as a bridge between Asia and Europe. Projects such as TANAP reinforce its role as an energy transit hub. Situated between energy-rich areas of the Middle East and energy-consuming Europe, it is no surprise that Turkey has become a crucial transit route for fossil fuels, especially natural gas. Furthermore, in recent decades, Turkey's leadership has pursued an ambitious plan to transform the country into a regional energy hub. Through TIKA, Turkey builds schools, hospitals, roads, and cultural centers (e.g., Yunus Emre Institutes), strengthening influence across Africa and the Balkans [26].

Japan invests strategically in transport and energy infrastructure in Asia and Africa, promoting stability and exporting technological standards (smart cities, water systems, energy efficiency).

UAE and Saudi Arabia invest heavily in megaprojects (e.g., NEOM), and through companies such as DP World, they purchase ports, logistics hubs, and airports worldwide—expanding global logistical and strategic influence [27].

3.2 Inter-State Actors—Alliances

The European Union uses regulatory power (e.g., GDPR, Green Deal standards) to shape global infrastructure development. Through the Global Gateway strategy, the EU invests in digital and energy projects in Africa, Latin America, and Asia. EU initiatives such as "Military Schengen" enhance internal and neighborhood infrastructural integration and military mobility.

Another manifestation of infrastructure weaponization is the geopolitical contest over 5G telecommunications networks. Western governments have treated control over 5G infrastructure as a national security priority, fearing that reliance on foreign (particularly Chinese) vendors could enable espionage or sabotage. Over the past few years, countries including the United States, Australia, Japan, the UK, and many EU states effectively banned Chinese firms (Huawei and ZTE) from their 5G rollouts. The outcome has been termed a "digital Iron Curtain," forcing nations to choose sides between US-allied technology ecosystems or China's. For China, the backlash against its flagship tech companies is seen as a strategic containment; for the West, it is pre-emptive defense of critical comms infrastructure. The 5G bans illustrate how infrastructure can be weaponized through standards and investments: by excluding Chinese gear, Western allies not only secure networks but also deny China global tech influence. In turn, China has pushed its own Belt and Road digital initiatives and telecom projects in friendly countries, to ensure it retains some strategic foothold. Thus, the battle over who builds and governs critical networks is itself a geopolitical confrontation [28].

3.3 Non-state Actors

Non-state actors are also weaponizing infrastructure, albeit with different motives and means. Cybercriminals and terrorist/extremist groups have identified CI as high-impact targets to advance their agendas. A notorious case was the 2021 ransomware attack on Colonial Pipeline in the U.S. (by a criminal gang) which forced a shutdown of fuel supplies along the East Coast, demonstrating how non-state hackers can inflict national-level disruptions. In 2023, U.S. authorities foiled a chilling domestic terror plot: two neo-Nazis were convicted of conspiring to physically attack power substations in Maryland with rifle fire and explosives, aiming to cause a massive blackout in the majority-Black city of Baltimore. Prosecutors revealed the attackers hoped to precipitate societal collapse and incite chaos by knocking out the electrical grid—a stark example of extremist ideology weaponizing CI to sow fear. Had the plot succeeded, it could have caused $70 million in damages and affected large portions of the city's infrastructure [29]. This incident, among others, prompted the FBI and homeland security agencies to warn of a rise in ideologically motivated sabotage against infrastructure, from power grids to telecom towers. Additionally, ransomware gangs and state-tolerated hacktivists have frequently struck hospitals, water systems, and government services with disruptive cyberattacks in 2023–2024. Though their aim may be profit or propaganda rather than classic geopolitics, the effect is to turn critical services into bargaining chips. A single coordinated cyber offensive by a non-state collective could impair banking networks or energy distribution in multiple countries, effectively wielding infrastructure disruption as a weapon.

One cannot overlook the role of the private sector in this strategic landscape. In some cases, private companies control infrastructure so vital that their decisions carry geopolitical weight. The most vivid illustration is SpaceX's Starlink satellite internet, which became lifeline communications infrastructure for Ukraine during the war. In 2022, as Ukraine undertook a major counteroffensive, SpaceX owner Elon Musk unilaterally ordered a shutdown of Starlink coverage over certain frontline areas—reportedly out of concern Russia might respond to Ukrainian advances with nuclear escalation [30]. The Starlink episode reveals how Big Tech proprietors can effectively act as geopolitical agents if they control critical digital infrastructure. Similarly, 2022 Western payment companies (Visa, Mastercard) and the SWIFT financial messaging network were leveraged to punish Russia for its aggression. Cutting off Russian banks from SWIFT was a coordinated act by governments and private financial actors, exemplifying infrastructure (the global banking network) used as a weapon of sanction [31]. In the inverse scenario, some Big Tech firms have become arenas of geopolitical tussle: e.g. U.S. pressures on TikTok (a Chinese-owned platform) citing data security, or Russia's throttling of Twitter and Facebook to control information flows. All these cases highlight that private sector platforms and services—from satellites to social media to banking networks—play an outsize role in critical infrastructure and can be co-opted or pressured as instruments of statecraft.

In summary, the period 2023–2025 has showcased a new reality: controlling or sabotaging infrastructure is now part of geopolitical competition. This heightens the need for resilience and protective measures. Democracies are responding by hardening infrastructure defenses and by tightening oversight of foreign involvement in vital sectors. Yet the offensive uses of infrastructure will likely persist, given the asymmetry they afford aggressors. Societal security planners must therefore anticipate that power plants, ports, wires, and chips can be turned into weapons, and devise both deterrence and rapid response strategies for such scenarios.

4 Interdependence of Infrastructure Systems and Its Impact on Regional Security Dynamics

Using the terminology of complex systems theory, CI represents a set of complex adaptive systems. They are highly connected, nonlinear, and capable of feedback loops that can amplify shocks. Research confirms that "interdependencies among infrastructures are a key factor in creating vulnerabilities in complex multi-infrastructure systems." [32]. The ability to co-evolve with external conditions enhances their resilience to threats originating both internally and externally. This is one of the key reasons why a systemic approach is essential for understanding infrastructure systems, given their influence on the environment and their dependence on environmental changes. These same dependencies also define their vulnerability to security challenges and threats. The systemic approach is best illustrated by examining the politization and securitization of energy supply, which opens a wide range of overlapping roles for numerous international and national actors involved in energy security governance. Recognizing and managing these linkages has become a priority in regional security frameworks (e.g. EU and NATO strategies) and in the application of systems-theory concepts to infrastructure resilience.

In the emerging multipolar order, with at least two global powers (the United States and China) and a significant number of strong regional hegemons (the EU without the UK, and influential actors such as Russia, India, Turkey, and Iran), as well as rapid technological development, we are witnessing major shifts in power vectors within regional geopolitical and security contexts.

4.1 European Union

The EU has overhauled its CI policy via the 2022 Directive on the Resilience of Critical Entities (CER). This directive explicitly mandates Member States to take a broad view of risks, conduct nationwide risk assessments covering all essential services, and build capacity for cross-sector risk oversight [33]. Under the CER Directive, countries must identify critical entities across 11 interlinked sectors and ensure those entities assess their own interdependencies and take resilience measures. The CER framework also establishes a Critical Entities Resilience Group to facilitate cooperation among states and the European Commission on these interconnected issues. In parallel, after sabotage incidents in 2022 (like Nord Stream), the EU Council adopted a recommendation for a "Union-wide coordinated approach" to infrastructure resilience, focusing on preparedness, response, and international cooperation as key pillars. This includes anticipating CER measures (e.g. stress-testing infrastructure) and using other instruments (like joint situational awareness tools) to respond to multi-country threats [33].

In line with the Joint Declaration on EU–NATO Cooperation (2023), the EU–NATO Task Force on the Resilience of Critical Infrastructure was established in March 2023. Its June 2023 report on CI resilience (covering energy, transport, digital infrastructure, and space) identifies three main threat categories: natural hazards, state actors (notably Russia, as the initiative was triggered by the alleged Nord Stream sabotage), and non-state actors (terrorist organizations). The report offers a general overview of challenges facing these four sectors due to the new geopolitical situation, especially the war in Ukraine and the perception of Russia as a hostile actor whose infrastructure remains closely interconnected with Europe [14].

The EU increasingly promotes the concept of *strategic autonomy*, defined as: reducing dependence on third countries in energy, digital infrastructure, and defense industries; investing in domestic production of chips, cables, batteries, 5G equipment, and critical raw materials (e.g., the *EU Chips Act*).

Screening mechanisms for foreign direct investments (FDI) allow blocking purchases of strategic infrastructure by foreign actors. For instance, Italy prevented the sale of its national electricity grid operator to China's State Grid using EU-level FDI screening tools. The EU launched Global Gateway, its strategic alternative to China's BRI, allocating more than €300 billion by 2027 to infrastructure projects across Africa, Asia, Latin America, and the Balkans. In the area of digital sovereignty, EU initiatives (e.g., Gaia-X) aim to reduce dependence on Chinese and American tech platforms. Regulations such as GDPR strengthen European control over data and restrict transfers of sensitive information to authoritarian regimes. Additionally, EU-driven public–private partnerships (PPPs) promote joint investments by domestic and European companies in telecommunications, railways, energy, and cybersecurity—further reducing dependence on non-EU actors.

4.2 NATO

Analyzing NATO's strategic concepts suggests that the Alliance seeks to regulate and protect its CI. The wide spectrum of possible scenarios for NATO's involvement in CI protection sometimes gives the impression that a militarized approach surpasses the Alliance's originally intended strategic role [34].

NATO has ramped up efforts acknowledging that infrastructure interdependence is integral to collective defense and regional stability. NATO's Article 3 encourages resilience of national infrastructure as a first line of defense. In the wake of hybrid attacks on pipelines and cables, NATO created a Critical Undersea Infrastructure Coordination Cell in early 2023 to improve Allied collaboration in safeguarding subsea energy and data links. By May 2024, NATO went further by convening a new Critical Undersea Infrastructure Network, bringing together experts and militaries to enhance info-sharing, surveillance, and joint protection of cross-border pipelines and cables [35].

NATO is also addressing cyber interdependence; since a cyberattack on one nation's grid or health system can have spillover effects (think of malware that jumps borders), NATO's Cooperative Cyber Defence Centre and information-sharing platforms enable a collective response to threats traversing the Alliance's networks.

4.3 Indicators of CI Resilience in the Western Balkans and Its Impact on Societal Security and Resilience

A key feature of infrastructure interdependence is how it ties into regional security dynamics, especially for smaller states or those in transition (e.g. EU candidate countries). Take the countries of Southeast Europe and the Western Balkans, many of these nations are not yet EU members or NATO allies, but their infrastructure networks (electric grids, gas pipelines, railways, digital links) are increasingly meshed with those of the EU. This interdependence can be a source of strength, enabling cross-border support in crises, but also a vulnerability if not managed. For example, the Balkan electric grid is synchronous with the continental European grid; disturbances in one corner can cascade. Recognizing this, initiatives supported by the EU's Instrument for Pre-Accession Assistance (IPA) have aimed to boost regional resilience. Such efforts acknowledge that a weak node in a regional network can endanger the whole, so raising the resilience of candidate and neighboring states is in the EU's direct interest. We see similar logic in NATO's partnerships (e.g. with Finland and Sweden before they joined, or programs with Moldova, Georgia): helping those countries protect their critical systems (energy, cyber, transport) not only stabilizes them but also shields the wider region from spillover disruptions.

Interdependent infrastructure systems are now recognized as a cornerstone of regional security. From a systems theory perspective, improving regional infrastructure resilience often means enhancing *redundancy, diversity,* and *coordination.* Redundancy ensures alternative supply lines or backup systems if one fails (for instance, having multiple interconnectors for electricity and gas among EU countries so no state is isolated). Diversity means not over-relying on a single type of infrastructure or supplier—a lesson Europe took to heart post-Ukraine war by diversifying gas sources to avoid a monoculture dependence on Russian pipelines. Coordination is probably the most human factor: it's about aligning policies, information flows, and response plans across jurisdictions.

Today's global drivers, cyber threats, geopolitical competition, climate pressures, digitalization, and energy transition, expose structural vulnerabilities in the Western Balkan region. Due to their small size, import dependence, and geopolitical contestation, these states face constrained infrastructure sovereignty. This results in: highly vulnerable supply chains, limited national control over critical technologies, external geopolitical leverage, and weak resilience during disruptions.

A growing share of CI critical to public welfare in the Western Balkans is in foreign ownership or under foreign operational control, particularly by entities that EU and NATO perceive as security risks, chiefly China and Russia. Russia is deeply embedded in the region's energy sector, while Chinese strategy is oriented toward dominance in transport, logistics, and mining. As geopolitical tensions escalate and future developments become increasingly uncertain, the pace and functionality of alignment with EU directives is questionable [14]. Challenges concern: interconnectedness-related vulnerabilities, centralization/decentralization trade-offs, privacy compromises, governance-related issues. According to Jovanović, instead of attempting to analyze unknown emerging risks directly, risks should be translated into *scenarios* and evaluated through the question of whether the "value to be protected" (ISO 31000) can "resist and adapt" (ISO 22316) under adverse conditions. Indicators serve as useful benchmarking tools, though their limitations must always be acknowledged. Resilience assessment examines the functionality of infrastructure through indicators that reflect the condition of individual functional elements throughout the scenario timeline [16].

Based on the above considerations, the subsections below provide a overview of some of the key CI resilience indicators in Western Balkan states (Serbia, Montenegro, Bosnia and Herzegovina, North Macedonia, Albania).

(a) **High Dependence on Foreign Technology and Suppliers**

Western Balkan states rely heavily on imported technology, energy equipment, digital systems, and industrial components. Foreign control over critical technologies (telecom, energy, transport) limits their autonomy during crises. Dependencies on suppliers from China, Russia, Turkey, and the U.S. create vulnerabilities stemming from sanctions, export controls, or political tensions. China has filled regional gaps by offering attractive financing and affordable construction services. The value of projects in the "16 + 1" framework exceeds €27 billion, with more than half concentrated in the Western Balkans. Between 75 and 85% of these projects are financed through Chinese loans. Debt exposure is significant: Montenegro—up to 18% of GDP, Serbia: 12%, Bosnia and Herzegovina: 10%, North Macedonia: 7%. High loan dependence correlates strongly with corruption risk [36]. Aging physical infrastructure magnifies these risks. Consequences include political conditionality, energy shocks, and economic destabilization during global crises.

(b) **Geopolitical Competition and Digitalization Fragmenting Supply Chains**

Major powers—EU, China, Russia, the U.S., compete for influence, creating fragmented infrastructure ecosystems. When states become "locked in" to a single supplier or geopolitical camp, their diversification capacity weakens. Sanctions or political pressure applied to one vendor can trigger national-level disruptions. As infrastructure digitalizes (smart grids, customs systems, transport corridors), cyberattacks grow more damaging. Breaches in logistics, customs, or electricity distribution can paralyze national systems—scenarios already observed in multiple Western Balkan countries.

(c) **Climate Stress on Import-Dependent Systems**

Rising climate pressures affect transport, agriculture, and energy systems, particularly in states that depend on imported fuels, food inputs, and high-voltage equipment.

The 2014 Serbian floods revealed deep fragility in regional logistics corridors.

Energy transition adds new dependencies, as Western Balkan nations import solar, battery, EV, and grid-modernization technologies. Critical mineral supply chains dominated by China, amplify this vulnerability.

(d) **Cross-Border Infrastructure Interdependence**

Western Balkan supply chains are integrated with EU transport and energy corridors (TEN-T, ENTSO-E, digital networks) [37]. Disruptions in neighboring states cascade regionally. Low levels of cross-border cooperation hinder crisis response, real-time data exchange, and cybersecurity coordination. Multi-hazard scenarios (simultaneous crises) remain a significant risk.

(e) **Public–Private Dependencies**

Ports, airports, telecom operators, and electricity distributors are frequently foreign-owned. This reduces strategic autonomy and limits governments' ability to control long-term infrastructure decisions (cybersecurity, modernization, crisis planning). When foreign owners come from mutually hostile geopolitical blocs, misalignment, lack of cooperation, and information withholding can severely undermine societal and economic stability. Sectors such as telecommunications and finance are particularly vulnerable to sabotage, espionage, and cyberattacks [14].

(f) **AI, Automation, Data Dependence and Skill Gaps**

As CI becomes increasingly AI-driven, resilience relies on: data integrity, model robustness, human oversight. The World Intellectual Property Organization (WIPO), in its analysis of technological trends in artificial intelligence, identifies several key application sectors that predominantly constitute CI: transportation, energy, telecommunications, medicine, agriculture, and public services [38]. Western Balkan countries also face shortages of skilled personnel in cybersecurity, energy systems, AI operations, and crisis communication. Dependence on foreign data centers and cloud services creates additional vulnerability during geopolitical or cyber disruptions.

(g) **Public Trust, Governance, and Regulatory Pressure**

Corruption, opacity in infrastructure contracts, and lack of public oversight undermine societal trust. Non-transparent agreements with foreign actors generate legal and political dependencies, "debt-colonialism," and erosion of democratic governance. Ukraine's experience underscores a central lesson: resilience arises not only from defensive measures but from eliminating vulnerabilities through integrated, multi-layered protection systems combining engineering, cyber, organizational, and emergency-response elements.

The list of indicators is based on an analysis of various scenarios in which the infrastructure sectors of small countries were compromised. Based of the best practice of national and CI resilience, proposed protective measures for Western Balkan states are:

- Transparency in infrastructure contracts (public disclosure, independent legal and security reviews).
- Diversification of partnerships (avoid over-reliance on China or the EU; include Japan, South Korea, Nordic states).
- Development of domestic expertise (cybersecurity, energy management, digital autonomy).

- Limiting high-risk foreign investments to technical services without ownership stakes.
- Alignment with EU standards (procurement, data protection, network security).
- Implementation of integrated, multi-layered protection models, following best practices demonstrated in Ukraine.

5 Critical Infrastructure Sovereignty

In today's world, CI lies at the core of what is known as "weaponized interdependence". Globalization has created dense networks that enable great powers to exploit their central positions within these networks as strategic assets—through energy transit and supply routes, 5G and telecommunications, cloud computing, digital platforms, and financial and payment systems.

Kallenborn introduces the concept of "globally critical infrastructure" to describe systems (e.g. undersea cables, semiconductor supply chains) so vital that their disruption would destabilize many nations [39]. He argues that traditional sovereignty based on territorial control is strained, since essential infrastructure often lies outside any single state's jurisdiction. This framework reframes CI sovereignty as a state's ability to maintain decisive influence over critical assets and avoid undue foreign leverage or control.

Digital sovereignty has been debated within control over data and digital infrastructure in cloud era. The European Commission tasked ENISA (the EU cybersecurity agency) with developing a common cloud security certification (the EUCS scheme) to regain some sovereignty, but this technical effort evolved into a heated political debates [40]. Member States diverged on how strict to be regarding non-EU providers. This case underscores that standards and certifications (often viewed as technical issues) are in fact geopolitical instruments: whoever sets the cloud security standards can indirectly shape who controls the data.

In the context of Extraterritorial access and data governance, European policymakers promote data localization, sovereign cloud initiatives (such as GAIA-X), and encryption standards that limit external surveillance. Academic works like Ganz et al. examine how submarine internet cables—carrying ~99% of international data—challenge digital sovereignty [41]. They conclude that improved international governance of internet infrastructure is needed to "defend states' digital sovereignty."

Thus, CI sovereignty becomes simultaneously: (a) a security issue (exposure to coercion, espionage, disruption); (b) a development issue (access to investment, capital, and technology), and (c) a status/autonomy issue (freedom to conduct independent foreign and security policy).

For small and medium-sized states, this often means navigating between competing infrastructure offers (EU vs. China vs. regional powers), while managing multi-vector dependence (e.g., energy from one geopolitical bloc, finance from another, digital technology from a third). At the same time, these states must maintain sufficient "room for maneuver", ensuring no single external actor can "hold them by the infrastructure."

CI sovereignty can be defined as the ability of a state (or political community) to determine, control, and govern the development, operation, data flows, and security of its essential infrastructure systems in accordance with its own strategic, legal, and societal priorities, without being structurally constrained, limited, or coerced by external actors. This sovereignty does not refer only to legal ownership or the "national flag" displayed on an asset. The central question is who actually has leverage over: energy flows, transport corridors, digital networks and platforms, financial and payment systems, health, water, food, and logistics infrastructure—and whether that leverage can be used to shape political or economic decisions.

5.1 Core Analytical Dimensions of CI Sovereignty

(a) **Ownership and Control**—Who owns the infrastructure: the state, domestic private entities, foreign companies, or foreign state-owned enterprises? Who controls strategic decisions such as: pricing, maintenance, expansion, shutdown policies, access to operational data? Formal ownership can be domestic, while *effective control* may remain external due to financing structures, proprietary technology, intellectual property, standards, or long-term contractual obligations.

(b) **Technological and Standard-Setting Dependence**—Which technology stack is used (hardware, software, protocols)? Is the system locked into proprietary foreign technology that makes replacement costly or impossible? Who sets interoperability requirements, certification rules, and technical standards? This dimension reflects *standard power*—the ability of a state or corporation to shape infrastructure evolution by controlling standards or intellectual property.

(c) **Data and Cyber Sovereignty**—Data sovereignty has become a central pillar of overall CI sovereignty. For digital and networked CI, key questions include: Where is data stored (data centers, cloud infrastructure, jurisdiction)? Which foreign intelligence services or regulators may legally or covertly access that data? Who manages cybersecurity, incident response, and updates?

(d) **Financial and Debt Dependence**—How is CI financed: via international financial institutions, EU funds, Chinese banks, Russian loans, or private equity? Are there political or economic conditionalities attached to financing? Does the structure of debt create leverage—for example, the possibility of taking over ports or energy assets in case of default? Here, CI becomes a tool of geoeconomics, where influence is exerted through credit, not military force.

(e) **Institutional and Regulatory Capacity**—Does the state have sufficient regulatory strength and expertise to govern foreign operators? Can it enforce transparency rules, security standards, environmental safeguards, and competition law? Where regulatory capacity is weak, sovereignty becomes nominal; rules exist on paper but not in practice.

CI sovereignty can be understood as a hybrid of *hard and soft power*. It is *hard* when disruption directly affects physical life, economic flows, and state functioning. It is *soft* when CI control allows subtle political influence without overt coercion. Thus, CI sovereignty is not only about protecting domestic infrastructure but also about assessing: (a) whether external actors can use infrastructure as an instrument of influence, and (b) whether the state itself can leverage its own CI in regional politics.

A state can be considered sovereign over its CI when it retains *decisive influence* over the strategic direction, security, and usage of its essential infrastructures—and when no external actor can unilaterally exploit infrastructural dependencies to force political or economic concessions contrary to the democratically defined interests of the population.

In policy practice, CI sovereignty requires: diversification of suppliers, routes, and technologies; screening of foreign direct investment; strong domestic regulation and cybersecurity; building local expertise and industrial capacity; aligning CI policy with broader foreign and security strategies rather than treating CI as purely economic.

5.2 Linking CI Sovereignty with Societal Security and Resilience

As infrastructures become more complex and interconnected, the risk increases that foreign states, corporations, or hybrid actors can exploit infrastructural dependencies for geoeconomic leverage, political influence, or strategic pressure. Consequently, *CI sovereignty emerges as a fundamental component of national and societal security.*

Societal security focuses on a community's ability to preserve its identity, essential functions, and cohesion when faced with external or internal pressures. In this context, CI forms the backbone of societal continuity:

- Energy powers households, industry, heating, and public services.
- Telecommunications maintain communication, information flows, and social integration.
- Transport and logistics ensure the supply of food, medicine, and essential goods.
- Digital infrastructure shapes public life, communication, and access to services.
- Healthcare infrastructure provides medical care and underpins social welfare.

When these infrastructures are externally owned, influenced, or weaponized, societal functions become vulnerable to manipulation, not only disruption. Failures in energy supply, telecom blackouts, or payment-system collapses can rapidly escalate into political instability, protests, and erosion of institutional legitimacy. External operators may influence society indirectly through: disinformation campaigns, cyberattacks, strategic infrastructure shutdowns, selective prioritization of their own national interests during crises.

CI sovereignty protects societal security during environmental disasters, pandemics, or infrastructure failures, ensuring faster response, clearer accountability and greater public trust. Societal trust and cohesion directly support crisis-management capacity. Reliable, sovereignly controlled infrastructures reduce the likelihood of social fragmentation during crises.

5.3 CI Sovereignty as a Buffer Against Systemic Risks

Modern risks are non-linear, unpredictable, and transboundary. The arena of uncertainty is shaped and expanded by global changes and interdependencies that interact at such speed that the shift can be described as a transition from Beck's "risk society" to a "crisis society." A crisis society can be characterized as the compression of a nonlinear dynamics of events which, through synergistic effects and the increasing amplitude of their disturbances, results in the transformation of systems into chaotic and seemingly ungovernable structures [42].

According to Keković, systemic risks represent a fundamental shift from a risk society to a crisis society, surpassing the ability of any single system or entity to control them effectively. The application of hierarchical decision-making patterns to complex, functionally diverse problems does not lead to problem-solving; on the contrary, it multiplies the problem. Given that every organization represents a system, the realization of events is often not the consequence of individual causes but rather the result of the emergence and development of a causal chain of events [42].

Policymakers and executives often become constrained by legacy patterns of action that cannot accommodate the cascading, cross-sectoral nature of systemic disruptions. Here, the strategic value of **CI sovereignty** becomes evident: only sovereign control over essential infrastructures provides the institutional and operational flexibility required to anticipate, absorb, and mitigate systemic shocks. For understanding these risks, it is essential to recognize that no subsystem or component of a broader system is more important than another, and that no hierarchy exists among them. This is especially evident in the example of a social community, whose functioning depends on each of its activities individually: economic, educational, cultural, security-related, etc.

Recognizing that no **subsystem within a broader infrastructure ecosystem is inherently superior to another** is essential for conceptualizing CI sovereignty. Energy reliability is contingent on digital infrastructure; digital infrastructure depends on transport and electricity; transport depends on fuel supply, logistics, and communication systems. This interdependence underscores why sovereignty must be systemic, not sector-specific.

The assumption that systemic risks are nonlinear means that beyond a certain threshold or tipping point, they may produce destructive effects. Without understanding this causal chain, neither risk nor the disruptive factor itself can be effectively managed. Disturbances of moderate intensity may be absorbed through redundancy, adaptation, or reconfiguration of the system. However, disturbances that exceed critical thresholds can trigger tipping points, driving infrastructures into states of disequilibrium. Crisis represents one such state—arising when systemic risks materialize across multiple infrastructural layers simultaneously.

Every crisis deepens social inequalities and income disparities, confirming that the key values that contribute to successfully containing them are inclusiveness, solidarity, and trust. Social trust is both a **precondition** and an **outcome** of CI sovereignty. External control undermines trust by creating: **uncertainty**, as citizens perceive dependency on foreign political dynamics; and **vulnerability**, since foreign actors may prioritize their own interests during crises.

Infrastructure disruptions and shocks amplify perceptions of societal vulnerability. CI sovereignty reduces this vulnerability and reassures citizens that societal welfare will not be subordinated to external interests. Sovereign governance strengthens resilience by reinforcing: public confidence in crisis management, legitimacy of institutional responses, coordinated decision-making, clear accountability among key actors.

For small and medium states such as those in the Western Balkans, where geopolitical competition intersects with infrastructural dependency, societal resilience must be not only **material** but also **structural**. CI sovereignty reduces exposure to systemic global shocks and limits the ability of external actors to exploit crises for political or economic leverage.

6 Conclusion

Our relationship toward CI is changing at an astonishing pace. These are no longer merely facilities, networks, and systems that are subject to criminal attacks in physical and cyber space—threats traditionally countered with physical, technical, and cybersecurity measures. The protection of CI has evolved into not only an economic and security priority, but also a first-order geopolitical issue and an increasingly important subject of diplomatic negotiations among major powers.

When the question of who controls CI is raised, it becomes clear that such infrastructure can be used as an instrument of soft power, an asset that great powers consistently employ to exert geopolitical influence, especially in politically and economically vulnerable regions. Control over critical resources and raw materials is likely to become one of the key drivers of geopolitical and strategic influence for major and regional powers, as well as a defining factor in contemporary geopolitical dynamics.

Infrastructure-driven great-power competition is today a reality and a strategic lever of influence and form of rivalry among investors. However, this does not necessarily imply a "pure conspiracy"; many investments have legitimate economic and developmental motivations and may significantly enhance regional infrastructure. Whether such investments become mechanisms of long-term dependence and control, or remain beneficial contributions, ultimately depends on how the projects are implemented, the degree of transparency, institutional strength, and regulatory oversight. The key lies in the strength of domestic institutions and their capacity to negotiate, regulate, and protect national and societal interests, preventing mechanisms of long-term dependency from taking hold.

CI sovereignty is not merely a technical or economic concern; it represents a core determinant of societal security and a structural foundation of societal resilience in contemporary geopolitics. In this paper it is presented by integrating the concepts: (a) ***CI sovereignty*** protects society from external coercion and structural vulnerabilities; (b) This ensures ***societal security***, preserving essential functions and social cohesion. (c) Secure, autonomous infrastructures form the foundation for ***societal resilience***, enabling shock absorption, adaptation, and long-term transformation.

To conclude: CI sovereignty is a structural determinant of societal security and an indispensable foundation of societal resilience in a geopolitical environment defined by interdependence, uncertainty, and complex systemic risks.

References

1. Keković, Z.: Security in New Corporate Environment, Security 6, VŠUP, Belgrade (2006)
2. NATO Parliamentary Assembly: Document 162 CDS 07 E rev 1—The Protection of Critical Infrastructures (2007)
3. Gordan, K., Dion, M.: Protection of Critical Infrastructure and the Role of Investment Policies Relating to National Security. OECD (2008)
4. NIPP 2013: Partnering for Critical Infrastructure Security and Resilience. U.S. Department of Homeland Security (2013)
5. National Security Strategy of the United Kingdom: Security in an Interdependent World. Cabinet Office (2010)
6. Keković, Z., Kešetović, Ž.: Societal security in Serbia. Serb. Polit. Though **15**(1–2), 181–200 (2008)
7. Cvrtila, V.: Security management and new threats. In: Keković, Đurić, Polović (eds.) Security Crises in the 21st Century and How to Manage Them, Social and Security Aspects, vol. 1, pp. 8–23. Center for Risk Analysis and Crisis Management (2020)
8. Hellmann, G.: Defending 'the west'? the transformation of national security in the European Union. In: Hellmann, G., Herborth, B. (eds.) Uses of the West. Security and the Politics of Order, pp. 231–254. Cambridge University Press (2017)
9. Kaufmann, M.: Emergent self-organisation in emergencies: resilience rationales in interconnected societies. Resilience **1**(1), 53–68 (2013)
10. Pavićević, O.: The resilience concept in sociology. Sociologija **58**(3), 432–449 (2016)
11. Ridley, G.: Resilience and national security. In: The Palgrave Handbook of Security. Risk and Intelligence, pp. 79–98. Palgrave Macmillan, London (2017)
12. Fjäder, C.: The nation-state, national security and resilience in the age of globalization. Resilience **2**(2), 114–129 (2014)
13. Holling, C.S.: Resilience and stability of ecological systems. Annu. Rev. Ecol. Syst. **4**, 1–23 (1973)
14. Keković, Z., Ninković, V.: Critical Infrastructure Protection: A Systemic Approach. Center for Risk Analysis and Crisis Management, Belgrade (2025)
15. Keković, Z., Dinić, J.: National security transformation: resilience within national security strategies of great powers in response to global threats. Zbornik Radova Pravnog Fakulteta Novi Sad **L**(4), 1141–1156 (2016)

16. Jovanović, A.: Managing risks and resilience of uncertain future(s): application to critical infrastructures. In: Keković, Đurić, Polović (eds.) Security Crises in the 21st Century and How to Manage Them, Social and Security Aspects, vol. 1, pp. 36–48, Belgrade (2020)
17. World Economic Forum: Global Risks Report 2025. https://reports.weforum.org/docs/WEF_Global_Risks_Report_2025.pdf
18. Basystyi, V.: Resilient by design: Ukraine's experience in protecting critical infrastructure, In: Keković (eds) Critical Infrastructure: Risk and Crisis Management in the New Reality, Conference Book of Abstracts, pp. 12–14. Center for Risk Analysis and Crisis Management (2025)
19. World Economic Forum, Global Risks Report 2024, on top emerging global risksweforum.orgweforum.org and expert views on AI and cyber threatsweforum.orgweforum.org. Last accessed 05 December 2025
20. Mosca, M., Dodson, D.: Cryptographic resilience (WEF, Dec 2024), on quantum computing risks to cryptography. weforum.orgweforum.org. Last accessed 05 December 2025
21. Barbe, A., Hunt, W.: Preserving the Chokepoints, Reducing the Risks of Offshoring Among U.S., Semiconductor Manufacturing Equipment Firms. cset.georgetown.educset.georgetown.edu. Last accessed 05 December 2025
22. Ocharo, S.: (SSRN, 2025), Africa's Role in Global Cybersecurity, on African CI cyber vulnerabilities vs. progress. papers.ssrn.compapers.ssrn.com. Last accessed 05 December 2025
23. Braw, E.: Atlantic Council Issue Brief, how the Baltic Sea nations have tackled suspicious cable cuts (2025). atlanticcouncil.orgatlanticcouncil.org. Last accessed 10 December 2025
24. Powers-Riggs, A., et al.: CSIS Brief, China's growing threat to gallium supply chains, on China weaponizing critical mineral exports (2025). csis.orgcsis.org. Last accessed 10 December 2025
25. GOV.UK. https://www.gov.uk/government/news/uk-announces-new-support-to-boost-british-exports-and-investment-in-the-western-balkans. Last accessed 05 December 2025
26. Novikau, A., Muhasilović, J.: Turkey's quest to become a regional energy hub: challenges and opportunities (2023). https://www.researchgate.net/publication/375093951_Turkey's_quest_to_become_a_regional_energy_hub_Challenges_and_opportunities. Last accessed 05 December 2025
27. Saudi Arabis's Vision 2030: Transforming Investment Strategies. https://chenjiazizhong.com/2025/08/14/saudi-arabias-vision-2030-transforming-investment-strategies/. Last accessed 05 December 2025
28. Council on Foreign Relations, Huawei Threat to U.S. Security? noting countries banning Huawei 5G (2023). cfr.org. Last accessed 05 December 2025
29. The Guardian: Neo-Nazi leader sentenced…plot to attack power grid, on foiled Baltimore substation plot (2025). theguardian.com. Last accessed 05 December 2025
30. Reuters Special Report: by Roulette, Bryan-Low, Balmforth, on Musk/Starlink's impact in Ukraine war (2025). reuters.comreuters.com. Last accessed 06 December 2025
31. EU Council Consilium: Timeline of Russia's aggression, noting EU excluding Russian banks from SWIFT (2022). consilium.europa.eu. Last accessed 06 December 2025
32. National Academies (U.S.) report, Complex and Interdependent Systems (as cited in unisdr.orgunisdr.org), on CI as system-of-systems and cascading failures. Last accessed 06 December 2025
33. European Commission DG Home: Critical infrastructure resilience at EU-level, on CER Directive provisions and Council Recommendation post-sabotage. home-affairs.ec.europa.euhome-affairs.ec.europa.eu. Last accessed 06 December 2025
34. Mitrevska, M., Mileski, T., Mikac, R.: Critical Infrastructure—Concept and Security Challenges. Friedrich Ebert Stiftung, Skopje (2019)
35. NATO News: First meeting of NATO Critical Undersea Infrastructure Network, Stoltenberg quote and NATO measures (2024). nato.intnato.int. Last accessed 16 December 2006
36. China in the Global Power Transition: Geopolitical Ambitions and the Transformation of the International Order. DFC, Podgorica (2025)
37. Methodology to Identify Regional Electricity Crisis Scenarios in accordance with Article 5 of the Regulation of The European Parliament and of The Council on Risk-Preparedness in the Electricity Sector and Repealing Directive 2005/89/EC

38. WIPO: WIPO Technology Trends 2019: Artificial Intelligence. World Intellectual Property Organization, Geneva (2019)
39. Kallenborn, Z.: Globally Critical Infrastructure: The Unique Risks and Challenges. RiskAnalysis. reviewedin emergingthreats.co.ukemergingthreats.co.ukemergingthreats.co.ukemergingthreats.co.uk. Last accessed 06 December 2025
40. Bômont, C.: Technical is political: cloud certification and Europe's digital sovereignty. EUISS Brief. iss.europa.euiss.europa.eu (2025)
41. Ganz, A., et al.: Submarine cables and the risks to digital sovereignty. Minds Mach. **34**(3), 31–57. papers.ssrn.compapers.ssrn.com (2024)
42. Keković, Z.: Effects of uncertainty in decision-making on systemic risks. Fac. Secur. Stud. Annu. 15–37 (2022)

The Theory of Neoclassical Realism and the Shaping of Critical Energy Infrastructure in the New International Reality

Boban Marjanović and Ema Kraktus

Abstract This paper analyzes critical energy infrastructure in Southeastern Europe through the theoretical framework of neoclassical realism in international relations. The new international reality is defined by great power politics and geopolitical rivalries that shape energy flows and patterns of dependency. In this context, energy transition in the region derives from strategic calculations aimed at reducing vulnerability, creating alternative corridors and strengthening alliances in line with the particular interests. The central hypothesis posits that international pressures from great powers determine the framework for the development and protection of critical energy infrastructure, while outcomes depend on domestic intervening factors such as institutional capacities and internal political-economic conditions. Neoclassical realism provides the tools to explain why countries exposed to similar external constraints adopt divergent strategies of energy security. The methodological approach combines analysis of Southeastern Europe with a comparative examination of state responses to systemic pressures. Particular focus is placed on the influence of the United States and Russia in shaping regional dynamics. The argument is twofold. It first shows how great power politics accelerate the transformation of energy systems in Southeastern Europe, and then explains how the interaction of external and domestic factors ultimately shapes the character of critical energy infrastructure.

Keywords Neoclassical realism · Critical energy infrastructure · Southeastern Europe

---

B. Marjanović (✉)
Institute for Political Studies, Belgrade, Serbia
e-mail: boban.marjanovic@ips.ac.rs

E. Kraktus
Ministry of Finance of the Republic of Serbia, Belgrade, Serbia

Z. Keković et al. (eds.), *Critical Infrastructure: Risk and Crisis Management in the New Reality*, Advanced Sciences and Technologies for Security Applications,
https://doi.org/10.1007/978-3-032-20948-1_2

1 Introduction

Global energy security in contemporary international relations constitutes a critical domain where diverse national interests converge. In the two decades following the major global oil crisis of the 1970s, energy politics was primarily approached within academic circles as an economic concern, with its security dimensions receiving comparatively less attention [1]. The increasing prominence of energy security in academic discourse became particularly evident after 2001, when then U.S. President George W. Bush publicly asserted that "energy security" should become a central priority of U.S. foreign policy and a key component of domestic policy. Soon thereafter, leaders of other industrialized nations began to underscore similar positions [2].

In the *Council Directive 2008/114/EC* adopted at the end of 2008, the Council of the European Union defined critical infrastructure as "an asset, system or part thereof located in Member States which is essential for the maintenance of vital societal functions, health, safety, security, economic or social well-being of people, and the disruption or destruction of which would have a significant impact in a Member State as a result of the failure to maintain those functions" [3, p. 345/77]. According to the Annex to this directive, critical energy infrastructure encompasses electricity generation, transmission, and distribution systems, as well as facilities and networks for the production, processing, storage, and transport of oil and natural gas, including liquefied natural gas terminals [3, p. 345/81]. In the current context of increasingly frequent disruptions in global energy markets, the reliability of energy supply and the protection of the infrastructure enabling its flow have become central to states' strategic policies. For the purposes of this paper, critical energy infrastructure includes the oil, natural gas, and electricity sectors as its key components [4].

In the context of contemporary international rivalry, understanding which actors shape the development of energy infrastructure and how they do so requires careful political and strategic analysis. Given that the protection of critical infrastructure is a core state priority, critical energy infrastructure can also be conceptualized as an instrument of foreign policy, since control over it allows governments to exert international political influence and strengthen their position in global affairs. In 2006, the European Commission highlighted that "Europe has entered into a new energy era" and warned that "rising import dependence from unstable regions and suppliers is a serious risk... [as] some major producers and consumers use energy as a political instrument" [5].

This topic has been selected in order to highlight the growing interlinkage between energy policy and security policy in a context where the structure of the international order remains insufficiently defined. In the period after 2022, the global energy system has faced profound changes which have altered diversification processes and generated new forms of dependency. These trends have been recognized by the European Union, which in its 2022 directive stressed the increasing interconnection of sectors and the impact of geopolitical crises on energy security, noting the existence of "a dynamic threat environment encompassing evolving hybrid and terrorist threats and a growing interdependence between infrastructures and sectors" [6].

This paper focuses on Serbia, Hungary, and Romania, as these countries represent distinct types of linkage between national energy policy and external influences, as will be demonstrated in the analysis. Within the context of regional energy infrastructure and transit routes, although geographically part of Central Europe, Hungary is often included in Southeast European energy corridor projects due to its integration into routes connecting the Caucasus and the Black Sea basin with the Balkans and Central Europe [7]. Furthermore, the manner in which energy infrastructure is controlled shapes relations among major powers and influences the behaviour of smaller states that seek to balance external pressures with their own national interests. For Southeast European countries situated at the intersection of the Euro-Atlantic and Eurasian energy spaces, these dynamics are particularly significant. The relevance of the subject stems from the fact that this region remains a focal point of great power competition, while the states concerned simultaneously strive to maintain energy stability.

The theoretical aim of this paper is to demonstrate how neoclassical realism, which links the systemic level of power distribution to intervening variables, can provide deeper insight into why states do not respond identically to similar external energy challenges, and how the interests of domestic elites, institutional frameworks, and decision-makers' perceptions of energy risks mediate between structural incentives and specific infrastructural outcomes [8]. The paper focuses on the application of neoclassical realism in international relations, as a foreign policy theory, to the process of shaping critical energy infrastructure in Southeast Europe. Accordingly, the main hypothesis is that international pressures and opportunities establish the general framework for the development of energy infrastructure, while its concrete and final configuration is shaped by domestic political and institutional factors.

Previous research on energy security has largely focused on global flows and the strategies of major powers such as the United States, Russia and China, as well as on the role of the European Union. Less attention has been devoted to the intervening factors that determine how smaller states respond to external political impulses originating in the international system. There is a lack of comparative analyses explaining how domestic political institutions, the structure of the energy sector, national interests, and the perceptions of decision-makers regarding threats and challenges shape strategies for protecting critical energy infrastructure. This paper seeks to address that gap by applying the theoretical framework of neoclassical realism, which links systemic-level political influences with domestic conditions.

The empirical section analyses three Southeast European states. Serbia is a transit country in the Balkans whose energy policy is significantly shaped by its dependence on Russian energy flows, while simultaneously maintaining an interest in diversifying its supply sources. Hungary, as a member of the European Union with a developed energy infrastructure, faces challenges arising from its geopolitical position between the West and Russia, which promotes a pragmatic approach to energy issues. Romania represents a relatively energy self-sufficient state whose strategy is oriented toward

NATO and the European system of energy security and which, unlike Serbia and Hungary, demonstrates the lowest level of dependence on Russian energy resources.

2 Neoclassical Realism and Critical Energy Infrastructure

Neoclassical realism builds upon the three levels of analysis that Kenneth Waltz popularized in the twentieth century: the international system, the state, and the individual [9, p. 18]. However, Waltz's neorealism emphasizes almost exclusively the anarchic character of the international system and the distribution of power among great powers as the principal determinant of foreign policy. Neoclassical realist theorists have observed that such an approach cannot account for numerous cases in which states behave differently under similar systemic conditions. As Vladimir Trapara notes, this new theoretical strand "supplements neorealism in order to enhance its explanatory power and render it suitable for the study of states' concrete foreign policies," with systemic factors operating indirectly, mediated through the foreign policy formulation process that depends on domestic factors as intervening variables [10].

In neoclassical realism, the international system remains the primary independent variable. The contemporary international system is marked by structural instability generated by geoeconomic and geopolitical pressures that are reconfiguring international relations. In her recent monograph, Emma Ashford conceptualizes these processes from a realist perspective, arguing that contemporary international politics unfolds through a process of structural transformation and asserting that the post-Cold War international system under U.S. leadership is eroding [11]. On this basis, one may analyse how the formation of a new international system affects the shaping of critical energy infrastructure and the behaviour of states and other actors.

Fundamentally, the structure of the international order still rests upon the anarchic principle that Kenneth Waltz identified as a core constant of world politics. Anarchy as a structural, immutable feature establishes the conditions in which states, as formally equal units of the system, must provide for their own survival. In an anarchic order, all states share the same formal role, as they are essentially equivalent units [9, p. 88]. The system is anarchic and defined by the distribution of power, but it does not by itself determine the foreign policy behaviour of individual states. Neoclassical realism emerged from the need to overcome the limits of structuralism and to explain why states with similar capabilities do not behave identically under comparable conditions. Gideon Rose articulated this claim in 1998 in *World Politics* journal, when he systematized this approach and argued that foreign policy cannot be understood without analysing the specific variables related to states' internal characteristics [8, p. 146].

Since states respond to international pressures through the prism of their own institutions and the perceptions of their leaderships, neoclassical realism emphasizes linking systemic and domestic variables. Rose stresses that foreign policy cannot be explained solely by the distribution of power in the international system, but also depends on how "state leaders interpret systemic imperatives and respond to the pressures arising from that system" [8, p. 146]. Over the last decade, this theory has been extended to issue areas beyond traditional foreign policy. Authors such as Norrin Ripsman, Jeffrey Taliaferro, and Steven Lobell, in their 2016 work *Neoclassical Realist Theory of International Politics*, highlight that foreign policy choices must be viewed in light of the state's capacity to articulate and implement appropriate strategies in response to the challenges it confronts. According to their argument, the selection of specific foreign policy options from the set of available alternatives depends on the particularities of the domestic political environment [12, p. 33].

Within the framework of foreign policy analysis articulated by neoclassical realism, energy infrastructures hold an important place. Starting from the assumption that the distribution of material capabilities in the international system sets the basic parameters of state behaviour, critical energy infrastructure can be conceptualized as a material resource that states may utilize to secure their external position. Research within the realist paradigm has demonstrated that energy resources play a significant role in shaping states' foreign policy power and their position in the international system [13, p. 30]. At the systemic level, the distribution of power and energy capacities remains central. States that control energy infrastructure are capable of structuring other states' dependencies and thereby enhancing their strategic importance. In the context of critical energy infrastructure, primary energy resources represent a fundamental component of state power. The greater the volume of resources at a state's disposal, the greater its potential power, assuming it possesses the capacity to efficiently extract and transport energy and operates under favourable conditions of global demand [13, p. 31].

Neoclassical realism, however, emphasizes that systemic stimuli arising from anarchy and from the distribution of energy capacities do not act directly. They are filtered through the unit level and depend on the institutional and political capacities of states to control available energy resources. In the context of critical energy infrastructure, this means that policy implementation and institutional performance are not only reactive responses to energy challenges, but also active factors shaping the energy system itself. As some authors note, "politics and policies actively drive, constrain and shape the material basis of the energy system and its geographic patterns," which leads to the conclusion that state interpretation of perceived international reality serves as the key intermediary between systemic pressures and foreign policy action [14].

Critical energy infrastructure thus reflects a fusion of material and institutional power, as control over it and the ability to protect it enable the transformation of material resources from their primary function into strategic political influence. Although often treated in international relations as merely a passive technical component, critical energy infrastructure can also be defined as "a mechanism that shapes power relations in international affairs and reconfigures spaces, policies and people" [15]. The translation of material power into political action lies at the core of the mediation between systemic pressures and state-level behaviour, since the state acts through institutional, technical, and political capacities that convert infrastructural resources into instruments of influence. In practical terms, states capable of safeguarding their energy systems from terrorist attacks, cyber threats or geopolitical sabotage, such as the destruction of the Nord Stream gas pipelines, are better positioned to preserve energy stability and maintain relevance in foreign policy, as highlighted by Italian scholars Gerlini and Indeo. Their work emphasizes that critical energy infrastructure constitutes a strategic resource in itself and that its protection is "a strategic priority for preserving the energy security of the European Union," a conclusion whose logic extends beyond the EU to other international actors [16].

Critical energy infrastructure thus becomes a conduit of state power through which systemic impulses and constraints are translated into concrete foreign policy outcomes. Its protection and management operate on two levels: the external level, where great powers act through sanctions, investments, or market control, and the internal level, where institutional stability and effectiveness as well as leaders' threat perceptions are decisive. When these domestic factors are misaligned, states are prone to what neoclassical realist theory defines as underbalancing, that is, an insufficient response to threats arising from vulnerability and limited capacity to utilize available resources, even when objective risks are evident [12, p. 145–146].

3 The Structure of the New International Reality

The outbreak of war in Ukraine in 2022 following the Russian invasion marked the beginning of a profound transformation in the structure of the international system. Some authors, such as Richard Haass, argue that this process of systemic change had begun earlier. Despite its status as the world's most powerful state, the United States had significantly less global leverage than in the immediate post-Cold War era [17]. The onset of systemic changes can be traced to Russia's annexation of Crimea in 2014, when the U.S. administration of President Barack Obama declined to provide Ukraine with lethal military assistance in its conflict with Russia, thereby avoiding direct confrontation while implicitly recognizing Russia as a great power [18].

These structural changes are visible not only in the military and political domains, but have also significantly reshaped international economic relations and the ways in which states define their core interests. The long-standing order rooted in economic interdependence has entered a phase of rivalry in which energy security became one of the central elements of state strategy. The significance of Russia's annexation of Crimea lies in the fact that, from that point onward, the United States and the European Union sought jointly to constrain the use of energy resources as an instrument of political coercion [19, p. 92.]. The key U.S. interest in supporting global energy security varies across regions, but is fundamentally linked to the protection of critical infrastructure with a view to long-term stabilization, particularly in cases such as Ukraine and Taiwan [20]. Since the annexation of Crimea, U.S. sanctions targeting the Russian energy sector have aimed at preventing Moscow from exerting political leverage on other states through its energy potential [19, p. 92].

Some analyses underscore that following the imposition of Western sanctions, Russia employed its energy exports as an instrument of political pressure by reducing natural gas deliveries to Europe by around 80% compared with previous levels, thereby severely jeopardizing European energy stability. Although many European states succeeded in securing alternative suppliers, energy prices increased sharply, partly due to the U.S. decision in March 2022 to ban imports of Russian oil, liquefied natural gas, and coal. The efforts of the European Union and the United States to increase the production of renewable energy have generated a new dependency on China, which controls a substantial portion of the manufacturing base and supply chains in this sector. At the same time, the 2022 U.S. National Security Strategy described Europe as "a fundamental partner in addressing global challenges," while emphasizing that Europe's energy stability directly affects U.S. security, prompting Washington to support European efforts to secure long-term supply arrangements [21].

The United States and the European Union seek to prevent the use of energy resources (primarily by Russia and China, author's note) as instruments of geopolitical pressure, recognizing energy security as a key strategic priority [19, p. 92]. A clear trend has emerged whereby the United States plays a central role in European energy security, both by increasing exports of crude oil and liquefied natural gas to Europe and by providing political and diplomatic support for diversification initiatives, especially after 2022 [19, p. 88]. Analysis of U.S. LNG exports to Europe up to 2023 shows that the United States has become a dominant supplier of energy to European markets [19, p. 93]. The United States and the European Union established a Joint Task Force on Energy Security to secure stable LNG deliveries from the United States to Europe and to facilitate the transition to renewable energy sources. As a result, U.S. LNG exports to Europe in 2022 increased by 141% compared with the previous year [21, p. 18]. At the same time, Europe reduced imports of Russian energy due to Western sanctions, so that U.S. supplies largely offset Russia's diminished access to European markets [19, p. 90].

According to a comparative assessment of the infrastructural agendas of China and the European Union, "the relationship between connectivity and geopolitics has evolved through several phases, gradually transforming connectivity into a geopolitical instrument" [22, p. 139]. The study argues that "connectivity infrastructure, reflected in the expansion of transport and energy networks and the enhancement of digital infrastructure, is essential for implementing both the EU connectivity agenda and China's Belt and Road Initiative in the Western Balkans" [22, p. 140]. Regarding China's approach to critical energy infrastructure, the analysis suggests that while both European and Chinese economic initiatives rely on similar infrastructural components, they are driven by different political and economic logics. The EU strategy is oriented toward integration and alignment with the European regulatory framework, whereas China's approach is based on expanding political influence through economic leverage [22, p. 136]. A similar conclusion is reached in the study *China's Belt and Road Initiative and its impact on the independence of the European Union*, which warns that "when financial support is tied to taking control of key energy infrastructure, whether through ownership of construction companies or their local subsidiaries, there is a risk of creating undesired dependence on an external political power, in this case China". In a hypothetical perspective, such dynamics may enable external actors to exert significant influence over national legislation, shaping future investment environments and regulatory frameworks in line with their own interests. This could generate heightened risks, including market isolation and reduced competitiveness in the energy sector. This issue is especially relevant given that Chinese energy projects constitute an integral component of the "Belt and Road Initiative" in Europe [23].

Russian energy policy functions to a considerable extent as a foreign policy instrument used by the political leadership in Moscow to preserve influence in Europe and other regions. According to several U.S. authors, "Russia has politicized its energy exports, cutting gas deliveries to Europe by 80%" after the introduction of Western sanctions in 2022, thereby transforming Europe's energy dependence into a mechanism of pressure [21, p. 13–14]. They note that "prior to the military intervention in Ukraine, Russia supplied about 33% of European gas imports," while "some countries, such as Poland, Finland, Slovakia, and Germany, were almost entirely dependent on Russian gas" [21, p. 15]. Germany was expected "to increase its gas imports from Russia to 60% with the completion of the second direct pipeline in 2022," which would have enabled Russia "to manipulate energy flows and cause serious economic harm" to European economies. At the same time, according to IMF estimates, "with high gas prices, European governments had to allocate up to six percent of their national income" to mitigate the effects of the energy crisis [21, p. 17]. These findings indicate that Russia, rather than acting as a conventional market-based supplier, employs energy policy as a geopolitical instrument to preserve its influence in Europe and safeguard its national interests.

The U.S. presidential administration of Donald Trump, which took office in 2025, formulated a new energy policy aimed at redefining U.S. energy power in the global context. According to an analytical report by the Institut français des relations internationales (*IFRI*), one of Europe's leading research institutes for international relations and energy policy, the main U.S. objectives in the energy field are directed toward supporting fossil fuels, the nuclear industry, and the critical minerals sector. Implementation relies on presidential executive orders concerning oil and gas projects, the coal industry, plans to strengthen the U.S. nuclear sector, and measures to develop projects related to critical metals [24, p. 3]. President Trump signed six executive orders in which he accused the previous administration of Joseph Biden of poor energy decisions and declared a national energy emergency, aimed at broadening U.S. room for manoeuvre and facilitating rapid decision-making through executive authority, bypassing Congress. In this context, the "National Council for Energy Dominance" was established as a new institution tasked with increasing energy production and achieving U.S. energy dominance, serving as the central institutional mechanism for implementing the new U.S. energy doctrine [24, p. 8–10].

The development of critical energy infrastructure occupies a prominent place in this framework. Nuclear energy is expected to play a key role as a primary source in U.S. energy policy under the new presidential administration. At the same time, the United States intends to become "the largest producer and processor of critical minerals within non-fuel energy resources" [24, p. 16]. The most concrete foreign policy implications will arise in the field of U.S. oil and gas exports. Natural gas production has been expanding more rapidly than oil production; in May 2025, gas output rose by 14% relative to May 2024, while oil production in the same period increased by 2.7%. The U.S. Energy Information Administration (*EIA*) projects that production will grow to 16 billion cubic feet per day by 2026, up from 12 billion cubic feet per day in 2024. Despite relatively low export prices, increasing export capacity continues to incentivize U.S. production [24, p. 11].

According to Yuliia Malohulko's analysis published in *Foreign Affairs* magazine, the principal objective of U.S. efforts is to strengthen Europe's energy security by diversifying supply sources and reducing dependence on Russian energy. Through legislation such as the *Countering America's Adversaries Through Sanctions Act* (2017) and the *Protecting Europe's Energy Security Act* (2020), which encourage diversification and constrain Moscow's ability to use energy as a political tool, the United States has employed diplomatic instruments and financial assistance to help European partners counter Russian energy influence [19, p. 89]. At the same time, the United States has become one of Europe's leading suppliers of oil and liquefied natural gas. EIA data show that in 2023 the United States exported an average of 4.1 million barrels of oil per day, with Europe as the principal destination [19, p. 91]. This policy has facilitated a significant reduction in the share of Russian energy in European imports. Malohulko argues that the United States has played a key role in transforming Europe's energy architecture, with 2022 representing a turning point and the United States emerging as the largest supplier of oil and LNG to Europe by 2023, supported by diplomatic and legislative initiatives that closely integrate energy and U.S. foreign policy strategy [19, p. 93].

An analysis of the aforementioned reports suggests that U.S. external energy policy seeks to exploit the emerging gap in global energy demand created by the reduction of Russian supplies due to Western economic sanctions. The new energy policy of Donald Trump signals a departure from the strategy of the previous administration, as it is oriented toward establishing U.S. energy dominance based on nuclear energy and the securing of critical minerals supplies. This policy emphasizes unilateralism, continuing the trajectory begun during Trump's first term (2017–2021), characterized by withdrawal from multilateral frameworks and the prioritization of U.S. national interests, including in the field of energy infrastructure, over earlier patterns of multilateral coordination [25].

4 Regional Framework and the Shaping of the Energy Policy of Southeast Europe

The case studies in this section focus on Serbia, Hungary, and Romania due to their specific domestic characteristics. Following the logic of neoclassical realism and its emphasis on intervening variables under systemic pressures, Southeast Europe can be understood as a complex region of overlapping influences. Systemic pressures generated by great power policies interact with domestic structural features and political decisions, producing distinct patterns of national energy policies.

This regional politics is characterized by simultaneous alignment with European Union norms and regulations, the enduring presence of Russian energy supplies as a long-term dependency, China's economic presence, and the U.S. external energy policy aimed at reducing Russian energy influence. This constellation generates dilemmas for states in protecting their critical energy infrastructure. They face the challenge of preserving autonomy amid overlapping market-regulatory standards and diversification efforts. Viewed through the lens of neoclassical realism, this dynamic requires careful analysis of how domestic characteristics shape responses to common external factors.

Many Southeast European states relied for decades on imports of Russian gas as a primary energy source, establishing a distinctive systemic feature in regional energy policy. Long before 2022 and stimulated by earlier Russo-Ukrainian disputes, these states began developing new terminals for energy imports and cross-border interconnections. Europe, and particularly Southeast Europe, now faces dual challenges: the need to diversify energy sources after Russia curtailed gas deliveries to much of Europe in 2022, and the broader imperative to reduce dependence on fossil fuels in order to tackle climate change. The combined challenge of resolving the consequences of the war in Ukraine and conducting the energy transition includes the construction of gas infrastructure to allow alternative suppliers to reach markets that were previously dependent on Russian gas [26, p. 5].

Since 2022, Europe has sought to reduce its dependence on Russian energy, especially gas. Russia responded to economic and financial sanctions by cutting gas supplies to several European states. Even countries that maintained relatively regular diplomatic relations with Russia, such as EU member Hungary and EU candidate Serbia, were significantly affected, since Ukraine's response to the invasion included halting Russian gas flows through key transit pipelines on its territory [26, p. 9].

At the same time, Russia has sought to diversify its export routes to Europe. Gazprom has maintained certain volumes of deliveries to Hungary and Serbia, and, by 2025, the only operational route for Russian gas pipeline exports to Europe was the Turkish Stream pipeline under the Black Sea [26, p. 14]. Russia may continue to use energy supplies as a geopolitical tool in Southeast Europe. Bulgaria, as a key transit country, plays a central role in this regard: a dispute between Bulgaria and Gazprom over contractual terms could trigger a serious energy security crisis, particularly affecting states dependent on Russian gas and lacking physical access to alternative sources. In the event of a transit disruption, Serbia, together with North Macedonia and Bosnia and Herzegovina, would be among the most severely affected, as all still rely on Russian gas imports [26, p. 20].

4.1 Serbia—Structural Dependence

According to the 2024 draft "Development Strategy of the Energy Sector of the Republic of Serbia," available on the official website of the Ministry of Mining and Energy, Serbia's energy sector is undergoing structural changes driven by "global and national circumstances" [27]. An analysis of Serbia's energy policy from late 2023 underscores that the country's energy sector faces a complex web of challenges which remained unresolved and had significantly deteriorated by the end of 2025, under the impact of U.S. sanctions on Naftna industrija Srbije (NIS). "Insufficient investment and inadequate management" of domestic energy systems coincided with the global energy crisis. The war in Ukraine further heightened concerns by exposing Serbia's vulnerability, as its energy supply depends heavily on companies under Russian ownership. These developments forced Serbia to incur unplanned expenditures in order to secure energy reserves and reopened fundamental questions about its strategic orientation and energy transition. Confronted with the energy crisis, Serbia has recognized the need to diversify gas suppliers and advance the shift toward renewable energy sources [28, p. 4].

Serbia's complete dependence on Russian gas has turned energy policy into a crucial national issue. Gas import and storage operations are managed by companies under Russian ownership, while NIS, majority-owned by Gazprom, is the only producer of crude oil and natural gas, the sole operator of an oil refinery, and the owner of the largest fuel retail network in the country [28, p. 7]. As the EU negotiated new sanctions against companies under majority Russian ownership and called on Serbia to align with them, the Serbian leadership decided to formally condemn the attack on Ukraine, but opted to maintain relatively stable diplomatic relations with Russia and refrain from joining sanctions targeting either Russian individuals or Russian-owned companies [28, p. 5].

In addition to the systemic shock generated by the war in Ukraine, EU policy functions both as a systemic stimulus and as a source of constraint on Serbia's energy strategy. Serbia has pledged, together with the EU, to work toward making Europe a "carbon–neutral" continent by 2050, with a commitment to reducing greenhouse gas emissions. As a *Contracting Party to the Energy Community Treaty*, whose objective is to extend the EU internal energy market to non-member states, Serbia is obliged to implement relevant EU legislation in the fields of energy, environmental protection, renewable energy, and energy efficiency. Through the *Stabilisation and Association Agreement*, Serbia has reaffirmed its commitment to the Energy Community agenda in energy and environmental policies [27].

From the systemic level, the EU's policy toward Serbia is articulated as a sustained expectation that Belgrade will pursue decarbonization and diversification, reflecting continued EU efforts to steer Serbia's energy policy. The EU is expected to "continue" providing Serbia with the conditions necessary to achieve security of gas supply and diversification through more and higher-capacity interconnectors, expanded underground gas storage, and joint gas procurement schemes [28, p. 3]. Serbia's institutional commitments as a party to the *Energy Community Treaty* function as an intervening variable between systemic pressures and domestic policy-making. They link the EU's energy transition agenda with Serbia's national energy strategy and influence Belgrade's choice of energy security instruments. In this sense, the Treaty, as an institutional framework between the EU and Southeast European states, directs EU influence on Serbia's strategic planning [27].

A particularly significant source of systemic pressure on Serbia's critical energy infrastructure is the economic sanctions imposed by the administration of U.S. president Donald Trump on NIS. After several postponements, these sanctions took effect in October 2025. Although formally targeting the Russian Federation and Russian companies operating in Serbia, given that NIS is majority-owned by Russian capital, Serbia has suffered substantial consequences despite the efforts of its leadership to resolve the situation [29]. The reactions of Serbian political authorities to the U.S. sanctions on Naftna industrija Srbije (NIS) show how external constraints acquire domestic significance only once they pass through the state's institutional capacities and elite perceptions of risk. When the Serbian government announced in February 2025 that NIS had officially requested a postponement of the sanctions, it stated that "the immediate imposition of sanctions would impair the company's ability to supply the citizens of Serbia with oil and oil derivatives" [30]. This formulation reveals that Serbian decision-makers interpreted the sanctions as a direct challenge to the functioning of essential energy infrastructure and to the stability of domestic supply. In the same month, Minister of Mining and Energy Dubravka Đedović Handanović emphasized that the U.S. administration "understands the importance of NIS for the Serbian economy and citizens, as well as for regional energy stability" [31]. Her statement illustrates the effort to soften systemic pressure through diplomatic communication, a pattern characteristic of smaller states balancing external constraints with domestic expectations.

By August 2025, the Serbian government publicly declared that "we have managed to preserve the country's energy stability thanks to the understanding shown by the U.S. administration" during the postponement of sanctions [32]. In the logic of neoclassical realism, this is precisely the point at which systemic pressure becomes politically meaningful because decision-makers evaluate its consequences for domestic welfare, credibility and social expectations. This view reflects the way political elites assess the relationship between external demands and internal stability. In November 2025, the situation intensified. The Serbian government announced that the United States had delivered "a very difficult message," requesting "the complete exit of Russian ownership from NIS," while stressing that "the functioning of NIS is of vital importance to the economy and citizens" [33]. Operational consequences of the sanctions were documented by Reuters. In late November 2025, it reported that the Pančevo refinery "will be idle until December 2 as it awaits a U.S. sanctions waiver" and that "banks will start to refuse processing payments for NIS due to the sanctions" which resulted in a temporary halt in refinery operations [34]. At the same time, the government attempted to reassure the public. According to Reuters, the government stated that "Serbia has sufficient fuel reserves" even as the refinery faced a potential sanctions-driven shutdown [35], as an attempt of the political elites to maintain between international obligations and domestic demand for continuity and predictability. President Aleksandar Vučić also noted publicly that "Russia is seeking a partner for the U.S.-sanctioned NIS," indicating that possible ownership restructuring was being discussed [36]. This acknowledgement reveals the search for additional maneuvering space within the constraints of the sanctions regime.

At the regional level Serbia continued work on gas interconnections necessary for diversification of supply. According to the Ministry of Mining and Energy, the Serbia–Bulgaria gas interconnector "ensures the diversification of supply sources, greater energy security and Serbia's important position on the gas map of Europe". The Serbia–Bulgaria gas interconnector "marks the first time Serbia has diversified its sources of natural gas supply, ensuring greater energy security for citizens and the economy," and is described as "a crucial international project" that will supply gas to Serbia and Central-Eastern Europe "from Azerbaijan and other sources, including the LNG terminal in Alexandroupolis, Greece". The Minister underlined that "the countries that have proven most resilient in the energy crisis and have weathered it more effectively are precisely those with diverse sources of supply" and noted that Serbia has already signed an interconnection agreement with Bulgartransgas and a commercial agreement to obtain "up to 400 million cubic meters of gas per year until 2026" from Azerbaijan, with the possibility that "from 2027, those quantities can increase up to three times" [37].

Taken together, the official statements of Serbian political leaders, the operational disruptions recorded by international agencies, and the institutional measures adopted in late 2025 indicate that the sanctions on NIS operated simultaneously as a source of external pressure and as a catalyst for domestic policy adjustments. It is precisely in this intersection of systemic constraint and internal political response that the dynamics described by neoclassical realism become visible. External pressures shape policy outcomes only through the perceptions, decisions, and capacities of domestic actors who manage the consequences for state stability.

Gazprom has retained its wholesale monopoly and a clear dominance in the retail market, leaving Serbia's dependence on Russian gas essentially unchanged since 2008 despite numerous challenges. The structural legacy of the Serbian-Russian arrangements from that period has ensured Serbia's long-term dependence, which persists in 2025 and limits the country's strategic room for manoeuvre. Lasting dependence on gas agreements with Russia has materialized as a pronounced external constraint in Serbia's foreign policy. At the same time, domestic infrastructural constraints operate as intervening variables that translate external pressures into specific national priorities. Until the construction of the interconnector with Bulgaria, which is a precondition for diversification, Serbia's internal limitations impede the development of critical energy infrastructure forcing the country to continue purchasing Russian gas [28, p. 11].

One explanation for Serbia's close linkage to Russian ownership stakes within its critical energy infrastructure, combined with parallel alignment with EU energy policy, can be found in the intervening variable of strategic culture as conceptualized in neoclassical realism. Taliaferro, Ripsman, and Lobell explicitly treat strategic culture as one of the four intervening variables that shape how states respond to systemic pressures and inform their foreign policy behaviour [12, p. 59]. The identity of the Serbian nation, rooted in the long-term preservation of statehood, is one of the foundations of contemporary Serbian strategic culture. Serbian geopolitical scholar Milomir Stepić describes Serbian strategic culture as a system of "comprehensive understanding, reflection and action in the military-security domain aimed at achieving national interests based on the Serbian identity code" [38, p. 17]. Within this framework, identity is understood as an intervening variable in the formation of Serbia's energy strategy. Geography, history, and religion are key elements shaping strategic culture in matters of security [38, p. 18]. Since these elements are central to Serbian perceptions of energy security and aspirations to strategic autonomy, Serbia is inclined neither to complete energy dependence on Russia nor to full integration into Western energy flows.

4.2 Romania—Euro-Atlantic Alignment

Romania, a strategically important state on NATO's eastern flank and within the European Union, faces a dual set of external influences and pressures in the new energy reality shaped by the war in Ukraine. On the one hand, Romania is exposed to structural pressures stemming from Russian revisionism in the Black Sea basin; on the other hand, systemic expectations emphasize Romania's Euro-Atlantic commitment and deepening transatlantic cooperation. Over the past two decades, Romania's foreign and security policy has thus been shaped by both structural and domestic variables [39, p. 1].

In the Black Sea region, the main structural driver is the power dynamics created by the strategic behaviour of two great powers. Since 2008, Russia has sought regional hegemony, gradually undermining the post-Soviet order and expanding its influence in the Black Sea and its surrounding neighbourhood. Concurrently, the United States, as the post-Cold War global hegemon, has increasingly viewed the Black Sea as a critical geostrategic pivot in the wider Euro-Atlantic security architecture [39, p. 13]. This is crucial for Romania's foreign and energy policy, since Romania's NATO accession in 2004, the Black Sea region has occupied a central place in nearly all major Romanian strategic documents [40, p. 39].

A core assumption of realist theory, including neoclassical realism, is that states operate in an anarchic international system where the power of great powers defines the environment for smaller actors. Under persistent uncertainty in recent years, Romania has oriented itself toward the hegemon capable of guaranteeing the stability of energy flows and security protection. Limited domestic resources that constrain full energy autonomy, along with high sensitivity to Russian policy in the gas and oil sectors, have reinforced dependence on Western integration. Within these systemic conditions, Romania's regional strategy has been twofold: facilitating its integration into the Euro-Atlantic security architecture and constraining Russian revisionism [39, p. 13]. Romanian foreign policy thus reflects an understanding of systemic pressures that shape the direction of its energy policy.

In balancing Russian influence in the energy domain, Romania's leadership has connected multilateral energy and geopolitical initiatives, such as participation in the *Three Seas Initiative*, alongside energy diversification and regional infrastructure projects aimed at reducing dependence on Russian imports. In this manner, Romania employs the multilateral energy framework as an instrument of strategic balancing against external threats. Early conceptions of the *Three Seas Initiative* emphasized infrastructure for "highways, high-speed railways and gas pipelines in order to bypass Ukraine," while subsequent interpretations, as highlighted by Alina Inayeh from German Marshall Fund, have linked its institutional design and infrastructural connectivity with a broader geopolitical framework [40, p. 52].

Following the logic of neoclassical realism, external influences are transmitted through intervening variables at the unit level, including leaders' beliefs and perceptions, strategic culture, domestic political dynamics, public opinion and institutions. While the broad orientation of foreign and security policy is constrained by systemic conditions, specific policy choices are shaped by these domestic factors [8, p. 147–152]. In Romania's case, a key intervening variable in energy security is the leadership's threat perception. Due to its geographic location and proximity to Russia, which political elites and most of the public view as a direct threat, Romania underscores that national security can be assured only through a pro-Western orientation and a strong strategic partnership with the United States and Western Europe [39, p. 13].

Another intervening variable shaping Romania's foreign and energy policy is strategic culture grounded in institutional continuity established during the country's post-socialist transition. As Zoltán Lechner notes, "during the first fifteen years after the Cold War, Romania paid little special attention to the Black Sea region [...] the only priority of Romanian foreign policy was participation in the liberal international order and accession to Euro-Atlantic institutions" [39, p. 3]. This early choice at the beginning of the twenty-first century has had a lasting impact on Romania's political culture and energy strategy, anchoring it in EU and NATO integration as the primary guarantees of security.

4.3 Hungary—Pragmatic Balancing

Hungary represents a typical example of a state whose energy system is shaped by the interaction of systemic pressures and domestic constraints. As a landlocked country with limited resources, it faces major challenges in securing stable energy supplies, which necessitate protecting and developing extensive gas, oil and electricity interconnectors with neighbouring states [41, p. 871]. According to IMF data, Hungary remains "heavily dependent" on imports of Russian energy: in 2024, approximately 74% of its imported gas and 86% of its imported oil came from Russia. Beyond high import dependence, energy flows into Hungary rely on several critical transit routes, particularly the Turkish Stream pipeline through Serbia and the Adriatic pipeline through Croatia, amplifying Hungary's vulnerability to systemic disruptions [42]. Furthermore, an analysis of Hungary's ongoing energy transition emphasizes that systemic impulses stemming from the war in Ukraine and from the EU decarbonization agenda have reshaped Hungary's approach to energy policy [41, p. 871].

In addition to geographical and resource-based factors, Hungary's internal energy sector structure, like systemic conditions, constrains its level of energy autonomy, especially in segments characterized by pronounced deficits. National data indicate strong dependence on foreign markets, most clearly in the procurement of hydrocarbons, with import dependence at 85%. This vulnerable domestic economic structure magnifies systemic pressures and narrows the space for independent action. Rather than reducing dependence, Hungary has been compelled to react pragmatically, managing energy risks through selective measures. This is confirmed by assessments showing that Hungary has focused on diversifying gas sources both by building physical infrastructure and by supporting a competitive wholesale market [41, p 872].

Hungary does not operate in isolation from the international environment. A report by the European Bank for Reconstruction and Development notes that new Hungarian "contracts with Turkey and Azerbaijan are expected to help ensure energy security despite potential disruptions in Russian gas transit through Ukraine" [43]. These agreements illustrate a diplomatic and economic balancing strategy whereby Hungary seeks to maintain stable relations with existing partners while responding to pressures from international institutions and markets. In this sense, Hungary employs energy policy as an instrument of strategic balancing, a pattern consistent with realist interpretations of smaller states' behaviour in a complex international system.

Hungary's national strategy is built around the so-called energy trilemma, which involves balancing energy access and affordability, environmental objectives and security. This framework guides its approach to critical energy infrastructure. Within this triangle, the essence of Hungary's balancing strategy between international norms, domestic interests and limited material capacities becomes evident, as the country continually adapts to changing structural conditions in the international system [41, p. 871].

Finally, the institutional framework of Hungary's energy policy clearly reflects an awareness of structural vulnerability and the need to build resilience amid conditions of high import dependence. The National Energy and Climate Plan (2024) emphasizes that "security reserve building, storage and stockpiling, source diversification and related route diversification options, and a reduction in the share of imports are of greater importance for all primary and secondary energy sources characterized by high import exposure. Accordingly, the aim is to reduce domestic consumption and, where possible, increase domestic production, expand the use of alternative, domestic energy sources and, finally, achieve a more balanced technological mix based on multiple solutions". This orientation underscores the role of institutionalized policy as an intervening variable that translates systemic constraints into concrete measures to strengthen critical energy infrastructure [44].

5 Conclusion

Starting from the definition of critical energy infrastructure and its place in contemporary international relations, this paper demonstrates that energy security is shaped through the interplay of material resources, infrastructural capacities and institutional frameworks that are directly linked to states' foreign policy power. The reliability of energy supply and control over energy flows have emerged as central issues, particularly in Southeast Europe, where Euro-Atlantic and Eurasian energy networks intersect.

At the theoretical level, the paper applies neoclassical realism to link the anarchic structure of the international system and the distribution of power with domestic intervening variables that guide specific foreign policy and energy choices. Neoclassical realism makes it possible to explain why states operating under similar structural conditions do not develop critical energy infrastructure in identical ways. The analysis highlights institutional capacities, strategic culture, and the internal structure of the energy sector as key mechanisms through which systemic pressures are translated into concrete infrastructural arrangements and foreign policy strategies.

The paper emphasized that developments associated with the Russian-Ukrainian conflict and the broader transformation of the post-Cold War order have intensified the use of energy resources as instruments of pressure and reshaped global energy flows. At the same time, efforts by the United States and EU to limit Russian influence in the energy sector have simultaneously generated new dependencies. Under these conditions, critical energy infrastructure remains one of the key domains through which the ongoing structural reconfiguration of the international system is unfolding.

Southeast Europe emerges as a region where systemic pressures from major powers and competition among international initiatives intersect directly with domestic institutional and infrastructural specificities. The analysis demonstrates that states in the region operate under the dual imperative of diversifying supplies following the reduction of Russian deliveries and advancing the energy transition. In such an environment, control over critical energy infrastructure becomes essential for preserving autonomy, yet the scope for action remains constrained by inherited arrangements, international commitments, and domestic limitations.

The cases of Serbia, Romania, and Hungary confirm the paper's central hypothesis. Serbia exhibits deep-seated structural dependence on Russian energy flows and ownership structures while remaining embedded within EU-related institutional frameworks and subject to U.S. sanctions targeting a key component of its energy infrastructure. This configuration generates a distinctive pattern in which systemic pressures intersect with domestic infrastructural constraints and a strategic culture oriented toward preserving flexibility between Eastern and Western alignments. Romania, by contrast, illustrates a state whose perception of threat from Russia and early commitment to Euro-Atlantic integration have anchored its energy infrastructure within a broader strategy to consolidate ties with NATO and the EU while limiting Russian influence in the Black Sea region. Hungary, characterized by high import dependence and reliance on narrow transit corridors, has pursued a pragmatic balancing strategy—maintaining cooperative relations with Russian suppliers while simultaneously institutionalizing alternative frameworks designed to enhance resilience amid structural vulnerability.

The paper concludes that the formation of critical energy infrastructure in the evolving international environment can be understood only through the simultaneous consideration of three analytical levels: the structure of the international system, regional dynamics and domestic intervening variables. While systemic pressures establish the overarching framework, the ultimate configuration of energy policies in Serbia, Romania and Hungary is determined by how domestic institutions and strategic cultures interpret, mediate and translate these external forces into national strategies.

References

1. Dyer, H., Trombetta, M.J.: The concept of energy security: broadening, deepening. In: Dyer, H., Trombetta, M.J. (eds.) International Handbook of Energy Security, pp. 3–15. Edward Elgar, Cheltenham Uk & Northampton
2. Klare, M.T.: Energy security. In: Williams, P.D. (ed.) Security Studies, An Introduction, pp. 602–618. Službeni glasnik, Beograd
3. Council of the European Union: COUNCIL DIRECTIVE 2008/114/EC of 8 December 2008 on the identification and designation of European critical infrastructures and the assessment of the need to improve their protection, 23. 12. 2008. Off. J. Eur. Union (2008). Available at: https://eur-lex.europa.eu/legal-content/EN/TXT/PDF/?uri=CELEX:32008L0114. Last accessed 9 November 2025
4. Mićović, D.M.: Specifičnosti kritične infrastrukture u Republici Srbiji. Kriminalističko-policijski Univerzitet, Beograd (2020)
5. Youngs, R.: Energy Security, Europe's New Foreign Policy Challenge. Routledge, London and New York (2009)
6. Council of the European Union: Directive (EU) 2022/2557 of the European Parliament and of the Council of 14 December 2022 on the resilience of critical entities and repealing Council Directive 2008/114/EC, 27. 12. 2022. Off. J. Eur. Union (2008). Available at: https://eur-lex.europa.eu/legal-content/EN/TXT/PDF/?uri=CELEX:32022L2557. Last accessed 9 November 2025
7. Hungarian Ministry of Innovation and Technology: National Energy and Climate Plan (2022). Available at: https://energy.ec.europa.eu/system/files/2022-08/hu_final_necp_main_en.pdf. Last accessed 9 November 2025
8. Rose, G.: Neoclassical realism and theories of foreign policy. World Politics **51**(1), 144–172 (1998)
9. Waltz, K.N.: Theory of International Politics. Addison-Wesley Publishing Company, Reading, Massachusetts (1979)
10. Trapara, V.: Neoklasični realizam – realizam za 21. vek. Međunarodni problemi **69**(2–3), 227–246 (2017)
11. Ashford, E.: First Among Equals, U.S. Foreign Policy in a Multipolar World. Yale University Press, New Haven & London (2025)
12. Ripsman, N.M., Taliaferro, J.W., Lobell, S.E.: Neoclassical Realist Theory of International Politics. Oxford University Press, New York (2016)
13. Česnakas, G.: Energy resources in foreign policy: a theoretical approach. Baltic J. Law Politics **3**(1), 30–52 (2010)

14. Kuzemko, C., Blondeel, M., Bradshaw, M., Bridge, G., Faigen, E., Fletcher, L.: Rethinking energy geopolitics: towards a geopolitical economy of global energy transformation. Geopolitics **30**(2), 1–35 (2024)
15. Bakonyi, J., Darwich,: Infrastructures and international relations: a critical reflection on materials and mobilities. Int. Stud. Rev. **26**(4), 1–26 (2024)
16. Gerlini, M., Indeo, F.: Critical energy infrastructures: geopolitical vulnerabilities and strategies of securitization. In: EasyChair Preprint, p. 10622 (2023). Available at: https://easychair.org/publications/preprint/p1tGd/open. Last accessed 9 November 2025
17. Haass, R.: A World in Disarray. Penguin Press, New York, American Foreign Policy and the Crisis of the Old Order (2017)
18. Kraktus, E.D.: Relations between the United States and the Russian Federation during Barack Obama's second presidential term: a case study of the Ukrainian crisis (2013–2017). Doctoral dissertation, Faculty of Political Science, Belgrade (2025)
19. Malohulko, Y.: Role of the USA in supporting energy security in Europe. Foreign Aff. **34**(3), 88–98 (2024)
20. Auth, K., Moss, T.: U.S. Energy Security Compacts**,** A Bipartisan Blueprint to Reinvigorate U.S. Influence Through Energy Investment. Energy For Growth Hub (2025). Available at: https://energyforgrowth.org/article/u-s-energy-security-compacts/. Last accessed 9 November 2025
21. Willis, S.M., Rizzuto, A.P., Ben-Itzhak, S.: US and European energy security amid great-power competition. J. Indo-Pac. Aff. **6**, 13–26 (2023)
22. Stekić, N., Nikolić, J.: Conflicting visions of connectivity: Eu's And China's infrastructural agendas for the Western Balkans. Natl. Interest **50**(1), 133–155 (2025)
23. Farah, P.D., Zoppolato, D.G., Ötvös, T.: China's belt and road initiative and its impact on the energy independence of the European union. Asian J. Law Soc. **12**, 274–303 (2025)
24. Michel, T.: The New US Energy Policy, Energy Dominance or Fallback? IFRI Studies, Center for Energy and Climate. IFRI, Paris (2025)
25. Restad, H.E.: What makes America great? Donald Trump, national identity, and U.S. foreign policy. Glob. Aff. **6**(1), 1–16 (2020)
26. Stambolis, C., Roberts, J., Mezartasoglou, D.: Fresh Challenges For Energy Security In Se Europe. Institute of Energy for South East Europe, Athens, Greece (2025)
27. Ministry of Mining and Energy of the Republic of Serbia: Energy Sector Development Strategy of the Republic of Serbia up to 2040 with Projections up to 2050 (Draft) (2024). Available at: https://www.mre.gov.rs/extfile/sr/5928/Draft%20-%20Energy%20Strategy%2015072024.pdf. Last accessed 9 November 2025
28. Ilić, V.: Serbia's energy dilemma: navigating geopolitical tensions, Russian ownership, and the path to renewable transition. BiEPAG (The Balkans in Europe Policy Advisory Group), 1–21 (2023)
29. Simić, D.R., Živojinović, D.: Pet znakova neuzvraćene Trampove ljubavi. Vreme, **1815**, 11–13 (2025)
30. The Ministry of Mining and Energy of the Republic of Serbia: Official request to US for postponement of sanctions on NIS (2025). Available at: https://www.srbija.gov.rs/vest/en/242590/official-request-to-us-for-postponement-of-sanctions-on-nis.php. Last accessed 29 November 2025
31. The Ministry of Mining and Energy of the Republic of Serbia: US sanctions on oil company NIS postponed (2025). Available at: https://www.srbija.gov.rs/vest/en/244255/us-sanctions-on-oil-company-nis-postponed.php. Accessed 29 November 2025
32. The Ministry of Mining and Energy of the Republic of Serbia: US sanctions on NIS postponed anew (2025). Available at: https://www.srbija.gov.rs/vest/en/256273/us-sanctions-on-nis-postponed-anew.php. Accessed 29 November 2025
33. The Ministry of Mining and Energy of the Republic of Serbia: United States sends request for Russian ownership to exit NIS (2025). Available at: https://www.srbija.gov.rs/vest/en/262582/united-states-sends-request-for-russian-ownership-to-exit-nis.php. Last accessed 29 November 2025

34. Reuters: Serbia's Russia-owned oil refinery faces shutdown as US sanctions bite, president says (2025). Available at: https://www.reuters.com/business/energy/serbias-nis-oil-refinery-halts-operations-us-sanctions-bite-report-2025-11-25/. Accessed 29 November 2025
35. Reuters: Serbia says it has enough fuel reserves as key refinery faces sanctions shutdown (2025). Available at: https://www.reuters.com/business/energy/serbia-says-it-has-sufficient-fuel-reserves-key-refinery-faces-sanctions-2025-11-24/. Last accessed 29 November 2025
36. Reuters: Serbia's president says Russia seeking partner for US-sanctioned NIS oil firm (2025). Available at: https://www.reuters.com/business/energy/serbias-president-says-russia-seeking-partner-us-sanctioned-nis-oil-firm-2025-11-05/. Last accessed 29 November 2025.
37. The Ministry of Mining and Energy of the Republic of Serbia: The Serbia-Bulgaria Gas Interconnector inaugurated today, ensuring the diversification of supply sources, greater energy security and Serbia's important position on the gas map of Europe (2025). Available at: https://www.mre.gov.rs/vest/en/339/the-serbia-bulgaria-gas-interconnector-inagurated-today-ensuring-the-diversification-of-supply-sources-greater-energy-security-and-serbias-important-position-on-the-gas-map-of-europe.php. Last accessed 29 November 2025
38. Blagojević, V., Gajić, A., Subotić, M., Igrutinović, M.: Savremena srpska strateška kultura. Institut za evropske studije, Beograd (2025)
39. Lechner, Z.: Anchoring NATO's eastern flank: a neoclassical realist analysis of Romania's Black Sea policy and its strategic interaction with Turkey. Secur. Defence Quart. **51**(3), 1–20 (2025)
40. Dumitrescu, L.: The three seas initiative and Romania's grand behaviour in the black sea area: change and continuity. Cent. Eur. J. Int. Secur. Stud. **18**(3), 39–60 (2024)
41. Gulyás, T., Palkovicsb, L., Gondola, C.: Analysis of potential energy transition schemes in Hungary: from natural gas to electricity. Chem. Eng. Trans. **114**, 871–876 (2024)
42. Malak, M., Panton, A., Rojas-Romagosa, H., Weller, A.: Promoting Energy Security in Hungary: A Model-based Analysis. International Monetary Fund, Washington, D.C. (2025)
43. European Bank for Reconstruction and Development (2024). Transition Report 2024–25: Navigating Industrial Policy. Country Assessments, Hungary. Available at: https://www.ebrd.com/content/dam/ebrd_dxp/assets/pdfs/office-of-the-chief-economist/transition-report-archive/transition-report-2024/country-assessments-2023-24/central-europe-and-the-baltic-states/Transition-Report-2024-25-Hungary.pdf. Last accessed 9 November 2025
44. Government of Hungary (2024). National Energy and Climate Plan (NECP) 2021–2030. Updated version. Available at: https://commission.europa.eu/document/download/0a2953f8-5789-4f6f-9714-03df3d4cbbab_en?filename=HU_FINAL%20UPDATED%20NECP%202021-2030%20(English).pdf. Last accessed 9 November 2025

Energy Relations of the European Union and the United States of America During the Second Trump Administration: New Energy (Inter)Dependence?

Olga Stevanović

Abstract This paper examines the subject of EU-US energy relations during the second Trump administration, with a particular focus on understanding their nature. A reflection of a broader transatlantic alignment in the energy sector should have contributed to a less conflict-prone environment in the context of the regional energy crisis. Additionally, friendshoring is usually expected to grant more autonomous foreign energy policy, in accordance with the de-risking energy strategy. Nevertheless, as we will argue in this paper, the relations seem to follow an old logic, rendering characteristic asymmetric energy (inter)dependence, with potential energy influence for the exporter, as well as competitive and/or conflictual behaviour. The classification of energy relations in this paper is based on the neorealist approach to energy relations, including interdependence. Accordingly, the supply security of nations without indigenous energy resources is critical for their national security, which results in the EU resorting to familiar patterns of achieving energy security, albeit with different partners. In addition, the new demand security paradigm in the US, coupled with the Trump administration's proactive foreign energy policy, contributes to the realization of a more traditional energy interdependence with the EU, with potential for using the energy power in achieving other foreign policy goals. The paper employs qualitative methodology, primarily a case study but entails both quantitative and qualitative within-case evidence.

Keywords Energy relations · Energy interdependence · Energy security · Supply security · Energy transition · Foreign policy · United States · EU

A previous version of this paper was presented at the *Critical Infrastructure: Risk and Crisis Management in the New Reality* conference, which took place in Belgrade, October 16–17, 2025, under the title "The Growing EU-US Energy (Inter)Dependence in the Context of the Energy Transition Crisis: Case Study of the Second Trump Administration". This paper was written within the research activity of the Institute for Political Studies, financed by the Ministry of Science, Innovation, and Technological Development of the Republic of Serbia.

O. Stevanović (✉)
Institute for Political Studies, Belgrade, Serbia
e-mail: olga.stevanovic@ips.ac.rs

Z. Keković et al. (eds.), *Critical Infrastructure: Risk and Crisis Management in the New Reality*, Advanced Sciences and Technologies for Security Applications,
https://doi.org/10.1007/978-3-032-20948-1_3

1 Introduction

Several energy relations emerged in the context of the regional energy crisis in Europe following the onset of the Russia-Ukraine war in 2022. Over the last three years, not only have the EU's energy trade patterns undergone significant changes, but there has also been a transformation in energy partners, the development of new energy infrastructure projects, and the institutionalization of energy interactions. Some of the nascent and growing energy relations reflect ideological alignment, sometimes referred to as "friendshoring" in literature, which, theoretically, should contribute to a less conflict-prone environment and more autonomy in energy. Friendshoring can be defined as "the rerouting of supply chains to countries perceived as politically and economically safe, or low-risk, to avoid disruption to the flow of business" [1]. It was introduced by the Secretary of the Treasury in the Biden administration, Janet Yellen, in 2022 [2], when she said "rather than being highly reliant on countries where we have geopolitical tensions and can't count on ongoing, reliable supplies, we need to really diversify our group of suppliers. And yes, friend-shoring means—and you've seen this in action—that we have a group of countries that have strong adherence to a set of norms and values about how to operate in the global economy and about how to run the global economic system, and we need to deepen our ties with those partners and to work together to make sure that we can supply our needs of critical materials" [3]. The term is usually associated with critical mineral strategies (e.g., [4, p. 2], [5, p. 1]); however, it can also refer to the oil and gas industry (e.g., [6]).

Considering the limited scope of this paper, we focused on one energy relation, which falls into the category of friendshoring based on "general trust", namely the US-EU energy relation.[1] In the aftermath of the Russia-Ukraine war and the EU's decision to phase out imports of Russian energy, the US has emerged as one of the EU's principal energy partners.[2] Since the initial decision, energy trade has surged, particularly in liquefied natural gas (LNG), as well as crude oil and petroleum products. It's worth noting that the US is currently among the primary suppliers of oil and LNG to Europe, with notable nuclear fuel exports [7] and coal shipments. During that time, energy trade was followed by the development of LNG infrastructure, especially import terminals within the EU.[3]

Recently, the energy relations between the US and the EU have come under scrutiny due to the Framework on an Agreement on Reciprocal, Fair, and Balanced Trade, particularly the provision on energy [8]. Under the key terms of this political agreement, the EU has agreed "[...] to procure US liquefied natural gas, oil, and

[1] Pesu and Sinkonnen differentiate between general and specific trust in the trans-Atlantic alliance, with special focus on NATO, and they point out that "Despite [...] periodic fluctuation in mistrust, several static elements maintain *general* trust within the trans-Atlantic alliance" [84, 214]).

[2] As stated on the EC website: "The U.S. has consistently been the EU's main LNG supplier since 2021, and played an important role in pursuit of the REPowerEU objective of fully phasing out Russian fossil fuels by 2027" [85].

[3] According to Gas Infrastructure Europe (GIE) data, the number of operational large-scale import terminals in the EU in 2024 was 33, while seven were under construction, and 16 were planned [38].

nuclear energy products with an expected offtake valued at $750 billion through 2028" [8]. The Agreement raises questions about both expected outcomes of friendshoring, as it reveals potentially conflictual elements of the relationship, and creates uncertainty regarding the EU's commitment to environmental protection. There are two points of contention regarding the US-EU trade deal in the context of energy: it may potentially compromise the EU's environmental policy goals, and it could further exacerbate the existing imbalance in energy trade, thereby creating leverage on the US's side. Accordingly, adverse reactions to the deal were reported in the EU public [9], and among certain members of the European Parliament who tried to initiate motions to redefine the energy provisions [10].

This points to the research question of the evolving nature of EU-US energy relations, and whether this relationship, based on friendshoring, is achieving the desired outcome of de-risking, or if it's exhibiting similar characteristics to the former energy (inter)dependencies. In other words, we ask whether this constellation of energy partners has a pacifying effect on their energy relation or if it produces a systemic outcome of an imbalanced energy relation. The central hypothesis of this paper is that the energy relations between the United States of America and the European Union can be categorized as asymmetric energy (inter)dependence, resulting in some level of instrumentalization of energy, despite the presumed broad ideological alignment between the two actors. Therefore, the objectives of the paper include classification and explanation of EU-US energy relations, with consideration of their implications on foreign policy goals. The classification and explanation are based on structural forces, i.e., different energy power of two actors, the EU's threats to supply security, and the US's opportunity for improving demand security, as well as their precedence over energy transition in the broader context of a regional energy crisis. Accordingly, they are considered within the neorealist approach to energy, including neorealist conceptualization of energy interdependence, as the central analytical framework used in this paper.

This paper will proceed as follows. We will review the existing literature regarding EU-US energy relations in the second section of the paper. The third section is dedicated to the conceptual and analytical framework of the paper, with special focus on the concept of energy interdependence and its operationalization in this paper. We will present empirical findings in line with the analytical framework in the fourth section of the paper, primarily using quantitative and qualitative indicators of EU-US energy interactions to classify and explain the relations accordingly. The fourth section will also encompass the empirical implications of such energy relations. Finally, the fifth section will entail our concluding remarks.

2 Literature Review

The research subject of this paper encompasses various strands of academic literature. First, this subject is situated within the research area of EU's energy relations after the Russia-Ukraine war. The impact of the Russia-Ukraine war on the EU's energy policy is a prominent area of research, particularly in terms of its implications for the energy transition. For instance, Kuzemko et al. [11] focus on the impacts of this war and the subsequent energy crisis on the EU's balancing of energy security, energy sustainability, energy equity, and social justice. They provide an essential analysis of the EU's energy policy as a response to energy supply securitization in the early days of the crisis [11]. The main conclusion of their analysis is that: "[...] responses to the current crisis represent a relative shift of political focus back to geopolitical supply security" [11, p. 6]. However, as they argue in the paper, "clean energy supply has emerged as the best solution to European energy security long-term", even though it created certain tensions with respect to affordability and justice [11, p. 6].

Francesco Albanese also writes about the impact of geopolitical risks, such as the Russia-Ukraine war, on sustainability goals, with special focus on parliamentary debates in the EU (see [12]). According to Albanese's model, geopolitical shocks, moderated by energy dependence, result either in reactive or stable legislative focus regarding renewables and sustainability [12, p. 4]. Results of his analysis of the parliamentary discourse in EU countries don't prove the hypothesis on "increased reliance on renewable energy technologies" [12, p. 4] in countries with high energy dependence during a crisis, although he doesn't rule out the possibility of changes in the framing of renewable energy [12, p. 6]. However, the author finds supportive evidence in his analysis regarding the second part of the hypothesis which states that: "Parliamentary engagement with sustainability issues tends to exhibit a more pronounced decline in countries with a higher energy dependency" [12, p. 7]. Similarly, Trushkina [13, p. 73] researches the impact of the 2022 energy crisis on the EU's "strategic calculus" between three imperatives—"short-term energy security", "long-term climate objectives", and "maintaining socio-political stability". Analyzing the progress of RePowerEU's targets in 2025, Trushkina [13, p. 79] concludes that the "EU succeeded in the immediate fossil-fuel-based components of its crisis response, but it is lagging in the more complex, long-term clean energy transition goals." Similarly to already mentioned inclusion of renewable energy into energy security, Trushkina [13, p. 82] notes "the unprecedented alignment of climate agenda and security agenda" in EU's energy policy, however she also identifies the danger of swapping the existing energy dependencies with dependencies on critical minerals essential for the green transition.[4]

This strand of literature identifies changes in the EU's energy policy goals, which could suggest an energy transition crisis, as they imply prioritizing the EU's supply security over energy transition, albeit in varying degrees, or at least an impact of supply security threats on the energy transition goals. There are also analyses with the opposite conclusions—for example, Steffen and Patt [14, pp. 4–5] explore public support for clean energy in Europe, in the context of the Russia-Ukraine war, with

[4] There are two schools of thought regarding the implications of energy transition on energy relations, or as Albanese [12, p. 2] puts it "'renewed conflict' and 'reduced conflict' camps". However, the belief that renewables will reproduce the existing patterns of interdependencies and conflicts in energy has become more salient recently (e.g., [86, p. 6], [87, p. 11], [88, p. 404]).

special focus on Switzerland, and conclude that there's higher support for phasing out fossil fuels and renewable energy after the war.

However, these articles primarily analyze the broader EU's energy policy as a response to the energy crisis, rather than specific energy relations. Even if they partially include energy relations, they are usually framed with reference to the EU-Russia energy interdependence (e.g., [11, 12, 15, 16], etc.), which is understandable considering that there are still remnants of this interdependence. However, it's important to analyze the evolving energy relations following the EU's decision to sever ties with Russia and how they relate to the EU's foreign energy policy, including the energy transition. This is still an underdeveloped area of research, although the idea is sometimes present in papers at a more general level. For instance, Jerzyniak's work is dedicated to the interconnection of energy dependence de-risking strategy and energy transition, as well as the possibility of creating a "new energy order" built around new energy partnerships [17, p. 3]. Shendrikova et al.'s [18] article on value compatibility in the EU's energy diplomacy is also an essential addition to this strand of literature. Their analysis of the EU's energy partnerships based on "like-mindedness" is invaluable for understanding the energy relation between the US and the EU, although the article has a broader scope than just the US [18, p. 2]. According to Shendrikova et al. [18, p. 2]: "In the energy supply, the strategy to cooperate with like-minded partners was translated into a drastic increase in the LNG imports from the US". The general idea behind this approach is "derisking through partner diversification" [18, p. 2], which implies avoiding interdependencies that can be instrumentalized or weaponized. They distinguish between three scenarios for the EU's partner diversification, normative-driven, pragmatic, and hybrid, and situate the US as a potential supplier within both normative and pragmatic approaches [18, p. 7].

We would also like to mention an important contribution to the research area of energy relations between the US and the EU, namely the 2025 research article authored by Michaela Hrabušajová and Matúš Mišík. Their article is dedicated to the analysis of the White House narratives on LNG exports to the EU from 2009 to 2024 to determine if the energy relation in the LNG sector was framed "as a market-driven business opportunity or as a geopolitical necessity for strengthening the US–EU alliance" [19, p. 1]. Hrabušajová and Mišík [19, p. 3] propose an analytical framework for explaining energy relations based on the elaboration of two common approaches to energy (realist and liberal). They provide a detailed description of the central tenets of both approaches, as well as the implications of each approach for energy relations. The results of the textual analysis of White House official statements on LNG, which they offer, show a prevalence of market-oriented narratives during three presidential terms [19, p. 5]. Proroković [20, p. 56] includes considerations of the Trump administration's approach to the Paris Climate Agreement, and to energy transition in general, in his analysis of the EU's energy security strategy. This analysis reveals significant divergence in the energy goals of the EU and the US, specifically regarding climate change and environmental protection [20, p. 60].

With this paper, we aim to contribute to the literature on energy relations that have been developing after the EU's decision to decouple from Russia. Namely, our goal is to build upon the research area of US-EU energy relations, using energy interdependence as an analytical framework. Moreover, we believe that this research offers a new perspective on energy interdependence between "friendly shores", with a consideration of whether it fosters the instrumentalization of energy interdependence or mitigates asymmetric interdependence.[5]

3 Conceptual and Analytical Framework

The primary research subject of this paper pertains to the nature of energy relations between the EU and the US in the first nine months of 2025, or the second administration of President Donald Trump. Considering that our general hypothesis revolves around proving that the energy relations between the two actors can be classified as energy (inter)dependence, it's important to define precisely what this concept entails. Therefore, the first part of this section is dedicated to explaining the conceptual framework behind the hypothesis, and the second part entails the operationalization of the concept.

Energy (inter)dependence is a commonly used term in this research area, although in practice it's usually associated with or equated to the quantitative indicator of net import dependency (e.g., [13, p. 76], [21]). For instance, Albanese [12, p. 1] defines energy dependence as "the extent to which a country relies on imported energy to meet its domestic needs". Recent literature encompasses both energy interdependence and energy dependence as key concepts, although the latter is more commonly discussed (e.g. [12, 22]) than the former (e.g., [19, 23]). Some authors distinguish between the terms (e.g. [23, p. 1]), while others use them interchangeably (e.g., [18, p. 5]). Some authors frame energy dependence as both the dependence on fossil fuels in general, within the context of the energy transition, and dependence on foreign sources (e.g., [21, 24]). Recent literature has also identified energy dependence associated with resources other than fossil fuels, primarily critical minerals (e.g., [25]) and more broadly, associated with supply-chains for various products (e.g., [26]).

This paper builds upon the existing literature on energy (inter)dependence attempting to conceptualize it as a type of energy relation that extends beyond just energy trade. Accordingly, energy interdependence in this paper is defined as an energy relation between two actors of different energy power, characterized by significant energy import dependency, and a potential for one or both actors to exert energy influence. The analytical framework for energy interdependence sometimes includes concepts in relation to interdependence developed by Nye and Keohane, primarily sensibility and vulnerability as costs of interdependence [27, p. 10], with vulnerability being more pronounced (e.g., [22, p. 2], [14, p. 2], [28, p. 4], [12, p. 1], [29, p. 68]) than sensitivity. Additionally, a few authors (e.g., [30, p. 534], [18, p. 1], [31, p. 346]) account for the interdependence symmetry, while others simply refer to

[5] According to Shendrikova et al. [18, p. 2]: "The second aspect of power balance in asymmetric interdependence suggests that authoritarian countries are more prone to instrumentalize the EU's energy dependence, which implies that like-minded countries would be less prone to do that."

energy interdependence. There is research dedicated to the systematic influence of energy trade on "bilateral relations between an importer and its exporter," particularly if energy dependence creates incentives for peaceful or conflictual energy relations (e.g., [22, pp. 1–2]). Both Gökçe et al. [22, p. 2] and Hrabušajová and Mišík [19, p. 3] differentiate between interdependence from a realist and liberal perspective, specifically around the idea of whether it contributes to conflict. According to Gökçe et al. [22, p. 2], even though there are differences depending on the energy resource and type of the market, their analysis "provide[s] evidence for the liberal argument and indicate[s] that the more dependent an importer is to an exporter of energy, the more pacific the importer will be in its foreign policy to that exporter". They also deem natural gas "the most pacifying energy resource" [22, p. 2]. However, the Russia-Ukraine war and the EU's response to the energy dimension of it, question the notion of interdependence being a pacifying factor in international relations (e.g., [18, p. 3]). Accordingly, interdependence has been deemed "liability" (e.g., [25, p. 3], [17, p. 1]), "weaponized" [15, p. 532] or associated with "risks" [13, p. 77] in recent literature.

The realist interpretation of interdependence appears more applicable to contemporary empirical findings, particularly when Nye and Keohane's distinction between the "realist condition" and "complex interdependence" is taken into account [27, p. 31]. Namely, energy can be a highly centralized policy area, whether the energy industry is nationalized, or there's high centralization in energy policy decision-making, especially during crises, which gives less space for transnational actors, or other channels, and prioritizes foreign energy policy at a high level. For instance, Kuzemko et al. [11, p. 2] note that securitization of energy and "emergency framings" can lead to "centralized, top-down, approaches".

Secondly, there is clear hierarchy of issues, especially in times of crisis when "security threats will set the agenda" [27, p. 31], and we believe that this includes matters within the energy area, such as energy security, which is an interest related to the actor's survival. This is based on the hierarchical order of national interests [32, p. 260], as well as Glaser's mechanisms connecting energy and national security (Glaser 2013 in [33, p. 101]). It also includes the assumption of "linkages between issues [...] made principally by strong states, using their power in one area of world politics (particularly their military power) to coerce other states on other issues" [27, p. 104]. Therefore, if the energy-related area is juxtaposed to an area of higher importance to survival, the resulting structure provides incentives for symmetrical interdependence. Conversely, if it's counterposed to an area of lower importance to survival, the resulting structure will offer incentives for asymmetrical interdependence. In situations of symmetrical energy interdependence, the energy influence of one actor is countered by the influence over territorial integrity and defense (or other interests in direct conjunction with the actor's survival in IR), which creates a relatively balanced energy relation between them and less potential for each of them individually to exert influence on the other. In situations of asymmetrical energy interdependence, one actor's energy influence would be countered by an opposing influence in areas not directly linked to their survival, thereby granting leverage for concessions in an unbalanced energy relation. A typical case in energy relations involves situations where one actor is dependent on energy supply, and the other on market access, on similar levels. However, in the odd case of extreme export dependency and export concentration, the resulting structure gives an incentive for symmetrical interdependence. There's also a possibility of interdependence only in energy, for example, if actors controlled different parts of a supply chain and had similar levels of energy trade for them.

Finally, dependence is another type of energy relation, and it's characterized by only one actor's susceptibility to the energy influence of the other, based on the concentrated reliance on the first actor for its foreign energy policy goals, especially energy security. With no influence of the importing actor in other areas, this would create a situation of unchecked energy influence on the exporting side. Additionally, an actor can choose whether and how to use energy power, so energy influence doesn't have to be always present; however, in its absence the energy relation cannot truly be deemed interdependent. Finally, convergence of interests in energy security (supply security on one side and demand security on the other) will create incentives for energy interdependence.

Therefore, the operationalization of energy interdependence in this paper is embedded in the neorealist perspective. Based on the energy power, as "the disposition of resources, within the country's territory at a given point in time, that can be extracted by existing technology and at acceptable costs" [33, p. 93], actors will have the potential to exert energy influence. This is embedded in the broader conceptualization of the global energy system as "totality of units on macro-level forming energy, and related political relations, limited by the power they have at their disposal", i.e., a dimension of the international system [33, p. 40].[6] Additionally, despite the differences between member states, we consider the entire EU as a single actor in the analyzed energy relation, based on growing integration in energy and foreign energy policy following the Russia-Ukraine war.[7]

Energy relations will also depend on the type of predominant energy resource/carrier they are built around. For instance, if most of the trade between two actors is related to fossil fuels, energy relations will mainly be organized on this basis,

[6] There are several ways of conceptualizing and operationalizing of the global energy system, as well as different terms, in contemporary literature. Prevalent approaches focus either on the predominant type of energy resources (especially in the context of energy transition), or the economic and political relations that stem from energy (see [33, p. 40]). For instance, Kuzemko et al. view the global energy system as: "comprised of a wide-ranging array of interconnected, material attributes that together deliver energy services to society" [30, p. 536], in line with the first approach. They, too, point out to the "transformation" of the global energy system, however, this concept takes on different meaning for them, in accordance with their definition. Criekemans [89, p. 21], writing about the "energy regimes", in line with the second approach, says: "there always was a close interconnection between the 'energy regime' of a specific era (coal, oil) and the actual 'power position' of specific countries and territories at a given time."

[7] For example, according to Anna Herranz-Surrallés [90, pp. 6–7]: "EU's response to the Russian gas dependency challenge sits between integration and reluctant integration because there has been a clear increase in the level and scope of EU policy making, most notably with the activation of a joint gas purchasing and coordinated phase-out of Russian gas imports, though with some important exceptions and implementation loopholes."

even though it doesn't have to be exclusively present. It's essential to distinguish between energy power which is material and quantifiable, and energy influence, which is relational and represents a way of utilizing energy power to achieve specific foreign policy objectives [33, p. 95]. Direct energy influence, which is proportionate to energy power, manifests as "supply or price control by the exporting country aimed to coerce, compel, or deter another country's foreign policy, economic, or military activity" [33, p. 95]. This is similar but not equivalent to the concepts of "energy weapon" and "weaponization of energy" [34, p. 68], [15, p. 532], which are usually associated with threats and coercion. Energy influence can be exerted through explicit or implicit threats to the importing country's energy security [33, p. 95], as well as through promises of support or support to the importing country's energy security. Accordingly, an exporting country can try to appeal to the importing country, and the energy influence needn't be used just as a coercive tool.[8] Another important concept related to energy influence is energy sovereignty, which La Belle [23, p. 4] defines as "sovereign control over *energy policy*" (italic in the original) rather than over resources. An iteration of this idea, presented in [33, p. 265, 349], takes the form of political independence or independence in foreign policy decisions related to energy. This essentially refers to the freedom of choice in foreign policy, which can be infringed upon by energy influence. Additionally, unlike Nye and Keohane's definition of power stemming from asymmetry [27, p. 15], we believe that energy influence is based on energy power and can be applied in situations of energy interdependence, but it's not limited to them. For example, a decision by an exporting country to alter its supply can be directed at foreign policies of other exporters, or it can indirectly affect decisions of other importing countries, even if there's no direct energy trade or relations between them.

In accordance with this analytical framework, we will proceed to classify and explain the US-EU energy relations. The central hypothesis is that US-EU energy relations during the second Trump administration can be classified as falling between asymmetrical interdependence and dependence, resulting from the differing energy power of the two actors and complementary goals in energy security. This type of energy relation will, therefore, include high levels of energy trade with a net import dependency, and other energy interactions with a potential energy influence from the US. The methodology used in this paper entails a case study of US-EU energy relations during the first nine months of the second Trump administration. Inferences are made based on both qualitative and quantitative within-case evidence, with some consideration of data prior to the timeframe of the case study used for comparison. Types of evidence used include statistical data from relevant national and international agencies, as well as institutional and individual public statements regarding the subject.

---

[8] For example, the US used energy influence based on its energy power to persuade other countries to impose sanctions on the Iranian oil sector during Obama's administration [33, 250].

4 US-EU Energy Relations

4.1 US-EU Energy Interactions

To operationalize the concept of energy interdependence for analytical purposes, we utilized several indicators of *energy interaction*. First, interdependence would involve established energy trade between actors, usually spanning over medium to long periods of time and encompassing medium to high quantities of energy goods. Secondly, energy trade can be accompanied by shared or tangent infrastructure (such as pipelines, interconnected electricity grids, etc.) or infrastructure designed specifically for a type of energy trade (such as LNG terminals, heavy oil and light oil refineries, hydrogen storage facilities, etc.). Thirdly, there's potential for institutionalization, such as long-term contracts, agreements, or different institutional arrangements, including bodies.

When it comes to energy trade between two actors, there is evidence that it is increasing, especially after the onset of the Russia-Ukraine war. As shown in Fig. 1, a growing petroleum trade has developed between the US and the EU, with significant quantities being exported to Belgium, France, Germany, Italy, the Netherlands, Spain, and Sweden. It has more than doubled in value compared to 2021 on the overall EU-27 level, and it has shown consistent exports over time. More importantly, the data on US net imports of crude oil and petroleum products by country suggest that there were negative net imports to all EU countries in 2024 (where there's energy trade), except for Lithuania, and overall, the US has a negative net import of crude oil and petroleum products to the EU (meaning that the US exported more than it imported from the EU) [35]. However, the leading energy resource traded between two actors is natural gas, especially LNG. As shown in Fig. 2, the US LNG exports to the EU have grown and doubled in volume from 2021 to 2024, with the highest export quantities in 2022 and 2023, coinciding with the energy crisis in the EU. Mainly, the US exported to the Netherlands, France, Spain, Germany, and Italy in the selected timeframe.

Secondly, there are indicators of infrastructure being built specifically for this type of energy trade. During this period, according to Gas Infrastructure Europe [38] new large scale LNG import terminals have been constructed in the EU in Croatia (Krk Island, 2021), Netherlands (EemsEnergyTerminal, 2022), Finland (Inkoo, 2023), France (Le Havre, 2023), Germany (Brunsbüttel, 2023; Wilhelmshaven 1, 2023; Mukran, 2024), Greece (Alexandroupolis, 2024), and Spain (El Musel, 2023). Simultaneously, notwithstanding the Biden administration's decision to impose a temporary ban on LNG export permits, two new export terminals were put into operation in the US in 2024 [39], and President Trump approved a new LNG project at the beginning of his second term [40].

The third indicator we used for classifying the relation is net energy import dependency on the one hand, and energy exports concentration on the other hand. As already mentioned, the US has a negative net import of crude oil and petroleum products to the EU [35]. Currently, the EU's net energy import dependency on the US is

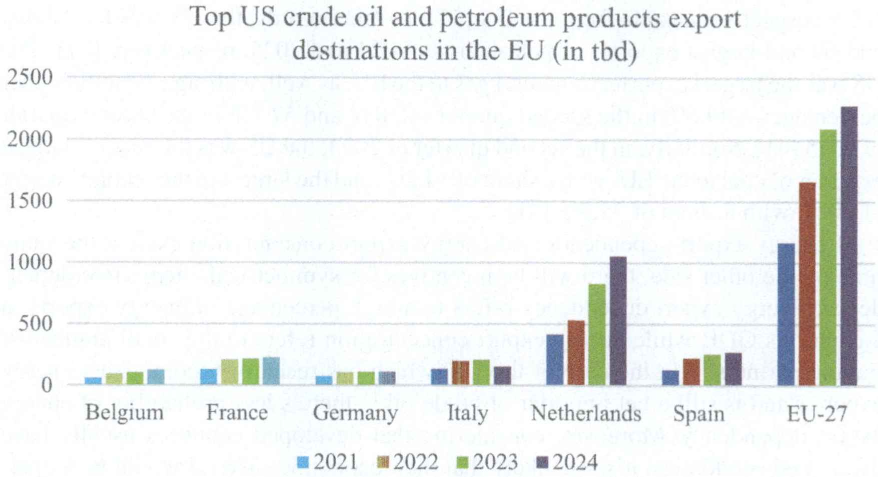

Fig. 1 US crude oil and petroleum products exports to the EU (in thousand barrels a day). *Source* Edited by the author based on EIA data [36]

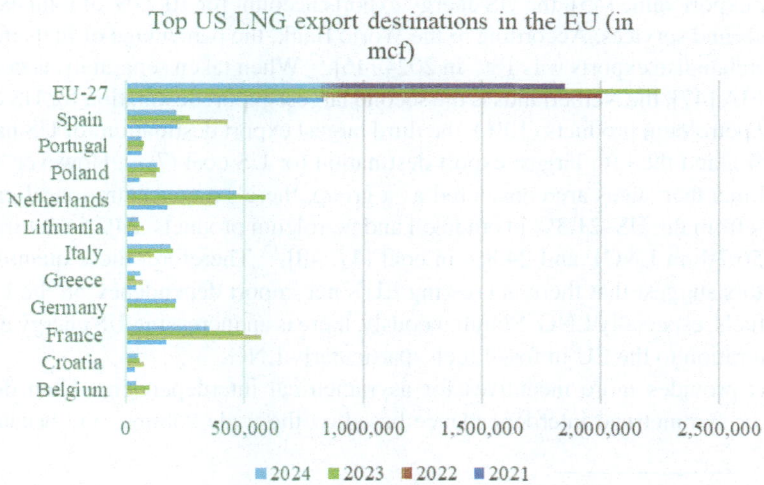

Fig. 2 US liquified natural gas exports to the EU (in million cubic ft). *Source* Edited by the author based on EIA data [37]

increasing in fossil fuels, particularly LNG.[9] According to the EC, 15.1% and 14.2% of imported petroleum oils in the second quarter of 2024, and the second quarter of

[9] The US imports of natural gas primarily come from Canada via pipeline, with no imports from any EU member state since 2020 [91]. On the other hand, the EU still exports and re-exports some refined petroleum products to the US. There's overall US negative net import of refined products from the EU, apart from Germany, Portugal, and Lithuania, which exported more petroleum products than they imported from the US in 2024 [92].

2025, respectively, came from the US [41]. In other words, the US was the largest and second-largest exporter of petroleum in 2024 and 2025, respectively [41]. The US was the largest exporter of natural gas to the EU, as well, with significantly higher percentages—49.6% in the second quarter of 2024 and 57.7% in the second quarter of 2025 [41]. Similarly, in the second quarter of 2024, the US was the second-largest exporter of coal to the EU, with a share of 34.2%, and the largest in the second quarter of 2025, with a share of 35.3% [41].

If energy export dependency and energy export concentration exist at the same time on the other side, there will be incentives for symmetrical energy interdependence. Energy export dependency refers to a high percentage of energy exports in a country's GDP, while energy export concentration refers to the small number of trading partners.[10] In the case of the US, which has recently become a net energy exporter and is still a net importer of crude oil,[11] there's less probability of energy export dependency. Moreover, considering that developed countries usually have diversified economies, it's less likely that their economic survival would be jeopardized, even if there's an energy export concentration. For instance, the percentage of total US goods and services exports in GDP was 10.9% in 2024 [42]. With $3.23 trillion in total export value of goods and services [43, 44], and $331.8 billion in total energy export value [45], the US energy exports account for 10.27% of total exports of goods and services. According to the World Bank, the percentage of fuels in total US merchandise exports was 18% in 2024 [46].[12] When taken separately, according to the EIA [47], the Netherlands is the second largest export destination for US crude oil and petroleum products (10%), the third largest export destination for US natural gas (5%), and the fifth largest export destination for US coal (7%). However, when all EU member states are considered as a group, there's a concentration of energy exports from the US–21.3% in crude oil and petroleum products, 24% in natural gas (with 56.7% in LNG), and 24.8% in coal [47, 48].[13] Therefore, these quantitative indicators suggest that there is growing EU's net import dependency on the US, in fossil fuels, especially LNG. Simultaneously, there is an increasing US energy export concentration to the EU in fossil fuels, particularly LNG.

This provides more incentives for asymmetrical interdependence than dependence or symmetrical interdependence based on the trade volume, and because in

[10] According to UNDP [93, p. 24]: "Export concentration reflects the degree to which a country's exports are concentrated on a small number of products or a small number of trading partners."

[11] The US has been a net total energy exporter since 2019, with the growing value of negative net imports, and total production surpassed total consumption of energy in 2022 [94, p. 3]. When it comes to oil imports, the US was a net exporter of total petroleum in 2023 and 2024 [94, p. 9, 11], a net exporter of petroleum products, natural gas, coal, biomass, and a net importer of crude oil and electricity [94, p. 13]).

[12] There are other indicators of fossil fuel export dependency, such as fossil fuel rents and resource revenue, as well as a dependency index which is a product of different indicators [95, p. 27]. According to Jensen's research, 40 countries are considered highly dependent on fossil fuels exports [95, p. 7], with the US not being one of them.

[13] The percentages for US exports are based on quantities of exported energy, not trade values, as in the case of the EU's imports, so they can't be compared.

the absence of energy trade with the US, the EU would suffer greater consequences than the US. The US has significant energy power, based on its high amounts of energy reserves, and subsequent increases in energy production, unlocked by the energy revolution (see [33]).[14] On the other hand, because it has limited indigenous hydrocarbon-based energy resources and production, with a concomitant high level of fossil fuels consumption, the EU is likely to import them.[15] Accordingly, the EU has a high rate of total energy import dependency, at 58.3% in 2023 [49]. When it comes to oil and petroleum import dependency, the average is 94.5%, with a median of 97.8%, which translates to all EU member states importing almost all of crude oil and petroleum products they use.[16] Similarly, the import dependency rate for natural gas is on average 90% in the EU, with a median of 98.6%.[17] Coal import dependency rates are the lowest (40.8% on average, with a median of 83.8%) compared to the other two.[18] Although renewable capacity is expected to increase in the EU over the next five years [50, p. 8], not all member states are equally represented in this forecast.[19] On the global level, green hydrogen development has also proven to be a more prolonged and more arduous process than initially believed, and according to the IEA green hydrogen will represent "4% of total hydrogen production in 2030" [50, p. 8]. Also, according to the IEA [50, p. 127]: "[...] despite their relative importance and increasing policy interest, none of the renewable fuels are on track with the Net Zero by 2050 Scenario". This is also recognized at the EU level as a need for acceleration in the clean energy transition [51, p. 12]. According to the 2025 Monitoring progress report, the goals of clean and affordable energy and climate action are qualified within the section "moderate progress" [51, p. 11]. It's worth noting that there is moderate progress in renewable energy share in final consumption, and in reducing the energy demand, including the demand for fossil fuels in the EU [51, p. 126].[20] Nevertheless, although renewables may replace fossil fuels in the future, and contribute to the EU's supply security, current foreign energy policy aimed at ensuring supply security is primarily based on fossil fuels imports. Consequently, current energy relations, including those with the US, are developed within this area.

---

[14] The US production of natural gas accounted for 25% of global production in 2024 [96, p. 37], crude oil production for 15.9% [96, p. 23], and oil production for 20.8% [96, p. 21].

[15] The average fossil fuel energy consumption in the EU was 66.3% in 2023 [97], and the energy supply consisted of five types of resources: "crude oil and petroleum products (37.7%), natural gas (20.4%), renewable energy (19.5%), solid fuels (10.6%), nuclear energy (11.8%)" [98]. The percentage of total global oil production located in the EU was 0.3% [96, p. 21], of crude oil production 0.4% [96, p. 23], and of natural gas production 0.8% [96, p. 37].

[16] The author's calculation based on Eurostat data [49].

[17] Ibid.

[18] Ibid.

[19] According to the IEA [50, p. 49]: "The majority (70%) of the expansion is concentrated in just seven countries, led by Germany, followed by the United Kingdom, Italy, Türkiye, France, Spain and the Netherlands."

[20] The EU's final energy consumption in 2023 had a 18.7% reduction compared to 2008, and 3% reduction compared to 2022 [51, p. 128] when the energy crisis began.

When it comes to the institutionalization of energy interactions, the US-EU energy relations encompass institutional arrangements, including the Task Force on Energy Security and the EU-US Energy Council. The meetings of the Task Force were more frequent during President Biden's term, and they primarily included the topic of the EU's energy security, although with mentions of both diversification of LNG imports and reducing demand [52], including through renewable energy [53] and clean energy transition [54]. Meetings of the EU-US Energy Council have included similar themes pertaining to energy security, US LNG as part of the EU's diversification strategy, the clean energy transition, and criticism of the weaponization of energy [55–57]. On the other hand, there aren't reports on meetings in these formats during Trump's second term. The new administration utilized the mentioned trade agreement to include energy, with a focus on fossil fuels and nuclear energy procurement, but without mentioning of renewables or energy transition.

Because the EU has chosen to phase out imports of Russian energy, and because LNG has typically been considered compatible with the EU's green energy policy, this resource has become the most attractive alternative, contributing to LNG trade as an area of energy influence for the US. Natural gas is commonly referred to as the "transition fuel" (e.g. [58, p. 18])[21]; however, some authors question or criticize the inclusion of natural gas, especially LNG, in low-carbon alternatives (e.g. [11, p. 4], [59, 60]). If this opinion is accepted, then expanding even LNG trade, and let alone fossil fuel imports to Europe, could be considered a form of energy influence, provided that the energy transition is defined as complete transition to renewable energy.[22]

As the "international agenda setter for climate policy" [25, p. 9], the European Union has framed its energy goals with a focus on energy transition, as already mentioned in the literature review. This remains the case for the EU, particularly since renewables are considered to contribute to both energy supply security (through domestic energy production) and the energy transition (e.g., [61, pp. 6–7], [62]). However, there's also a focus on supply security in the fossil fuels sector, specifically in the context of substituting imports from Russia. Therefore, the RePowerEU Roadmap is almost entirely dedicated to the diversification of fossil fuel imports from Russia, with LNG having a prominent place, and the Action Plan for Affordable Energy contains sections on natural gas as an important part of the EU's supply security [63, p. 17], [64, p. 9]. There are references to supply security in von der Leyen's public statements, mostly along three themes—reducing dependencies that can be weaponized, investing in domestic supply of renewable energy, and lowering the energy prices (e.g. [62]). However, there also are references to the conventional fossil fuel supply, which are framed as an unwanted necessity, also implying potential susceptibility to influence based on fossil fuels imports. For instance, von der Leyen stated: "To be clear: I know of course that Europe will continue to use oil and gas for years to come as a basis without any question. But that will never be a source of

---

[21] For an analysis of the discourse on national gas as a transitional fuel, and its compatibility with climate action see [99].

[22] For a list of definitions of energy transition see [100], 2.

advantage for Europe's industry, or its consumers, or our planet" [65]. A pronounced theme in von der Leyen's speeches is independence, including energy independence as a goal (e.g. [66, 67]) which can be understood as energy sovereignty—"We must take control over the sensitive technologies and energy sources that will power our economies" [66]. Therefore, even though energy transition goals remain high on the EU's agenda, and are incorporated into the goal of supply security, the EU still has a prominent goal of fossil fuels supply security, which is compatible with the US's goal of demand security, thus creating incentives for energy interdependence.

Unlike the previous administration, which shared energy transition goals with the EU, US energy policy during President Trump's second term prioritizes goals of supply and demand security, with less interest in environmental goals. The second Trump administration has immediately put energy high on the agenda, as reflected in the declaration of the national energy emergency and several executive orders, which essentially constitute its domestic and foreign energy strategy. One of the central ideas of the Trump administration's energy policy revolves around domestic production of energy, used for achieving both security of supply domestically and demand security internationally—"An affordable and reliable domestic energy supply is essential to the national and economic security of the United States, as well as our foreign policy" [68]. Therefore, domestic production of resources is linked to economic prosperity and national security of the US, including "military preparedness" [69]. Accordingly, a stated policy goal is "to encourage energy exploration and production on Federal lands and waters, including on the Outer Continental Shelf, in order to meet the needs of our citizens and solidify the United States as a global energy leader long into the future" [69]. Energy dominance, including "mineral dominance", is a prominent theme in the documents [68, 69]. The documents also reflect administration's position on energy transition, renewable and sustainable energy with "terminating the New Green Deal" [69], review of regulations impeding domestic production of "oil, natural gas, coal, hydropower, biofuels, critical mineral, and nuclear energy resources" [69, 70], the critique of "burdensome and ideologically motivated 'climate change' or energy policies that threaten American energy dominance and our economic and national security" [68] and the critique of "social cost of carbon" which is viewed as "rendering the United States economy internationally uncompetitive, encourages a greater human impact on the environment by affording less efficient foreign energy producers a greater share of the global energy and natural resource market" [69].[23]

[23] A lesser interest in environmental goals can also be inferred from the second withdrawal from the Paris Climate Agreement, pause on permitting of wind farms in the US [40], as well as President Trump's critique of climate action and renewable energy during his address at the UN GA in September 2025 [82].

4.2 *Energy Influence Potentially Stemming from US-EU Energy Interactions*

To understand the empirical manifestation of the US's energy influence over the EU in areas beyond energy policy, it's important to determine the goals related to energy demand security in US energy policy, as well as the intent to instrumentalize energy for achieving those goals. The Trump administration views the global energy system as a net-zero game, or "[...] increasingly crucial theater of global competition" [71], which would contribute to using energy as a lever. Accordingly, there's evidence of the second Trump administration's intent to utilize energy exports to achieve economic prosperity, particularly to lower the trade deficit with the EU. As stated in the declaration, energy exports of domestic energy, besides providing economic prosperity in the US, can be used to "improve the United States' trade balance, help our country compete with hostile foreign powers, strengthen relations with allies and partners, and support international peace and security" [71].[24] It's important to note that LNG export projects are expedited in accordance with the Trump administration's energy policy, and they are associated with US national security [69]. A general propensity for using energy for achieving economic prosperity is also present in the inaugural speech of President Trump: "We will bring prices down, fill our strategic reserves up again right to the top, and export American energy all over the world. We will be a rich nation again, and it is that liquid gold under our feet that will help to do it" [72].

In his public statements, Trump himself linked exporting US energy products with lowering of trade deficit: "So, you know, we have, like, a $300 billion deficit with the European Union. So we'll straighten that out with either tariffs, or they have to buy our oil. The one thing they can do—or oil or gas. The one thing they can do to catch up quickly: buy our oil or gas" [73]. Another example of a similar public statement also contains a dimension of the EU's supply security—"We have a deficit with the European Union of $350 billion, and it's going to disappear fast. And one of the reasons and one of the ways that that can disappear easily and quickly is, they're going to have to buy our energy from us, because they need it. *They're going to have to buy it from us.* They can buy it. We can knock off $350 billion in 1 week. They have to buy and commit to buy a like amount of energy, and we have that" [74] (italic by the author of this paper). President Trump also said that he would guarantee the security of LNG supply to the EU.[25] Statements made during the news conference after Trump and Von der Leyen's meeting in July signal a connection between areas of energy security

---

[24] During Trump's first time in office, there were also indications of using energy as leverage in his administration's approach to energy security of Poland and the Baltic countries [102, p. 54]. However, it was primarily framed as a means to help allies reduce their energy dependence [102, p. 48, 55] by diversification. Expectedly, in accordance with the expansion of trade, and the context of energy security threats to the EU during his second term, this leverage is much more pronounced and generally more associated with US interests.

[25] "[...] yes, I would. I would make sure that you get it. If we make a deal, we make a deal; you'll get it. Because a lot of people do have that problem. They make a deal, and then it can't get supplied because of war-type problems and other problems. So we would absolutely do that" [101].

and the economy [75]. Ursula von der Leyen spoke about the imbalance of trade, with a US trade deficit,[26] while Donald Trump emphasized the importance of the energy that the US is providing [75].[27] These statements are dated prior to the Framework on an Agreement meaning that the energy provisions were probably included in the negotiations. The energy provisions, and especially the high trade volume included in the Agreement, can therefore be considered an outcome resulting from the energy influence of the US. This is also important when considering the quantity of energy to be traded until 2028, as it would require either a substantial increase in fossil fuel imports and consumption (e.g., [76]), or a higher concentration of US exports. According to one analysis: "the bloc would have to essentially triple its American imports over the next three years—in the process shunning other providers, such as Norway, which provides cheaper gas via pipeline" [77]. This could potentially jeopardize not only energy transition goals but the current EU's energy security strategy, impeding further import diversification and energy demand reduction. It could also result in higher asymmetric interdependence.

Finally, following the Joint Statement on the Framework on an Agreement in August, public statements from EU officials, including von der Leyen, reaffirm that the energy part of the deal is connected to the EU's supply (e.g. [78–80]). Simultaneously, the Agreement was deemed "a massive trade deal" and "historical" in the US, with the White House promoting it as a way to "[…] strengthen the United States' energy dominance, reduce European reliance on adversarial sources, and narrow our trade deficit with the EU" [81].

When it comes to the impact of the Agreement on the energy transition, the EU has addressed this issue, publicly stating that it remains committed to achieving the goals of decarbonization and climate-neutrality by 2050 [7], although the 2030 goals aren't mentioned in this statement. A particularly interesting piece of evidence regarding the potential US energy influence over the EU pertains to President Trump's address to the UN GA in September 2025. A significant part of his speech is dedicated to critiquing the green energy transition, with several mentions of the EU in this context [82]. For instance, during this address, President Trump stated: "Europe, on the other hand, has a long way to go with many countries being on the brink of destruction because of the green energy agenda.", also adding costs of renewable energy and green energy transition, as well as energy poverty into the critique, and displaying overall suspicion regarding climate change [82]. The speech included calls to action for Europe to "take control strongly and immediately of the unmitigated immigration disaster and the fake energy catastrophe before it's too late", and commended Germany's energy policy for including fossil fuels and nuclear energy in its energy supply [82]. Judging by this speech, the Trump administration's approach to the

[26] "*President von der Leyen.* So, as we—the starting point was an imbalance, a surplus on our side and a deficit on the U.S. side. And we wanted to rebalance the trade relation, and we wanted to do it in a way that trade goes on between the two of us, across the Atlantic, because the two biggest economies should have a good trade flow between us" [75].

[27] "*President Trump.* And, again, the energy is a very important component because we have more energy than anybody else in that sense. And I think it was very wise that they buy a lot of energy, and it's great stuff, and it solves a lot of problems" [75].

EU's goals of a green energy transition appears to be one of substantial divergence, including some engagement in efforts to change it. It's too early to draw definite conclusions if this will produce a real change in EU's energy policy; however, besides Trump's speech, there are other indications of possible energy influence over EU's energy transition goals, for example, within US' and Qatar's reported critique of the EU's Corporate Sustainability Due Diligence Directive proposal, especially when it comes to LNG exports [83].

5 Conclusion

We argued that the growing energy relations between the US and the EU during the second Trump administration exhibit similar characteristics of a typical energy (inter)dependence in this paper. To test this hypothesis, we operationalized the concept of energy interdependence from a neorealist perspective and utilized quantitative and qualitative evidence pertaining to the case. First, there is growing energy trade in fossil fuels, especially LNG, between the US and the EU, with energy infrastructure on the EU's side designed specifically for importing LNG. Secondly, the initiatives to organize the relations through institutional arrangements were replaced by a political agreement, which included the procurement of US energy. Thirdly, there has been a simultaneous net import dependency on the EU's side in all fossil fuel categories and export concentration on the US side, although LNG is the area with the most potential for interdependence, based on quantitative indicators. Based on this indicator, energy interdependence is more fitting than dependence in classifying the energy relation. Complementary goals in energy security also corroborate this conclusion. Namely, notwithstanding the enduring EU's goals of energy transition and framing renewable energy as part of the solution for the supply security, there is recognition of fossil fuel imports in supply security. On the other hand, the US goal of demand security is prominent in the second Trump administration's foreign energy policy, which provides an opportunity for a higher level of interdependence between the two actors.

Additionally, documents and individual public statements pertaining to both actors disclose an association between energy security and the economy, with a prominent theme of using energy to lower the trade deficit in President Trump's speeches. The energy provisions of the Framework on an Agreement on Reciprocal, Fair, and Balanced Trade, along with the trade volume of the US exports accepted under the Agreement, suggest that this type of energy influence has produced an outcome. When it comes to using energy for other goals within the energy area, there are indications that the Trump administration would favor fossil fuels over renewables in the EU, mostly within his UN GA address. Still, it's early to tell if this produced any outcome regarding the EU's energy transition beyond the indirect effects of the agreed procurement of fossil fuels under the Framework Agreement.

References

1. Ellerbeck, S.: What's the difference between 'friend-shoring' and other global trade buzzwords? The World Economic Forum, (2023). https://www.weforum.org/stories/2023/02/friendshoring-global-trade-buzzwords/. Last accessed 21 Oct 2025
2. Benson, E., Kapstein, E.B.: The limits of 'Friend-shoring'. Center for Strategic and International Studies, (2023). https://www.csis.org/analysis/limits-friend-shoring. Last accessed 21 Oct 2025
3. Yellen, J.: Transcript: US Treasury Secretary Janet Yellen on the next steps for Russia sanctions and 'friend-shoring' supply chains. Atlantic Council, (2022). https://www.atlanticcouncil.org/blogs/new-atlanticist/transcript-us-treasury-secretary-janet-yellen-on-the-next-steps-for-russia-sanctions-and-friend-shoring-supply-chains/. Last accessed 21 Oct 2025
4. Deberdt, R., Smith, N.M., Calderon, J.L., McCall, S.K.: Critical mineral lists for low-carbon transitions: Reviewing their structure, objectives, and limitations. Energy Res. & Soc. Sci. **127**, 1–18 (2025). https://doi.org/10.1016/j.erss.2025.104252
5. Braunstein, J., Chuchko, M.: Resource curse in the age of critical minerals: Geopolitical forces and market maturity. Energy Res. Soc. Sci. **127**(2025), 1–12 (2025). https://doi.org/10.1016/j.erss.2025.104247
6. Łoskot-Strachota, A., Keliauskaitė, U., Zachmann, G.: Future European Union gas imports: balancing different objectives. Bruegel (2024). https://www.bruegel.org/analysis/future-european-union-gas-imports-balancing-different-objectives. Last accessed 22 Oct 2025
7. European Commission [EC].: EU-US trade deal explained—energy aspects. (2025). https://ec.europa.eu/commission/presscorner/detail/en/qanda_25_1935. Last accessed 21 Oct 2025
8. European Commission [EC].: Joint Statement on a United States—European Union framework on an agreement on reciprocal, fair and balanced trade. Directorate-General for Trade and Economic Security, (2025). https://policy.trade.ec.europa.eu/news/joint-statement-united-states-european-union-framework-agreement-reciprocal-fair-and-balanced-trade-2025-08-21_en. Last accessed 21 Oct 2025
9. Liboreiro, J.: Majority of Europeans think EU-US trade deal is a 'humiliation', new poll shows. Euronews, (2025). https://www.euronews.com/my-europe/2025/09/09/majority-of-europeans-think-eu-us-trade-deal-is-a-humiliation-new-poll-shows. Last accessed 22 Oct 2025
10. Corlin, P.: MEPs call on European Commission to drop energy purchase promise in EU-US deal. Euronews, (2025). https://www.euronews.com/business/2025/09/15/meps-call-on-commission-to-drop-energy-purchase-promise-in-eu-us-trade-deal. Last accessed 22 Oct 2025
11. Kuzemko, C., Blondeel, M., Dupont, C., Brisbois, M.C.: Russia's war on Ukraine, European energy policy responses & implications for sustainable transformations. Energy Res. & Soc. Sci. **93**, 1–8 (2022). https://doi.org/10.1016/j.erss.2022.102842
12. Albanese, F.: Navigating trade-offs: Energy dependency, geopolitical shocks, and sustainability in European parliamentary debates. Energy Res. Soc. Sci. **127**(2025), 1–14 (2025). https://doi.org/10.1016/j.erss.2025.104198
13. Trushkina, N.: Energy security as a factor of political stability in the European Union. Politics & Security **12**(2), 72–87 (2025). https://doi.org/10.54658/ps.28153324.2025.12.2
14. Steffen, B., Patt, A.: A historical turning point? Early evidence on how the Russia-Ukraine war changes public support for clean energy policies. Energy Res. Soc. Sci. **91**(2022), 1–10 (2022). https://doi.org/10.1016/j.erss.2022.102758
15. La Belle, M.C.: Energy as a weapon of war: Lessons from 50 years of energy interdependence. Global Pol. **14**(2023), 531–547 (2023). https://doi.org/10.1111/1758-5899.13235
16. Erdogan, S., Pata, U.K., Alola, A.A., Kartal, M.T., Depren, S.K.: How does geopolitical risk affect clean energy generation? Daily evidence from five highly Russia-reliant European countries. Energy Strat. Rev. **55**, 1–25 (2024). https://doi.org/10.1016/j.esr.2024.101492
17. Jerzyniak, T.: The EU de-risking of energy dependencies: Towards a new clean energy geopolitical order? Polit. Gov., **12**, 1–18 (2024). https://doi.org/10.17645/pag.8285

18. Shendrikova, D., Rinaldi, L., Colombo, E.: Measuring value compatibility in European Union energy diplomacy: Conceptual framework and application to natural gas sector. Energy Res. & Soc. Sci. **127**, 1–11 (2025). https://doi.org/10.1016/j.erss.2025.104254
19. Hrabušajová, M., Mišík, M.: The market will save the day: White House narratives on liquefied natural gas exports to the European Union. Energy Res. Soc. Sci. **127**(2025), 1–9 (2025). https://doi.org/10.1016/j.erss.2025.104197
20. Proroković, D.: Europe's energy security and the conclusions of the UN climate change conferences: Between desire and possibilities. Serbian Pol. Thought. **90**(2), 51–72 (2025). https://doi.org/10.5937/spm90-57227
21. Rodríguez-Fernández, L., Fernández Carvajal, A.B., Fernández de Tejada, V.: Improving the concept of energy security in an energy transition environment: Application to the gas sector of the European Union. Extr. Ind. Soc. **9**, 1–9 (2022). https://doi.org/10.1016/j.exis.2022.101045
22. Gökçe, O.Z., Hatipoglu, E., Soytas, M.A.: The pacifying effect of energy dependence on interstate conflict: A Large-N analysis. Energy Res. & Soc. Sci. **78**, 1–12 (2021). https://doi.org/10.1016/j.erss.2021.102133
23. La Belle, M.C.: Breaking the era of energy interdependence in Europe: A multidimensional reframing of energy security, sovereignty, and solidarity. Energ. Strat. Rev. **52**(2024), 1–13 (2024). https://doi.org/10.1016/j.esr.2024.101314
24. Peréz, M. de la E.M., Scholten, D., Smith Stegen, K.: The multi-speed energy transition in Europe: Opportunities and challenges for EU energy security. Energy Strat. Rev. **26**, 1–6 (2019). https://doi.org/10.1016/j.esr.2019.100415
25. Kalantzakos, S., Overland, I., Vakulchuk R.: Decarbonisation and critical materials in the context of fraught geopolitics: Europe's distinctive approach to a net zero future. Int. Spect. **58**(1), 3–22 (2023). https://doi.org/10.1080/03932729.2022.2157090
26. Monsees, L.: The paradox of semiconductors—EU governance between sovereignty and interdependence. Camb. Rev. Int. Aff. **38**(1), 3–21 (2025). https://doi.org/10.1080/09557571.2024.2405915
27. Nye, J.S., Jr., Keohane, R.O.: Power and Interdependence, 4th edn. Longman (2012)
28. Vezzoni, R.: Green growth for whom, how and why? The REPowerEU Plan and the inconsistencies of European Union energy policy. Energy Res. Soc. Sci. **101**(2023), 1–15 (2023). https://doi.org/10.1016/j.erss.2023.103134
29. Pavković, O.: Politika Sjedinjenih Američkih Država prema Kraljevini Saudijskoj Arabiji od 2009. do 2017. godine. Srpska Politička Misao. **65**(3), 63–90 (2019)
30. Kuzemko, C., Blondeel, M., Bradshaw, M., Bridge, G., Faigen, E., Fletcher, L.: Rethinking energy geopolitics: Towards a geopolitical economy of global energy transformation. Geopolitics, **30**(2), 531–565 (2025). https://doi.org/10.1080/14650045.2024.2351075
31. Oana, I., Moise A.D., Truchlewski, Z.: Demand-side constraints on European solidarity for energy sanctions: Experimental evidence from seven EU countries. Eur. Union Polit. **26**(2), 344–368 (2025). https://doi.org/10.1177/14651165251318955
32. Živojinović, D.: Razumevanje pojma nacionalni interes u realističkim teorijama. In: Jović, D. (ed.) Teorije međunarodnih odnosa: REALIZAM, pp. 249–263. Politička kultura, Zagreb (2013)
33. Stevanović, O.: Politika Sjedinjenih Američkih Država prema Persijskom zalivu u izmenjenom globalnom energetskom sistemu od 2013. do 2021. Godine. Doctoral Dissertation. Faculty of Political Science, University of Belgrade, (2025)
34. Bordoff, J., O'Sullivan, M.L.: The return of the energy weapon: An old tool creating new dangers, Foreign Affairs, November/December, pp. 56–71 (2025)
35. The Energy Information Administration [EIA].: Petroleum and other liquids. Data: U.S. Net Imports by Country. (2025). https://www.eia.gov/dnav/pet/pet_move_neti_a_ep00_IMN_mbblpd_a.htm. Last accessed 2 Oct 2025
36. The Energy Information Administration [EIA].: Petroleum and other liquids. Data: Exports by destination. (2025). https://www.eia.gov/dnav/pet/pet_move_expc_a_EP00_EEX_mbblpd_a.htm. Last accessed 2 Oct 2025

37. The Energy Information Administration [EIA].: Natural gas. Data: U. S. Natural gas exports and re-exports by Country. (2025). https://www.eia.gov/dnav/ng/ng_move_expc_s1_a.htm. Last accessed 3 Oct 2025
38. Gas Infrastructure Europe [GIE].: GIE LNG map 2024. (2024). https://www.gie.eu/publications/maps/gie-lng-map/. Last accessed 21 Oct 2025
39. The Energy Information Administration [EIA].: The eighth U.S. liquefied natural gas export terminal, Plaquemines LNG, ships first cargo. Principal contributor: Victoria Zaretskaya. Energy Information Administration, (2025). https://www.eia.gov/todayinenergy/detail.php?id=64224. Last accessed 2 Nov 2025
40. The White House.: President Trump is unleashing American energy. (2025). https://www.whitehouse.gov/articles/2025/03/president-trump-is-unleashing-american-energy/. Last accessed 10 Oct 2025
41. Eurostat.: EU imports of energy products—latest developments. (2025). https://ec.europa.eu/eurostat/statistics-explained/index.php?title=EU_imports_of_energy_products_-_latest_developments. Last accessed 3 Nov 2025
42. The World Bank.: Exports of goods and services (% GDP). (n.d.). https://data.worldbank.org/indicator/NE.EXP.GNFS.ZS. Last accessed 3 Nov 2025
43. Bureau of Economic Analysis [BEA].: Table 2.1. U.S. international trade in goods. (2025). https://apps.bea.gov/iTable/?ReqID=62&step=2&_gl=1*biz8is*_ga*MjEyMTg3MzEzOC4xNzYyMjUxMjQ1*_ga_J4698JNNFT*czE3NjIyNTEyNDUkbzEkZzEkdDE3NjIyNTE0NDQkajYkbDAkaDA.#eyJhcHBpZCI6NjIsInN0ZXBzIjpbMiw2XSwiZGF0YSI6W1siVGFibGVMaXN0IiwiNDUiXV19. Last accessed 3 Nov 2025
44. Bureau of Economic Analysis [BEA].: Table 3.1. U.S. International trade in services. (2025). https://apps.bea.gov/iTable/?ReqID=62&step=2&_gl=1*biz8is*_ga*MjEyMTg3MzEzOC4xNzYyMjUxMjQ1*_ga_J4698JNNFT*czE3NjIyNTEyNDUkbzEkZzEkdDE3NjIyNTE0NDQkajYkbDAkaDA.#eyJhcHBpZCI6NjIsInN0ZXBzIjpbMiw2XSwiZGF0YSI6W1siVGFibGVMaXN0IiwiNTEiXV19. Last accessed 2 Nov 2025
45. United States International Trade Commission [USITC].: Energy-related products. (n.d.). https://www.usitc.gov/research_and_analysis/tradeshifts/2024/energy. Last accessed 4 Nov 2025
46. The World Bank.: Fuels exports (% of merchandise exports). (n.d.). https://data.worldbank.org/indicator/TX.VAL.FUEL.ZS.UN. Last accessed 4 Nov 2025
47. The Energy Information Administration [EIA].: United States exported 30% of the energy it produced in 2024. Principal contributor: Mickey Francis. Energy Information Administration, (2025). https://www.eia.gov/todayinenergy/detail.php?id=65924. Last accessed 4 Nov 2025
48. The Energy Information Administration [EIA].: Coal data browser. (n.d.). https://www.eia.gov/coal/data/browser/#/topic/41?agg=0,2,1&rank=g&freq=A&start=2001&end=2023&ctype=map&ltype=pin&rtype=s&maptype=0&rse=0&pin. Last accessed 4 Nov 2025
49. Eurostat.: Shedding light on energy in Europe–2025 edition. Energy Sources: Energy Imports Depend., (2025). https://ec.europa.eu/eurostat/web/interactive-publications/energy-2025#energy-imports-dependency. Last accessed 28 Sep 2025
50. International Energy Agency [IEA].: Renewables 2024: Analysis and forecast to 2030. International Energy Agency, (2024)
51. Eurostat.: Sustainable development in the European Union: Monitoring report on progress towards the SDGs in an EU context, 2025 edn. Eurostat, (2025)
52. European Commission [EC].: Joint statement between the United States and the European Commission on European energy security. Statement, (2022). https://ec.europa.eu/commission/presscorner/detail/en/STATEMENT_22_2750. Last accessed 05 Nov 2025
53. European Commission [EC].: Readout of EU-US task force on European energy security meeting. Statement, (2022). https://ec.europa.eu/commission/presscorner/detail/en/STATEMENT_22_3980. Last accessed 05 Nov 2025
54. European Commission [EC].: Joint statement following the latest meeting of the EU-US task force on energy security. Statement, (2022). https://ec.europa.eu/commission/presscorner/detail/en/statement_22_6582. Last accessed 05 Nov 2025

55. European Commission [EC].: The ninth U.S.-EU Energy council. News Article. (2022). https://commission.europa.eu/news-and-media/news/ninth-us-eu-energy-council-2022-02-07_en. Last accessed 5 Nov 2025
56. European Commission [EC].: Joint statement by the EU and the US following the 10th EU-US Energy Council. Statement, (2023). https://ec.europa.eu/commission/presscorner/detail/en/statement_23_2121. Last accessed 06 Nov 2025
57. European Commission [EC].: Statement by the EU and the US following the 11th EU-US energy council. Statement, (2024). https://ec.europa.eu/commission/presscorner/detail/en/statement_24_1516. Last accessed 06 Nov 2025
58. World Energy Council.: World energy trilemma 2024: Evolving with resilience and justice. WEC, (2024).
59. Howarth, R.W.: The greenhouse gas footprint of liquefied natural gas (LNG) exported from the United States. Energy Sci. Eng. **12**(2024), 4843–4859 (2024). https://doi.org/10.1002/ese3.1934
60. Brauers, H.: Natural gas as a barrier to sustainability transitions? A systematic mapping of the risks and challenges. Energy Res. Soc. Sci. **89**(2022), 1–18 (2022). https://doi.org/10.1016/j.erss.2022.102538
61. European Commission [EC].: Report from the Commission to the European Parliament, the Council, the European Economic and Social Committee and the Committee of the Regions: State of the Energy Union Report 2025 (pursuant to Regulation (EU)2018/1999 on the Governance of the Energy Union and Climate Action). Brussels, 6.11.2025 COM (2025) 667 final (2025)
62. European Commission [EC].: Speech by President von der Leyen at the Copenhagen competitiveness summit. Press Corner, (2025). https://ec.europa.eu/commission/presscorner/detail/en/speech_25_2272. Last accessed 07 Nov 2025
63. European Commission [EC].: Communication from the Commission to the European Parliament, the Council, the European Economic and Social Committee and the Committee of the Regions: Action plan for affordable energy unlocking the true value of our Energy Union to secure affordable, efficient and clean energy for all Europeans. Brussels, 26.2.2025 COM (2025) 79 final (2025)
64. European Commission [EC].: Communication from the Commission to the European Parliament, the Council, the European economic and social committee and the committee of the regions: Roadmap towards ending Russian energy imports. Brussels, 12.5.2025 COM (2025) 440 final/2
65. European Commission [EC].: Speech by President von der Leyen at the European Parliament plenary ahead of the European Council of 23 October 2025. Press Corner, (2025). https://ec.europa.eu/commission/presscorner/detail/en/speech_25_2470. Last accessed 07 Nov 2025
66. European Commission [EC].: Speech by President von der Leyen in the European Parliament plenary debate on the Commission work programme 2026. Press Corner, (2025). https://ec.europa.eu/commission/presscorner/detail/en/speech_25_2462. Last accessed 06 Nov 2025
67. European Commission [EC].: 2025 State of the Union Address by President von der Leyen. Speech, (2025). https://ec.europa.eu/commission/presscorner/detail/ov/SPEECH_25_2053. Last accessed 08 Nov 2025
68. The White House.: Protecting American energy from state overreach. Executive Orders, (2025). https://www.whitehouse.gov/presidential-actions/2025/04/protecting-american-energy-from-state-overreach/. Last accessed 06 Nov 2025
69. The White House.: Unleashing American energy. Executive Orders, (2025). https://www.whitehouse.gov/presidential-actions/2025/01/unleashing-american-energy/. Last accessed 06 Nov 2025
70. The White House.: Zero-based regulatory budgeting to unleash American energy. Executive Orders, (2025). https://www.whitehouse.gov/presidential-actions/2025/04/zero-based-regulatory-budgeting-to-unleash-american-energy/. Last accessed 06 Nov 2025
71. The White House.: Declaring national energy emergency, (2025). https://www.whitehouse.gov/presidential-actions/2025/01/declaring-a-national-energy-emergency/. Last accessed 25 Oct 2025

72. Trump, D.: Inaugural address. The American Presidency Project, (2025). https://www.presidency.ucsb.edu/documents/inaugural-address-54. Last accessed 25 Oct 2025
73. Trump, D.: Remarks at a document signing ceremony and an exchange with reporters. The American Presidency Project, (2025). https://www.presidency.ucsb.edu/documents/remarks-document-signing-ceremony-and-exchange-with-reporters. Last accessed 25 Oct 2025
74. Trump, D.: Remarks prior to a meeting with prime Minister Benjamin Netanyahu of Israel and an exchange with reporters. (2025). https://www.presidency.ucsb.edu/documents/remarks-prior-meeting-with-prime-minister-benjamin-netanyahu-israel-and-exchange-with-8. Last accessed 3 Nov 2025
75. Trump, D.: Remarks following a meeting with President Ursula von der Leyen of the European commission and an exchange with reporters in South Ayrshire, United Kingdom. The American Presidency Project, (2025). https://www.presidency.ucsb.edu/documents/remarks-following-meeting-with-president-ursula-von-der-leyen-the-european-commission-0. Last accessed 3 Nov 2025
76. Galgóczi, B.: The 'dirty' US-EU trade deal. European Trade Union Institute, (2025). https://www.etui.org/news/dirty-us-eu-trade-deal. Last accessed 25 Oct 2025
77. Jack, V.: The EU's 'fantasy' $750B energy promise to Trump. Politico, (2025). https://www.politico.eu/article/eu-trade-war-donald-trump-us-oil-gas-energy/. Last accessed 1 Nov 2025
78. European Commission [EC].: Remarks by Commissioner Šefčovič on EU-US trade relations. Press Corner, (2025). https://ec.europa.eu/commission/presscorner/detail/en/speech_25_1977. Last accessed 28 Sep 2025
79. European Commission [EC].: Questions and answers on the EU-US joint statement on transatlantic trade and investment. Press Corner, (2025). https://ec.europa.eu/commission/presscorner/detail/en/qanda_25_1974. Last accessed 11 Jan 2025
80. European Commission [EC].: Statement by President von der Leyen on the deal on tariffs and trade with the United States. Press Corner, (2025). https://ec.europa.eu/commission/presscorner/detail/da/statement_25_1915. Last accessed 11 Jan 2025
81. The White House.: The United States and European Union reach massive trade deal. (2025). https://www.whitehouse.gov/fact-sheets/2025/07/fact-sheet-the-united-states-and-european-union-reach-massive-trade-deal/. Last accessed 25 Oct 2025
82. Trump, D.: Remarks to the United Nations General Assembly in New York city. The American Presidency Project, (2025). https://www.presidency.ucsb.edu/documents/remarks-the-united-nations-general-assembly-new-york-city-20. Last accessed 5 Nov 2025
83. Smith, J., Hancock, A.: US and Qatar issue energy and trade threats to EU over climate rules. Financial Times. (2025). https://www.ft.com/content/05ecdd19-a779-4b69-8609-c7f04ca49314. Last accessed 24 Oct 2025
84. Pesu, M., Sinkkonen, V.: Trans-atlantic (mis)trust in perspective: Asymmetry, abandonment and alliance cohesion. Camb. Rev. Int. Aff. **37**(2), 206–225 (2024). https://doi.org/10.1080/09557571.2023.2225650
85. European Commission [EC].: United States of America. Energy, climate change, environment. (n.d.). https://energy.ec.europa.eu/topics/international-cooperation/key-partner-countries-and-regions/united-states-america_en. Last accessed 21 Oct 2025
86. Sanderson, H.: Volt rush: the winners and losers in the race to go green. Oneworld publication, London (2022)
87. Nakanwagi, S.: Critical minerals, sustainability, and the energy transition in the Global South: A justice perspective. Hart Publishing, London (2024)
88. Giuli, M., Oberthür, S.: The EU's external energy governance in the age of the energy transition. In: Scholten, D. (ed.) Handbook of on the geopolitics of the energy transition, pp. 404–419. Edward Elgar Publishing, Cheltenham, UK/Northampton, MA, USA (2023)
89. Criekemans, D.: Geopolitics, geoeconomics, and energy security in an age of transition towards renewables. In: Scholten, D. (ed.) Handbook of on the geopolitics of the energy transition, pp. 21–43. Edward Elgar Publishing, Cheltenham, UK/Northampton, MA, USA (2023)

90. Herranz-Surrallés, A.: External pressures from 'enemies', 'rivals' and 'allies': EU energy policy integration in response to geopolitical challenges. J. Eur. Publ. Policy (2025). https://doi.org/10.1080/13501763.2025.2569576
91. The Energy Information Administration [EIA].: U.S. natural gas imports by country. (2025). https://www.eia.gov/dnav/ng/ng_move_impc_s1_a.htm. Last accessed 5 Dec 2025
92. The Energy Information Administration [EIA].: U.S. net imports by Country (products), (2025). https://www.eia.gov/dnav/pet/pet_move_neti_a_EPP0_IMN_mbblpd_a.htm. Last accessed 5 Dec 2025
93. United Nations Development Programme [UNDP].: Towards human resilience: Sustaining MDG progress in an age of economic uncertainty. UNDP, (2011)
94. The Energy Information Administration [EIA].: Monthly Energy Review August 2025. U.S. Energy Information Administration, 8/25/2025. DOE/EIA-0035(2025/8), (2025)
95. Jensen, L.: Global decarbonization in fossil fuel export-dependent economies: Fiscal and economic transition costs. UNDP, (2023)
96. Energy Institute.: Statistical review of the world energy 2025, 74th edn. Energy Institute in collaboration with Kearny, (2025)
97. The World Bank.: Fossil fuel energy consumption (% of total). (n.d.). https://data.worldbank.org/indicator/EG.USE.COMM.FO.ZS. Last accessed 28 Sep 2025
98. Eurostat.: Shedding light on energy in Europe–2025 edition. Energy sources, (2025). https://ec.europa.eu/eurostat/web/interactive-publications/energy-2025#energy-sources. Last accessed 28 Sep 2025
99. Szabo, J.: Energy transition or transformation? Power and politics in the European natural gas industry's trasformismo. Energy Res. Soc. Sci. **84**(2021), 1–14 (2021). https://doi.org/10.1016/j.erss.2021.102391
100. Yang, Y., Xia, S., Huang, P., Qian, J.: Energy transition: Connotations, mechanisms, and effects. Energy Strat. Rev. **52**, 1–13 (2024). https://doi.org/10.1016/j.esr.2024.101320
101. Trump, D.: Remarks to the world economic forum and a question-and-answer session with business leaders. The American Presidency Project, (2025). https://www.presidency.ucsb.edu/documents/remarks-the-world-economic-forum-and-question-and-answer-session-with-business-leaders. Last accessed 11 July 2025
102. Stevanović, O.: The Trump administration's approach to energy security in Poland and the Baltic States. Polit. Nac. Bezb. **18**(1), 33–67 (2020)

Energy and Technological Aspects of Electric Power Systems' Security in Transition Trajectories

Vojin Grković

Abstract Security is a key requirement that must be met when designing and operating power systems (PS). The ultimate consequence of the insecurity of the PS is manifested in its technological disintegration, or colloquially, "blackout" followed by the energy supply break. The paper presents the concept of energy and technological security of power systems in transition, structured in three levels. The first, basic level is the level of the portfolio for electricity generation. The second level is the level of individual equipment within the portfolio, and the third level is equipment maintenance and management of the electricity generation process with available equipment and with the available portfolio. The energy security of the PS is correlated with dispatchability, or the ability of the system to respond to a request for the power increase. Appropriate indicators can be used to quantitatively define security. At the first level, these are dispatchability indicators at the second level, indicators based on the rate of change of a power plant's load, and at the third level, the performance indicators can be used. The paper provides examples of the impact of the trajectory of a hypothetical energy or technological transition on the dispatchability of power systems in nine European countries. The paper also presents the effects of the real energy transition on the dispatchability of the power system. It is concluded that the chosen transition trajectory, or the chosen final technological structure of the electricity generation portfolio, has a major impact on the numerical value of the dispatchability indicators and, through them, on the energy supply security of the power system, in terms of the risk of the system's technological disintegration. Insufficient security achieved at the first, basic level, cannot be compensated by the security improvements at the second and third levels.

Keywords Security of power systems · Dispatchability · Energy transition · Technological transition

V. Grković (✉)
Faculty of Technical Sciences, University of Novi Sad, Novi Sad, Serbia
e-mail: vojingr@uns.ac.rs

Z. Keković et al. (eds.), *Critical Infrastructure: Risk and Crisis Management in the New Reality*, Advanced Sciences and Technologies for Security Applications,
https://doi.org/10.1007/978-3-032-20948-1_4

1 Introduction

Today, the main goals of the energy transition are the drastic reduction, or rather the elimination, of carbon dioxide emissions during energy transformation and consumption, and, as a second goal, ensuring sustainable development. Energy security is one of the pillars of the energy transition. In the system of displaying and analyzing the periodic and final results of the energy transition, a complex indicator known as the energy trilemma is used [1]. Based on this indicator, the World Energy Council has been publishing annual reports on the ranking of countries since 2010. One of the three dimensions of this indicator refers to energy security [1]. The countries gathered in the BRICS group emphasize the "fundamental role of energy security in achieving the Sustainable Development Goals" [2]. In this sense, we can say that the energy transition is inseparable from sustainable development, in which energy security plays an essential role.

Within the framework of energy security at the level of states, or groups of states, the security of power systems is of particular importance for the implementation of energy transition tasks. Due to the unique and very demanding technology of power systems' operation, their energy security is a very complex issue. The main aspects of this issue will be given due attention in the following wording.

2 Power System

The power system represents a unique energy and technological entity. It includes all sources of electricity, the transmission system and electricity distribution systems, all belonging electricity consumers, and all connections with neighboring power systems. Each of these parts of the power system must be dimensioned for the appropriate operation load. The basic technological requirement for the safe operation of the power system is that the power of all energy sources at any time must be equal to the sum of the power of all connected consumers and all losses in transmission and distribution. If the power of all connected consumers (including all losses) is less than the sum of the powers of all sources, then the turbogenerators accelerate, so it is necessary power plant's control system to reduce the power of the turbogenerators. Technically, this power reduction occurs, as a rule, not in all, but only in a certain number of generators, so that the total power of all generators is equal to the sum of the powers of all consumers and all losses. The problem arises when the power of all connected consumers and all losses is greater than the power of all sources of electrical energy. Then there is a slowdown, or a certain reduction in the number of revolutions of all generators. This reduction in the number of revolutions is accompanied by an increase in torsional stresses in the rotors of the turbogenerators. If the system cannot increase the power (all generators operate at maximum power), there are two options: to increase the power of electricity import from neighboring power systems, and/or to disconnect a part of the consumers from the network. If these measures are not sufficient, or if they are not implemented quickly enough, the protective system of one turbogenerator automatically operates and disconnects the generator from the network. Since at that moment the imbalance between the power of all consumers and all generators is even greater, in a very short time, all remaining generators on the network are automatically disconnected, so the consumers lose their energy supply almost instantly. This scenario represents technological disintegration, or colloquially, the "blackout" of the power system.

In the case that one or more power plants suddenly "fall out" of operation due to a failure, for example on a transmission line, or due to natural conditions (wind cessation), an imbalance immediately occurs between the power of all remaining power plants in operation and the power of all currently connected consumers plus losses, which, in the case of large imbalances, can lead to the already described scenario of a power system blackout.

The largest and the most consequential blackout of a power system occurred in the year 1965 in the New York and some east states of the USA [3]. The last in a series of major power system collapses occurred on April 28, 2025 and involved the power systems of Spain, Portugal, and parts of southern France [4].

3 Security Structure of a Power System

There are three levels of implementation of power systems' energy and technological security in the energy transition. The first level refers to the structure of applied technologies for electricity generation in the system. The second level refers to the level of special technological characteristics of individual technologies, i.e., thermal power plants. Finally, the third level refers to the quality of equipment maintenance and management of the electricity generation process in the power plants, Fig. 1.

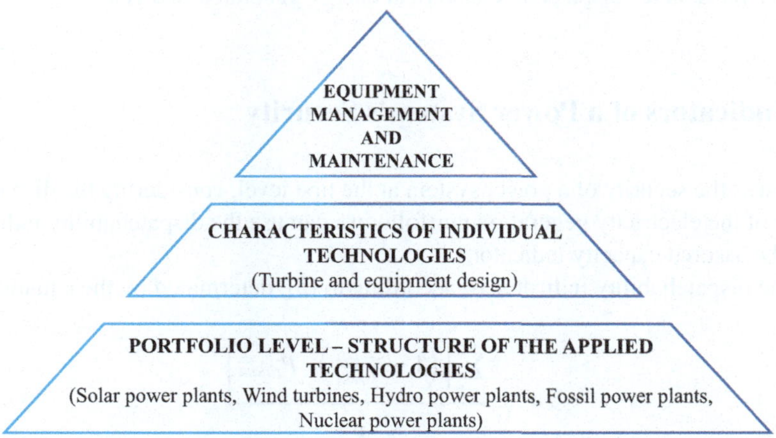

Fig. 1 Three levels of the power system's energy and technological security

The ability of the power system to respond to a demand for the power increase, i.e., its ability to increase the power of the power sources in the timely manner to the power level of all connected consumers plus losses, is called the dispatchability of the power system. We can directly influence the dispatchability of the power system by choosing the appropriate structure of electricity generation technologies—the first level, then by choosing equipment with appropriate characteristics—the second level, and indirectly through the availability of the equipment, which can be obtained by equipment's' good maintenance, as well as by good management of the electricity generation process—the third level.

The sources of electricity in the power system can be different technologies, namely: fossil fueled (coal, oil, or natural gas), nuclear, hydroelectric, wind, solar, and biomass power plants.

There is a significant difference between individual technologies for generating electricity in terms of their ability to first respond and then to respond quickly enough to the system's demand for the power increase. For example, fossil or nuclear power plants with steam or gas turbines can safely change power in the range between their minimum and maximum power. On the other hand, solar and wind power plants, when there is no wind, i.e., the sun, cannot increase power. Therefore, the technological structure, i.e., the structure of electricity sources in the power system's portfolio, is very important for the power system's security. We define the technological structure as the participation of individual technologies in the total electricity production of the observed system for a year. Particularly indicative is the ratio of electricity produced by non-dispatchable sources (wind and sun) in a year ($E_{i\text{-}RES}$) to the annual production of all electricity sources in the observed system (E_{tot}), which is analytically presented by Eq. (1), according to [5].

$$\lambda_{i-RES} = \frac{E_{i-RES}}{E_{i-RES} + E_{DIS}} = \frac{E_{i-RES}}{E_{tot}} \tag{1}$$

Here, E_{DIS} denotes dispatchable electrical energy generated in a year.

4 Indicators of a Power System's Security

To assess the security of a power system at the first level, considering the dispatchability of the electricity generation portfolio, we can use the dispatchability indicator and the assured capacity indicator.

The dispatchability indicator is, according to [5], determined by the equation:

$$DI = \frac{\sum_{i=1}^{n} \left[\left(1 - \frac{P_{timin}}{P_{tinom}} \right) \cdot P_{tinom} \right]}{\sum_{i=1}^{n} P_{tinom}} \tag{2}$$

where P_{tinom} (MW) denotes the nominal power of the i-th power plant, P_{timin} (MW) denotes the minimum power of the i-th power plant, while n is the number of power plants in the observed power system. We can see that this indicator is based on the sum of the available power change ranges for each of the power plants in the portfolio.

The assured capacity indicator at the power system level is based on the concept of assured capacity at the technology level, defined by Urlich and Schiffer [6]. The assured capacity indicator at the power system level is, according to [7], determined by the equation:

$$ACT = \sum_{i=1}^{n} \frac{IC_i}{\sum_{i=1}^{n} IC_i} \cdot AC \tag{3}$$

Here, AC denotes the assured capacity at the technology level in percentage, according to Urlich and Schiffer [6], while IC_i denotes the installed capacity of the considered technology in the power system. The index i denotes the i-th technology, i.e., the power plant, while n is the number of power plants in the observed power system.

In general, the numerical values of both indicators depend on the technological configuration of the considered power system. It is very useful to express the configuration by the participation of intermittent renewable energy sources (i–RES) in the total electricity generation $\lambda_{i\text{-}RES}$, according to Eq. (1).

Bothe dispatchability indicators—*DI* and *ACT*—quantify the technical ability of the power system to respond to a demand for power change. However, the extent to which this ability will be used in practice depends on the realistically achievable rate of the power change in the system, and this is the subject of analysis in the second level of the power system security. Low values of *DI* and *ACT* indicators cannot be significantly compensated by the choice of power plant design in the second level. However, inadequate power plant design, selected in the second level, can significantly weaken the effects of dispatchability achieved in the first level, i.e., the overall dispatchability of the power system. This conclusion applies, first of all, to systems with a large share of i-RES in total electricity generation.

At the second level of energy and technological security of the power system, it is necessary to take into account the detailed characteristics and technical solutions of the equipment of each individual source, such as, for example: what speed of power change is allowed, or enabled by the specific turbine design, then, whether the power control is carried out at a fixed or sliding pressure, etc. In addition, it is necessary to assume a reference speed of electricity consumption power change and then compare the values of the required and possible speed of power change. All in all, this is a very detailed analysis for which it is necessary to have a fully defined specific portfolio of the power system in which all technological characteristics of each individual electricity source are known, on the one hand, and when we have a fully defined electricity consumption, on the other hand. The result of this analysis should be an indicator of the security of the power system at the second level, where the speed of power change of individual plants is key. In that respect, an important indicator of dispatchability at the second level is the duration of power changes in the power system. This indicator, according to Grković et al. [5], is determined by the equation:

$$DPC = \frac{\sum_{i=1}^{n} \left[\left(1 - \frac{P_{timin}}{P_{tinom}} \right) \cdot P_{tinom} \right]}{\sum_{i=1}^{n} \dot{p}_{t\,i}} \tag{4}$$

Here $\dot{p}_{t\,i}$ (MW/min) denotes the rate of load change of the i-th power plant.

In the first and second levels, the energy and technological security of the power system are defined in relation to the technical factor—the technological structure of the electricity generating portfolio and the applied equipment. In the third level, the human factor is decisive, that is, the people who manage the production and maintenance of the equipment. Three things are key here. First, each employee must have the appropriate education to perform the tasks entrusted to them. Second, the hierarchy of production and maintenance management must coincide with the hierarchy of knowledge and education. The one who should define and/or give instructions for the implementation of tasks and control how the tasks are performed must have an appropriate higher education and knowledge than those to whom he gives instructions and whom he controls. This applies to all levels in the organization of the power system.

Third, it is necessary to implement the best equipment maintenance system. In this respect, the development of the energy sector has advanced significantly, so that current maintenance systems provide a significantly higher level of security of the power system compared to previous maintenance systems. Here, unlike the first two levels, we can objectively determine the level of security only by monitoring the results of electricity production in a specific system, or, possibly, by comparing it with other similar systems. An exhaustive overview of over forty different performance indicators for monitoring electricity production facilities was provided by IEEE [8]. Here we will point out three, generally the most frequently used, indicators of operational performance, i.e. availability indicator, equivalent availability, and forced outage rate. In addition to performance indicators, energy transformation efficiency indicators can also be used to assess the quality of work of personnel responsible for production and maintenance in a power plant. Inadequate plant management and equipment maintenance lead to equipment degradation, which is then manifested by a deterioration in energy transformation efficiency indicators.

5 Application Examples and Discussion of Results

Different transition trajectories can lead to the set goal: transformation and use of energy without or with drastically reduced carbon dioxide emissions. These different transition trajectories have different effects on the security of the power system, as well as on the overall investments. To prove this hypothesis, we will observe nine European countries in a hypothetical energy transition along two transition trajectories: (1) replacement of coal-fired power plants (including lignite) with wind power plants and (2) replacement of coal-fired power plants (including lignite) with nuclear power plants. Of the nine countries considered, eight are members of the European Union, and one—Serbia—is not. For each country, numerical values of the *DI* indicator were calculated after the hypothetical transition to nuclear power plants (a), after the hypothetical transition to wind power plants (b), and before the beginning of the transition (c). The results of the analysis are shown in Fig. 2, according to Grković et al. [9], where we can see two groups of countries. The first group includes Belgium, France, and Austria, where the numerical values of the dispatchability indicators differ rather little between the transition trajectories to wind power plants and nuclear power plants. These small differences can be explained by the small share of coal-fired power plants at the beginning of the transition in these countries. In the countries of the second group, both transition trajectories lead to a decrease in the dispatchability of the power system compared to the initial level. In addition, in the countries of this group, the transition to wind power plants results in a significantly lower dispatchability of the power system, which practically means its lower security, expressed as resistance to disintegration, and thus the resistance to the loss of energy supply, than the transition to nuclear power plants. In this regard, of the countries considered, Serbia finds itself in the most difficult situation, because the transition to wind power plants would result in the dispatchability of the power system being nearly five times lower than in the conditions before the transition began. Similar relationships between these two transition trajectories, in the nine European countries observed, are also obtained by applying the *ACT* indicator [7].

In addition to these basic ones, other, specifically designed transition trajectories are also very interesting from the energy and technological security point of view. For example, there is the possibility of coupling the ecological and energy transitions. This solution is based on the use of forests and hydro flora to store carbon dioxide and the use of fossil fuel power plants to produce electricity, according to Grković [10].

Now, we will consider an example of the energy and technological security of the power system in the actual energy transition that is taking place in line with the previously developed project. In the part related to the power system, the transition includes the replacement of solid fossil fuel power plants and the remaining nuclear power plants in Germany with wind power plants and solar power plants, according to Kai van de Loo [11]. Figure 3 shows calculated dependences of the values of the *ACT* indicator, and the $\lambda_{i\text{-}RES}$ indicator, for the actual energy transition in Germany. The calculation was carried out based on the data published by Kai van de Loo in

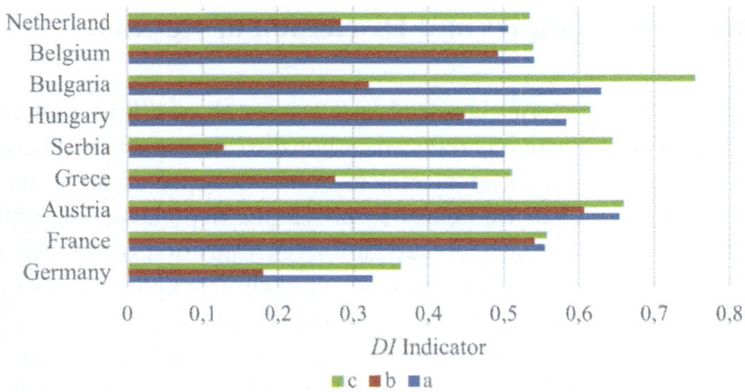

Fig. 2 Dispatchability indicator of power systems in nine European countries, **a** after transition to nuclear power plants, **b** after transition to i–RES, **c** before the transition, according to [9]

[11]. From Fig. 3, we see that during the implementation of this transition, there will be a significant increase in participation of i–RES followed by significant decrease in dispatchability. This decrease in dispatchability leads to a significant reduction of the power system's technological and energy security pointing out the possibility of the power system disintegration. Potential improvements in dispatchability that could be achieved by building energy storage capacities are not included here.

We can also mention the predictions of Goldman Sax [12] that the volume of investments in the power systems in Europe over the next ten years will amount 3000 bn€ (3 trillion €). A large part of these investments is planned for improving the security of electricity supply including increasing the dispatchability of the power systems. Another part of the investments is planned to modernize outdated transmission lines, and to reconfigure them to better suit the new configuration of electricity sources.

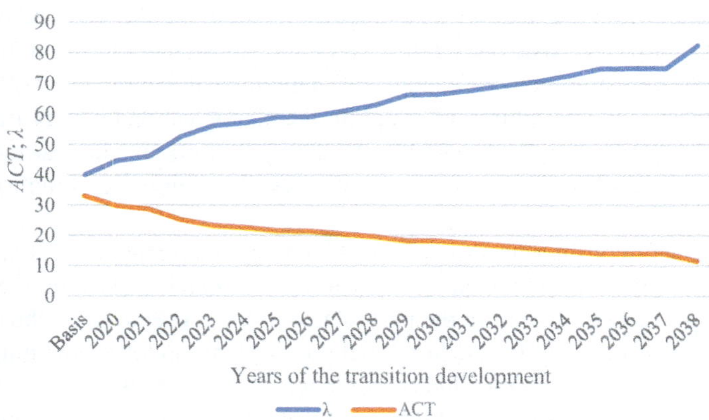

Fig. 3 Values of the *ACT* indicator and $\lambda_{i\text{–}RES}$ for the actual energy transition in the German power system, calculated according to data from [11]

For improving the dispatchability and thus the security of energy supply in the power system in the energy transition on the trajectory from fossil fuel power plants to wind and solar power plants, it is necessary to build certain systems for storing the electricity generated by i–RES. The construction of these systems can significantly improve the dispatchability of the power system. If all capacities of energy storage technologies amount to 40% of all wind and solar capacities, the dispatchability of such a unit could be at the level of fossil fueled power plants' dispatchability [13]. A very useful technology for storing energy is the technology of pumped-hydro storage power plants. There are possibilities to improve this technology by combining it with a compressed air energy storage system, according to Huanran et al. [14]. Batteries are of particular interest for solar power plants, for which there are several different variants. The most promising are the so-called organic flow batteries, whose development is ongoing [15]. Among other batteries, we can single out ferro-chromium batteries, which can have about six times smaller cost of stored electricity than lithium-ion batteries, and sodium-sulfide batteries, which can have about three times smaller cost of stored electricity than lithium-ion batteries, according to Eliett [16]. The construction of energy storage technologies is a very complex undertaking, and their intensive development is in full swing [17].

6 Conclusions

The break of energy supply can be caused by serious technological misfunctioning within the power system. Therefore, energy security in the power system is closely connected to the technological operations security. In other words, energy supply security as the top-level request is conditioned with corresponding technological security of the power system, which is in this paper considered as the system's ability to avoid its disintegration, or colloquially, blackout.

For ensuring energy supply security, the power system's technological security is conceptualized in three levels in this paper. The first level represents the technology portfolio level, the second level represents the individual technology level, and the third level represents the power plant's maintenance and operation.

In the first level the energy supply security in the form of technological security, can be quantified with the dispatchability indicators, in the second level with the indicators based on the speed of power plant's load change, while in the third level the security can be indirectly indicated with performance indicators. If sufficient security is not achieved at the first, basic level, it cannot be compensated by the security improvements at the second and third levels. This is of great importance in defining the energy transition trajectory.

The transition trajectory from fossil fueled power plants to nuclear power plants enables good dispatchability and thus the good technological, as well as good energy supply security. In the case of transition trajectory from fossil fueled power plants to i-RES, the energy storage technologies must be applied to improve the dispatchability, and thus the technological and finally the energy supply security of the power system.

References

1. Marti, L., Puertas, R.: Sustainable energy development analysis: Energy trilemma. Sustain. Technol. Entrep. **1**, (2022). https://doi.org/10.1016/j.stae.2022.100007
2. XIV BRICS Summit.: (2022). http://brics2022.mfa.gov.cn/eng/dtxw/202206/t20220624_10709295.html (16.08.2022)
3. https://www.baruch.cuny.edu/nycdata/disasters/blackouts-northeast.html
4. https://www.entsoe.eu/publications/blackout/28-april-2025-iberian-blackout/
5. Grković, V., Čantrak, D.J., Benisek, M., Petrović, D., Doder, D.J.: Assured capacity, total assured capacity and dispatchability of Serbian power system and power plants—a contribution to the research. IEEP 2022, Belgrade, Srbija (2022)
6. Ulrich, S., Schiffer, H.-W.: Prospects for development of power generation in Europe. VGB PowerTech **12**(2019), 43–50 (2019)
7. Grković, V., Doder, D.J.: Indicating power system's dispatchability. Acad. Lett., Article 2231, (2021). https://doi.org/10.20935/AL2231
8. IEEE standard definitions for use in reporting electric generating unit reliability, availability, and productivity. IEEE 3 Park Avenue, New York (2007)
9. Grković, V., Kljajić, M., Doder, D.J., Živković, V.: Integral approach to competitiveness analysis of electric power systems in technological transition and post transition, SIMTERM 2022, Niš, pp. 139–149 (2022)
10. Grković, V.: Coupled environmental and energy transition for power systems' security—the Relevance of Serbia for BRI. In: Jovanović, A., Junkai, L., Hong, H. (eds.), Innovation for enhanced resilience and sustainability along the belt & road initiative. Springer Nature, pp. 241–256 (2025) (in press)
11. Van de Loo.: Suspend coal phase-out—keep existing plants in operation and reactivate available capacities, give transition more time. Vgbe Energy J., **3**, 73–84 (2023)
12. Goldman Sachs: Europe needs $3.5 trillion of power investment through 2035, (2025). https://www.goldmansachs.com/insights/articles/
13. Grković, V.: Transition and post-transition: energy and technology. Society of engineers in Zrenjanin and City national library in Zrenjanin, Zrenjanin, (2024) (in Serbian)
14. Huanran, W., Guang, X., Ruixiong, L., Erren, Y., Zhongguo, S., Zhiheng, W., Hansen, Z., Lanning, L., Xin, H., Yufei, Z.: A study on the combined pumped-hydro and compressed air energy storage system. School of Energy and Power Engineering, Xi'an Jiaotong University, Xi'an 710049, China (2024)
15. Ferreira, H.L., Garde, R., Fulli, G., Kling, W., Lopes, J.P.: Characterisation of electrical energy storage technologies. Energy **53**(2013), 288–298 (2013)
16. Eliett, R.: Renewable energies and their challenges: Uncertainty and variability of the wind and solar resource. Acad. Lett., Article 2341 (2021). https://doi.org/10.20935/AL2341
17. Grković, V.: Problems and controverses of energy storing from the perspective of energy transition. DIT, Society of Engineers in Zrenjanin, no. 43, pp. 41–52 (2025). (in Serbian)

Enhancing Critical Infrastructure Resilience: An Integrated Framework for Hazardous Material Management of Electric Vehicle Batteries

Dragan Vukmirović, Stefan Radojičić, Nebojša Stanojeveć, Nebojša Dragović, and Tijana Čomić

Abstract This paper addresses the "electrification paradox": while electric vehicles (EVs) are vital for decarbonising transport, the rapidly growing wave of end-of-life (EoL) EV batteries creates a new, high-stakes hazardous waste stream that threatens the resilience of storage, transport and recycling infrastructures. We first synthesise the regulatory and scientific basis for treating EoL lithium-ion batteries as hazardous materials, highlighting the combined risks of toxic emissions, long-term environmental contamination, and acute thermal runaway events. Building on this risk profile, we propose an integrated technology framework that combines the Internet of Things (IoT), high-fidelity Digital Twins, and Generative Artificial Intelligence (AI) to move from reactive, compliance-driven management to predictive, adaptive and resilient operations. Three detailed use cases—proactive thermal-runaway prevention in warehouses, dynamic rerouting of high-risk loads, and AI-assisted emergency response—illustrate how the framework senses, simulates, and responds to evolving hazards in real time. We then examine key implementation challenges, including cybersecurity, scalability, interoperability, explainability, human-in-the-loop governance and algorithmic fairness. Finally, we show how the same data backbone underpins a Digital Battery Passport that enables second-life markets, optimised recycling and alignment with emerging policy, positioning EoL batteries as traceable resources within a secure circular economy. We conclude by outlining future research directions in multi-agent control, quantum optimisation and end-to-end digital threads for critical materials.

D. Vukmirović (✉) · S. Radojičić · N. Stanojeveć · N. Dragović
Faculty of Organizational Sciences, University of Belgrade, Belgrade, Serbia
e-mail: dragan.vukmirovic@fon.bg.ac.rs

T. Čomić
The Faculty of Organizational Studies "Eduka", University Business Academy in Novi Sad, Belgrade, Serbia
e-mail: tijana.comic@vos.edu.rs

Z. Keković et al. (eds.), *Critical Infrastructure: Risk and Crisis Management in the New Reality*, Advanced Sciences and Technologies for Security Applications,
https://doi.org/10.1007/978-3-032-20948-1_5

Keywords Electric vehicle batteries · Hazardous materials · Critical infrastructure resilience · Internet of Things · Digital Twin · Generative artificial intelligence · Circular economy

1 Introduction

The global transition toward electric mobility represents a critical step in decarbonising the transportation sector and mitigating the impacts of climate change [1, 2]. However, this progress introduces a significant and escalating challenge for critical infrastructure: managing a massive, imminent wave of end-of-life (EoL) electric vehicle (EV) batteries. This "electrification paradox" posits that addressing one environmental crisis by electrifying energy systems creates a new and formidable hazardous waste stream that threatens the resilience of waste management facilities, transportation networks, and recycling centres [2, 3, 5]. Spanning from electric scooters to commercial trucks, these lithium-ion batteries are not benign; they are classified as hazardous materials due to their inherent chemical, fire, and environmental risks, which persist and even increase at the end of their useful life [2–4]. Current risk management paradigms, which are predominantly reactive and rooted in regulatory compliance, are proving fundamentally inadequate for the unique complexities of EoL EV batteries. The sheer volume of batteries entering the reverse supply chain, combined with their diverse and often proprietary chemistries and the high-consequence nature of potential failures, overwhelms traditional methodologies [2, 3, 5]. The most acute of these risks is thermal runaway—a rapid, self-sustaining, and uncontrollable exothermic reaction that can lead to intense fires, the release of toxic gases, and explosions [4]. This danger is exacerbated in EoL batteries, which are more susceptible to the physical damage and internal degradation that can trigger such events. The prevailing linear economic model of "take–make–waste" further compounds these risks by treating EoL batteries as a disposal liability rather than a valuable resource, thereby failing to incentivise the development of robust, safe, and sustainable EoL pathways [3, 5].

This paper posits that a paradigm shift from a reactive to a predictive, adaptive, and resilient management model is not only necessary but imminently achievable through the synergistic integration of key Industry 4.0 technologies. We propose a novel, integrated framework that leverages the Internet of Things (IoT), Digital Twins, and Generative Artificial Intelligence (AI) to revolutionise the lifecycle management of EoL EV batteries. In this framework, IoT serves as a distributed sensory layer, providing a constant stream of real-time data on the condition and location of every battery. This data feeds into a high-fidelity Digital Twin, a dynamic, virtual simulation environment that contextualises real-time data and enables predictive "what-if" analysis. At the core of the system, Generative AI functions as the cognitive engine, analysing data from the IoT network and running simulations on the Digital Twin to predict failures, optimise logistics, and dynamically generate context-aware safety protocols and emergency response plans. This integrated approach is designed not only to mitigate the immediate safety and logistical risks but also to provide the foundational data infrastructure required to unlock a secure, resilient, and sustainable circular economy for EV batteries.

The main contributions of this paper are fourfold:

1. **Conceptual contribution**: We introduce a holistic, integrated framework that combines IoT, high-fidelity Digital Twins, and Generative AI specifically for the hazardous material management of EoL EV batteries, with a focus on critical infrastructure resilience.
2. **Operational contribution**: We develop detailed operational workflows and use-case scenarios—covering proactive thermal runaway prevention, resilient reverse logistics, and intelligent emergency response—that demonstrate how the framework can be deployed in real-world storage, transport, and crisis-management settings.
3. **Governance and policy contribution**: We align the proposed framework with emerging regulatory and critical-infrastructure resilience agendas, including hazardous materials regulations and the EU Batteries Regulation, highlighting how IoT–DT–GenAI integration can support compliance, explainability, and human-in-the-loop decision-making.
4. **Circular-economy contribution**: We show how the same data infrastructure can underpin a Digital Battery Passport, enabling safe second-life applications, efficient recycling and material recovery, and a data-driven circular economy for EV batteries.

Together, these contributions provide both a conceptual blueprint and a practical roadmap for enhancing the resilience of critical infrastructure in the face of the rapidly growing EoL EV battery challenge.

2 The End-of-Life EV Battery: A High-Stakes Hazardous Material

To fully appreciate the necessity of an advanced management framework, it is essential first to establish the precise nature of the threat. EoL EV batteries are not merely waste; they are a complex, high-stakes category of hazardous materials, governed by stringent regulations that reflect their significant inherent risks. This section provides a detailed examination of the regulatory landscape and the scientific underpinnings of the chemical, environmental, and thermal hazards associated with these batteries.

2.1 Methodology

This study employs a conceptual, **design-oriented research methodology** to develop an integrated operational framework for managing hazardous materials associated with EoL EV batteries. Given the systemic, interdisciplinary, and emerging nature of the problem—situated at the intersection of hazardous materials management, critical infrastructure resilience, digital technologies, and regulatory compliance—the research does not rely on primary empirical data; instead, it synthesises established knowledge, regulatory frameworks, and emerging technological paradigms.

The methodological approach is structured around **four complementary components**.

First, a **systematic review and synthesis of secondary sources** is conducted, encompassing international and national regulatory documents, technical standards, scientific literature, and industry reports related to lithium-ion battery hazards, transport and waste classification, and circular economy policies. This component establishes the regulatory and risk baseline that defines EoL EV batteries as a high-stakes hazardous material category.

Second, the study applies a **conceptual integration methodology**, combining established digital paradigms—including the Internet of Things (IoT), Digital Twin technology, Generative Artificial Intelligence, and distributed ledger technologies—into a unified architecture tailored specifically to hazardous material management. Rather than proposing novel technologies in isolation, the research focuses on their **functional integration within a safety–critical,** regulation-intensive context.

Third, a **design science logic** is employed to develop the proposed framework. This involves the iterative construction of a system architecture, supported by risk–technology mapping and use-case reasoning, with particular attention to emergency response, logistics operations, and EoL battery traceability. The framework is evaluated through **theoretical validation and practical plausibility**, assessing its capacity to address known shortcomings in current hazardous material management systems.

Finally, the research incorporates a **scenario-based analytical perspective**, examining how the proposed framework could operate under real-world conditions, including thermal runaway incidents, transport disruptions, and regulatory non-compliance risks. This approach enables critical assessment of the model's strengths, limitations, and implementation challenges without overstating empirical claims.

Through this methodology, the paper aims to contribute a **coherent, operationally grounded, and policy-relevant framework** that bridges the gap between regulatory requirements, technological capabilities, and the practical realities of managing EoL EV batteries as hazardous materials.

2.2 Regulatory Classification and Compliance Burden

The classification of EoL lithium-ion batteries as hazardous material is not a matter of administrative convenience but a direct codification of their intrinsic physical and chemical properties. In the United States, multiple federal agencies have established a complex regulatory framework governing the handling, transportation, and disposal of these materials. The U.S. Environmental Protection Agency (EPA), under the Resource Conservation and Recovery Act (RCRA), classifies most lithium-ion batteries disposed of as hazardous waste due to their ignitability (waste code D001) and reactivity (waste code D003) characteristics [7, 8]. These classifications are a direct response to the scientifically established risk profile of lithium-ion batteries: high energy density, flammable organic electrolytes, and the potential for uncontrolled exothermic reactions that can culminate in thermal runaway [6]. Thus, the regulations themselves act as a lagging indicator of the underlying physical hazards. A truly resilient management system must therefore address the root scientific causes of these risks—the potential for uncontrolled thermal and chemical feedback loops—rather than merely aiming for post-facto regulatory compliance [6, 9].

Complementing the EPA's framework, the U.S. Department of Transportation (DOT) regulates lithium batteries as a Class 9 hazardous material under its Hazardous Materials Regulations (HMR; 49 C.F.R., Parts 171–180) [9]. These regulations apply to any material that the DOT determines can pose an unreasonable risk to health, safety, and property during transport in commerce, and they mandate strict requirements for UN-certified packaging, labelling, hazard communication, and employee training across all modes of transport [9, 11].

To ease the compliance burden on the wide variety of commercial entities that handle ordinary hazardous wastes, the EPA allows sure lithium-ion batteries to be managed under the streamlined Universal Waste Rule (40 C.F.R. Part 273) [10]. This framework reduces specific administrative requirements, such as the need for a hazardous waste manifest during shipment, while maintaining baseline controls on labelling, accumulation time, and transport [10]. However, a critical and often overlooked limitation is that this streamlined pathway does not apply to Damaged, Defective, or Recalled (DDR) batteries, which are subject to more stringent handling and packaging guidance because of their elevated risk of fire and venting [11, 12]. This is a significant constraint, as EoL batteries—especially those from vehicles involved in collisions or improperly handled—are disproportionately likely to be in a DDR state. Consequently, a substantial portion of the EoL battery stream may not qualify for the simplified universal waste pathway, forcing handlers back into the whole hazardous-waste management regime [7, 8, 10].

This regulatory nuance underscores the need for a system capable of accurately assessing and tracking the physical integrity and risk status of each battery pack throughout the reverse logistics chain, and of aligning that dynamic risk assessment with appropriate regulatory categories and emergency response guidance [11–13].

Beyond the United States, similar risk-based classifications exist in other major jurisdictions. In the European Union, lithium-ion batteries used in EVs are generally classified as dangerous goods under UN numbers 3480 (lithium-ion batteries) and 3481 (lithium-ion batteries contained in equipment or packed with equipment). They are subject to multimodal transport regulations, including ADR, RID, and the IMDG Code [14, 15]. At the waste stage, EoL EV batteries fall under the broader EU hazardous waste framework. At the same time, the new EU Batteries Regulation (EU) 2023/1542 introduces a lifecycle-oriented approach that tightens requirements on safety, collection, and traceability [16]. This regulation, discussed in more detail in Sect. 6.3, further reinforces the need for data-driven systems capable of tracking the chemistry, state of health, and integrity of each battery pack throughout its logistics chain [16].

2.3 The Spectrum of Risks

The regulatory framework is built on a scientific understanding of a multifaceted risk profile encompassing both long-term environmental contamination and acute, high-consequence safety events.

Chemical and Environmental Hazards

EV batteries contain a cocktail of materials that are both valuable and hazardous. The cathodes are typically composed of lithium-containing compounds such as lithium cobalt oxide and related layered oxides. At the same time, the anodes are typically graphite-based, and the electrolyte is a flammable organic solvent containing lithium salts, such as lithium hexafluorophosphate ($LiPF_6$), dissolved in carbonate solvents [17, 18].

Improper disposal of these batteries—for example, in municipal landfills or through informal dumping—poses a significant environmental threat. Over time, battery casings can corrode and degrade, leading to the leaching of heavy metals and other toxic species into soil and groundwater, causing long-term contamination of terrestrial and aquatic ecosystems [19]. In addition, studies of field incidents and laboratory simulations have shown that fires involving spent lithium-ion batteries can release highly toxic and corrosive gases, including hydrogen fluoride (HF) and, under certain conditions, hydrogen cyanide (HCN), as well as metal-containing particulates [19–21]. These emissions pose an immediate threat to human health and the environment, particularly in enclosed facilities or densely populated areas.

Taken together, these environmental risks underscore the imperative to shift from a linear disposal model to a circular economy, in which valuable metals such as lithium, cobalt, nickel, and manganese are safely recovered and reintroduced into the supply chain [2, 19]. Such circular approaches not only reduce the environmental footprint and ethical concerns associated with primary resource extraction but also help secure the critical supply of raw materials for future generations of EV batteries [2].

The Acute Threat of Thermal Runaway

The most immediate and dangerous risk associated with lithium-ion batteries is thermal runaway. This is not simply a conventional fire; it is a rapid, self-sustaining, and violent exothermic chain reaction within a battery cell that becomes extremely difficult to control once initiated [22]. The process begins when a cell overheats beyond a critical threshold. For fresh, non-aged cells, the onset temperature for runaway is often around 200–250 °C, but ageing-related degradation processes—such as increased internal resistance and growth of unstable interphases—can reduce this threshold to values on the order of ~150 °C in heavily cycled cells, substantially narrowing the safety margin [22, 26].

Several categories of abuse can trigger this initial overheating:

- **Mechanical abuse**—Physical damage such as crushing, penetration, or severe vibration can compromise the thin polymer separator that keeps the anode and cathode apart, leading to internal short circuits [22].
- **Electrical abuse**—Overcharging, deep over-discharging, or excessively rapid charging/discharging can promote lithium plating, dendrite growth, and localised hot spots, which in turn can pierce the separator and trigger internal shorts [22].
- **Thermal abuse**—Exposure to external heat sources (e.g., nearby fires, hot surfaces, or poor thermal management) can raise the internal temperature of the cell to its decomposition point [22].
- **Internal defects**—Manufacturing flaws, contamination, or latent defects in aged cells can evolve into internal short circuits without any apparent external trigger [22].

Once thermal runaway is initiated, the reaction becomes self-perpetuating. Exothermic decomposition of the solid electrolyte interphase (SEI), electrodes, and electrolyte releases heat faster than it can be dissipated, driving cell temperatures to well above 800–1,000 °C in extreme cases [22, 24]. At these temperatures, the flammable organic electrolyte vaporises and decomposes, producing a mixture of toxic, highly flammable gases, including hydrogen, carbon monoxide, methane, and various fluorinated species [20, 23]. The rapidly increasing internal pressure eventually causes the cell casing to rupture in a process known as venting, releasing a dark cloud rich in metal-containing particles, followed by a white or grey vapour cloud of flammable gases [20, 23]. If this vapour cloud ignites—which can occur seconds to minutes after venting—it can result in an explosion and a jet-like flame that can melt through metal containers.

In battery packs and modules, this jet-like flame and intense heat can easily propagate to adjacent cells, triggering a cascading failure throughout the entire assembly if no practical thermal barriers or mitigation strategies are in place [22, 25]. Experimental work on module-level thermal runaway has demonstrated that propagation speed and severity are strongly influenced by cell spacing, pack design, state of charge, and cooling strategy [25]. This risk is markedly elevated in the EoL battery population, which exhibits a higher prevalence of damaged, defective, or recalled units and more pronounced ageing-induced degradation [19, 26]. As a result, the prediction and prevention of thermal runaway—particularly in clusters of heterogeneous, partially degraded cells—are paramount concerns for the safe management of EoL EV batteries.

3 An Integrated Technology Framework for Proactive Resilience

To counter the multifaceted risks posed by EoL EV batteries, a reactive, compliance-driven approach is insufficient. A paradigm shift is required toward a proactive, data-driven, and intelligent system capable of anticipating, mitigating, and responding to threats in real time. This paper proposes such a system: an integrated technology framework built on three synergistic pillars—the Internet of Things (IoT), Digital Twins, and Generative Artificial Intelligence (AI). This architecture transforms the management of hazardous materials from a series of disconnected, manual processes into a cohesive, continuously learning ecosystem, as shown in Fig. 1.

The framework is organised into three tightly coupled layers. The sensing layer uses an IoT-enabled "nervous system" to capture high-frequency data from EoL EV batteries and logistics assets across the reverse supply chain. The virtualisation layer aggregates these data into high-fidelity Digital Twins that provide real-time situational awareness and a sandbox for risk-free simulation. At the top, the cognitive core combines predictive machine learning and prescriptive Generative AI to anticipate failures, optimise logistics, and generate context-aware emergency protocols. The following Sects. 3.1–3.3 elaborate on each of these layers in turn.

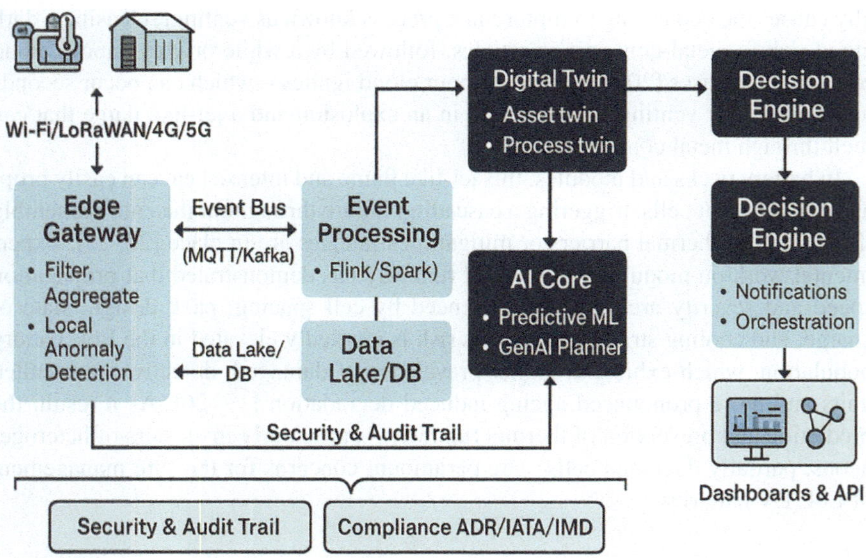

Fig. 1 Integrated technology framework for proactive resilience

3.1 The Sensing Layer: An IoT-Enabled Nervous System

The framework's foundation is a robust IoT sensor network that serves as a distributed nervous system, providing continuous, real-time data from every asset in the reverse logistics chain [27, 31]. This granular visibility is the first step in moving from assumption-based to evidence-based risk management. Sensors are embedded within individual battery packs, specialised transport containers, and storage facilities to monitor a suite of critical parameters.

- **Sensor Types and Monitored Parameters**
  - **Thermal Monitoring**. Arrays of thermocouples or thermistors are deployed to provide continuous temperature data at the cell, module, and pack level. This is the most direct method for detecting the early stages of thermal excursion, a primary precursor to thermal runaway [27–29].
  - **State of Health (SoH) and State of Charge (SoC) Monitoring**. Voltage and current sensors—such as shunt-based or integrated current-sense ICs—continuously measure electrical parameters during residual charging or discharging. This data is crucial for algorithms that estimate the battery's SoC (its current energy level) and, more importantly, its SoH—a measure of degradation and remaining capacity relative to the original state [27, 28, 32, 33]. A deteriorating SoH is a key indicator of increased internal resistance and a heightened risk profile.
  - **Physical Integrity Monitoring**. Micro-Electro-Mechanical Systems (MEMS) accelerometers and gyroscopes are integrated to detect and record physical events such as shocks, impacts, and excessive vibration during handling and transport. These sensors can measure acceleration over a range of ± 2 to ± 16 g and often include embedded features such as free-fall detection, providing an immutable record of mechanical abuse that may lead to latent internal damage and subsequent failure [31].
- **Data Architecture and Transmission**
  The data generated by this sensor array is transmitted wirelessly via protocols such as Wi-Fi, LoRaWAN, or cellular (4G/5G) to a local gateway or directly to a cloud platform [27, 31]. To manage the immense volume of data and ensure real-time responsiveness, a hybrid data processing architecture is employed. Edge computing devices located near the assets perform initial data filtering, aggregation, and anomaly detection, thereby reducing latency and bandwidth requirements while maintaining core safety functions even under degraded connectivity [30, 31]. This processed data is then streamed to a central cloud platform for more complex analysis, simulation, and long-term storage, where it can be fused with contextual information in the Digital Twin and consumed by AI-driven decision-support services (Fig. 2) [27–30].

Sensor data are ingested at the edge, streamed via an event bus to cloud-based real-time processing and batch analytics, and finally consumed by the decision engine

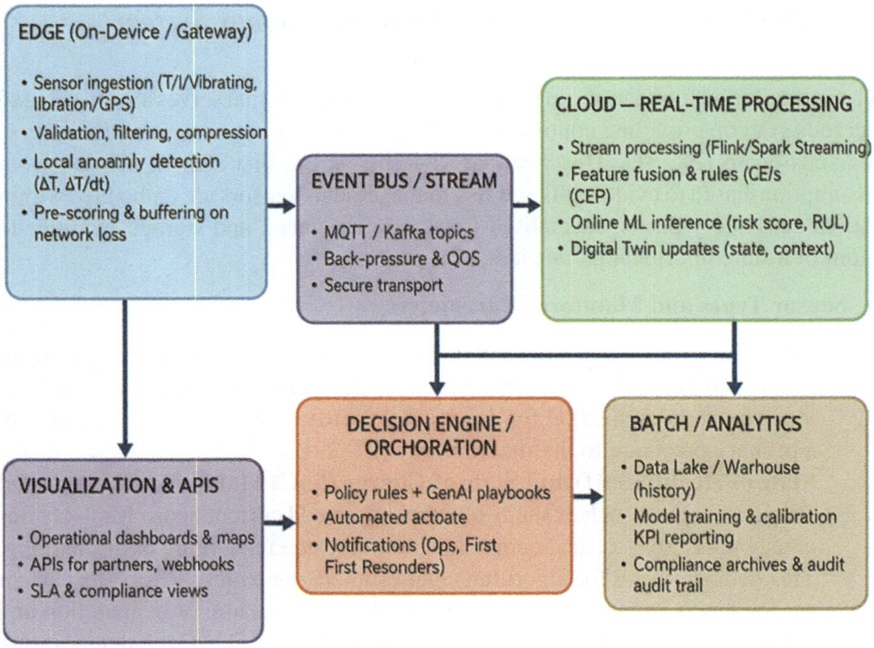

Fig. 2 Hybrid data processing architecture for the proposed IoT–Digital Twin–Generative AI framework for EoL EV battery management

and operational dashboards. This architecture enables low-latency risk detection with deep analytics and directly supports the use cases described in Sect. 4. The same data plane underpins predictive mitigation, dynamic rerouting, and intelligent emergency-response workflows in Sects. 4.1–4.3.

3.2 *The Virtualisation Layer: High-Fidelity Digital Twins*

The continuous stream of data from the IoT layer feeds the second pillar of the framework: a high-fidelity Digital Twin. This is not merely a 3D model or a dashboard; it is a dynamic, virtual replica of the entire EoL battery ecosystem, synchronised with its physical counterpart in real time [27–30]. The Digital Twin provides the context necessary to transform raw sensor data into actionable intelligence. Its architecture is multi-layered to reflect system complexity:

- **Asset-Level Twins**. A unique Digital Twin is associated with each battery pack. This virtual entity serves as a dynamic record containing manufacturing specifications (chemistry, capacity), complete operational history (charge–discharge cycles, temperature exposure), and current real-time state as reported by dedicated IoT sensors (SoH, temperature, location, physical integrity status) [27, 28, 32, 33].

- **Process-Level Twins**. A higher-level Digital Twin provides a comprehensive, holistic model of the entire reverse supply chain, including virtual representations of all physical locations (collection points, warehouses, recycling plants), transportation assets (trucks, railcars), and the logistical routes connecting them [27, 29, 30].

This multi-layered Digital Twin serves two primary, interconnected functions:

1. **Real-time mirroring and situational awareness**
   The Digital Twin provide a single, unified interface for stakeholders to visualise the exact location, status, and condition of every battery and logistical asset in the network. This eliminates information silos and provides end-to-end visibility, enabling managers to monitor operations, detect anomalies, and identify emerging risks in real time [27–30].
2. **Predictive simulation and scenario analysis**
   The Digital Twin acts as a risk-free virtual environment—a "sandbox"—to simulate the consequences of potential failures and test the efficacy of mitigation strategies. For instance, if the IoT layer reports a rising temperature in a battery, the Digital Twin can simulate the progression of a thermal runaway event within its virtual storage facility, model heat transfer to adjacent batteries, estimate time to cascading failure, and visualise the potential scale of the resulting fire [27–29]. Similarly, it can simulate the impact of a transportation accident, model container breach and hazardous-material release, and evaluate alternative emergency-response plans to determine the most effective course of action [29, 30].

3.3 The Cognitive Core: Generative AI for Predictive Intelligence and Dynamic Response

The cognitive core of the framework is a sophisticated AI engine that synthesises real-time data from the IoT layer and contextual simulations from the Digital Twin layer to generate predictions, optimisations, and dynamic responses. This engine employs a synergistic combination of predictive machine learning (ML) and prescriptive Generative AI [29, 32–36].

Predictive analytics with machine learning

The system utilises established ML models, such as Long Short-Term Memory (LSTM) networks and Random Forest classifiers, which are particularly adept at analysing time-series data. Trained on historical datasets of battery performance and failure modes, these models analyse incoming sensor streams (voltage, current, temperature, vibration) to support predictive maintenance and probabilistic risk assessment [29, 32, 33]. They can forecast a battery's Remaining Useful Life (RUL) and, critically, estimate the probability of imminent failure events such as thermal runaway. Empirical studies have shown that LSTM- and ensemble-based approaches can achieve high prediction accuracies—often exceeding 95%—for specific datasets and operating regimes when properly trained and validated [32, 33]. This predictive capability transforms risk management from a reactive exercise into a proactive process.

Dynamic optimisation and protocol generation with Generative AI

While ML models excel at prediction, Generative AI excels at creation and prescription. It leverages predictive insights from ML models and contextual understanding from Digital Twin simulations to generate novel, optimised solutions in real time [34–36].

- **Logistics optimisation**. Instead of following static routes, Generative AI can dynamically propose the safest and most efficient transportation plans. It analyses a multitude of real-time variables—traffic conditions, weather forecasts, road closures, and dynamically updated geographic risk profiles (e.g., avoiding tunnels or densely populated zones when transporting a high-risk battery load)—to generate risk-aware routing instructions that balance safety, cost, and timeliness [34].
- **Automated protocol generation**. In the event of an incident, Generative AI does not simply retrieve a static, pre-written emergency plan. Instead, it synthesises all available real-time data—the precise incident location, the specific battery chemistry involved, the number of units, current weather conditions, and proximity to vulnerable infrastructure—to generate a dynamic, context-aware emergency-response protocol. Recent work on large language models and generative decision-support systems in supply chain risk management and emergency response demonstrates their potential to draft scenario-specific action plans, warnings, and operating procedures that can be rapidly reviewed and deployed by human experts [34–36]. These bespoke plans can be disseminated instantly to facility managers, transport operators, and first responders, providing clear, actionable instructions tailored to the crisis's unique circumstances.

This synergy between predictive ML and prescriptive Generative AI is crucial. The ML models answer the question, "What is likely to happen?" The Digital Twin answers the question, "What will be the consequences if it does?" And Generative AI answers the ultimate question, "Given the prediction and its consequences, what is the best possible thing we can do right now?" This closes the loop from data collection to intelligent, semi-automated action, elevating the system from a passive monitoring tool to an active, adaptive risk-management agent. Table 1 operationalises this logic by mapping each stage of the reverse-logistics chain—Collection and Sorting, Transportation, Storage, and Repurposing/Recycling—to its dominant risk drivers, the corresponding IoT sensing modalities, the Digital Twin functions, and the AI/Generative-AI applications that support risk-informed decision-making across the EoL battery lifecycle.

Table 1 Integrated framework component mapping to lifecycle risks

Stage of reverse logistics	Key risks	and	Digital Twin function	AI/generative AI application
Collection and sorting	Misidentification of battery chemistry/SoH; handling damage causing internal shorts	Voltage/current sensors for SoH estimation; MEMS accelerometers for shock detection	Asset-level twin maintains a unique digital record of each battery's history and condition	AI-driven sorting algorithms to automatically classify batteries by risk level; generate safe handling protocols for damaged units
Transportation	Physical shock from road hazards; vibration-induced component failure; thermal runaway in transit	GPS for real-time location; MEMS accelerometers for shock/ vibration data; thermal sensors for temperature monitoring	Real-time traffic and route simulation; modelling of collision impacts and potential container breach	Dynamic, risk-aware rerouting to avoid hazards; predictive alerts for drivers; automated emergency protocol generation for first responders
Storage	Cascading thermal runaway in high-density storage; gradual degradation leading to unexpected failure	Continuous cell-level temperature monitoring; gas sensors for early detection of electrolyte venting	Thermal propagation simulation to model fire spread; facility-level twin for optimising layout and fire suppression systems	Predictive failure alerts for specific batteries; generates optimised segregation plans to isolate high-risk units; and dynamic evacuation plans
Repurposing/ recycling	Handling of unknown/ damaged batteries; exposure of workers to chemical/ electrical hazards	SoH and internal resistance data; record of physical damage events	The Digital Twin provides a comprehensive "Battery Passport" that includes historical and real-time data	AI recommends the optimal EoL pathway (reuse vs. recycle) and generates specific, safe disassembly instructions based on the battery model and condition

Taken together, these mappings show how the cognitive core translates raw sensor data and virtual simulations into concrete, stage-specific interventions across the reverse-logistics chain. Rather than treating collection, transport, storage, and repurposing/recycling as isolated operations, the framework links them through a common data and decision layer, providing a blueprint for systematic risk reduction and operational optimisation, as further illustrated in the use cases of Sect. 4.

4 Framework in Operation: From Predictive Mitigation to Automated Crisis Management

The theoretical architecture of the integrated framework gains practical significance when applied to real-world operational scenarios. By examining specific use cases, we can illustrate how the synergistic interaction among IoT, Digital Twins, and Generative AI yields tangible improvements in safety, efficiency, and resilience throughout the EoL battery lifecycle.

4.1 Use Case: Proactive Thermal Runaway Prevention in a Storage Facility

The operational steps of this scenario are summarised in Fig. 3.

This scenario demonstrates the framework's ability to shift from reactive incident response to proactive hazard mitigation.

- **Scenario**. Within a large-scale warehouse housing hundreds of EoL battery packs, the IoT sensor network detects a subtle yet persistent increase in the

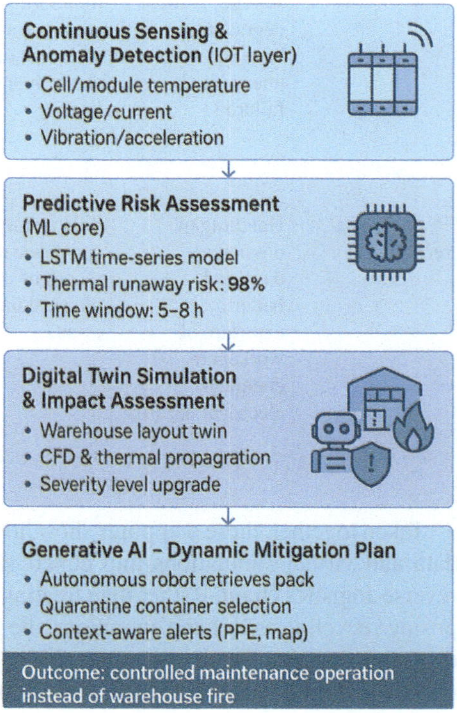

Fig. 3 Operational workflow of the integrated framework for proactive thermal runaway prevention in a storage facility

internal temperature of a single battery pack, accompanied by anomalous voltage fluctuations.

- **Framework response**:
  1. **Prediction**. The raw sensor data are streamed to the AI core. A data-driven time-series model, such as a deep recurrent neural network (e.g., LSTM/GRU variant) trained on coupled thermal and electrical degradation patterns, analyses the data stream. It identifies the observed pattern as a precursor to thermal runaway and flags the specific battery with a high probability (e.g., 98%) of triggering a critical event within a 5–8-h predictive window [37, 38]. An immediate, non-critical alert is sent to the facility manager.
  2. **Simulation and impact assessment**. The Digital Twin, which maintains a precise virtual replica of the warehouse layout, ingests this predictive alert. It creates a virtual clone of the facility and initiates a high-fidelity simulation of the predicted thermal runaway event. Using computational fluid dynamics (CFD) and thermal-propagation models, it visualises the likely path of the fire, calculates the time until adjacent battery packs reach their critical temperature thresholds, and estimates the potential for a catastrophic, cascading failure across the entire storage bay [37, 39]. The simulation output automatically elevates the incident's severity level.
  3. **Generated action plan**. The Generative AI synthesises predictions, simulated impacts, and the facility's real-time layout to generate an optimised, step-by-step mitigation plan. This is not a generic checklist but a bespoke set of instructions. It may include: (a) dispatching an autonomous robotic unit along the safest, pre-calculated route to retrieve the identified battery pack; (b) designating a specific, pre-cooled fire-resistant quarantine container at a safe distance from other assets; and (c) automatically generating and disseminating context-aware alerts to on-site safety personnel, complete with detailed handling protocols, required personal protective equipment (PPE), and a map to the designated quarantine area [40].

The entire process transforms a potential warehouse-scale disaster into a controlled, manageable maintenance operation, illustrating how the proposed framework operationalises predictive mitigation in a high-risk, real-world setting.

4.2 *Use Case: Resilient Reverse Logistics and Dynamic Rerouting*

This use case highlights the framework's capacity to adapt to unforeseen disruptions in the transportation network, thereby ensuring the continuous, safe flow of hazardous materials along the reverse logistics chain [41–43]. Dynamic, risk-aware routing has long been recognised as a key lever for reducing population exposure and infrastructure risk in hazardous materials transportation [41, 44].

- **Scenario**
  A truck is transporting a pallet of Damaged, Defective, or Recalled (DDR) batteries—classified as high-risk—along a primary highway route. An unrelated multi-vehicle accident forces a complete closure of the highway, leaving the truck stationary in a traffic jam.
- **Framework response**

  1. **Real-time detection**
     The system's Digital Twin, which continuously monitors external data feeds, immediately detects highway closures via real-time traffic APIs. At the same time, the truck's onboard IoT sensors confirm its stationary GPS location and prolonged idle time, flagging an abnormal stop duration for a high-risk load.
  2. **Dynamic risk analysis**
     The AI core updates the risk profile in real time. A stationary truck carrying hazardous materials in an uncontrolled public environment for an unknown duration poses a significantly higher risk than a vehicle in controlled transit. Similar risk metrics are used in the hazmat-routing literature, where accident probability and exposed population are jointly considered [41, 44, 45].
  3. **Generative rerouting**
     The Generative AI immediately begins generating and evaluating multiple alternative routes. Crucially, its optimisation algorithm is not configured to find the fastest alternative. Instead, it prioritises the safest route by weighing variables such as population density, proximity to critical infrastructure (e.g. schools and hospitals), and the avoidance of tunnels or major bridges where an incident would have disproportionate consequences [41, 44, 45]. The AI selects the route that minimises overall risk and transmits it directly to the driver's in-cab navigation system. It simultaneously provides updated ETAs to the receiving facility and logs the deviation and its justification for compliance and post-incident review [42, 43].

This use case illustrates how integrating IoT, Digital Twins, and Generative AI operationalises long-standing principles from the hazmat-routing literature—risk equity, population exposure minimisation, and infrastructure protection—within a real-time, data-driven decision-support workflow [41–45].

4.3 Use Case: Dynamic and Intelligent Emergency Response

This scenario illustrates the framework's role during an active crisis, demonstrating how it can dramatically enhance the effectiveness and safety of emergency responders. Recent work on AI-enabled emergency management systems highlights the potential of combining sensor networks, simulation environments, and AI decision support to enhance situational awareness and response coordination during complex incidents [46–48].

- **Scenario**

  Despite preventive measures, a battery pack on a transport truck experiences a critical internal failure and enters thermal runaway.
- **Framework response**:

  1. **Instantaneous alert**

     Onboard IoT sensors (thermal, gas, and acoustic) detect the rapid temperature spike, venting of electrolyte gases, and characteristic acoustic signatures, triggering an immediate, high-priority automated alert. The incident is classified as a suspected thermal runaway event involving DDR lithium-ion batteries, with initial risk estimates generated by pre-trained ML models.
  2. **Automated information dissemination**

     The system does not wait for manual intervention. It automatically compiles and transmits a critical data packet to the local emergency services dispatch centre (e.g., 911). This packet contains the truck's precise real-time GPS coordinates, a manifest detailing the number, type, and chemistry of all batteries on board, a live stream of telemetry data (such as temperatures in adjacent packs), and a secure link to the live Digital Twin visualisation of the truck and its cargo [46, 47]. This integrated view provides dispatchers and incident commanders with a high-fidelity picture of the evolving situation before the first units arrive on scene.
  3. **Generative response protocol**

     Simultaneously, the Generative AI synthesises a tailored, initial emergency-response plan based on the specific incident parameters and a curated knowledge base of hazardous-materials protocols and best practices [46–48]. The generated protocol is transmitted alongside the data packet and provides actionable guidance for the first units arriving on scene. It may include:

     - **Specific suppressant recommendation**

       For example: "Incident involves 24 NMC-811 battery packs. Use large volumes of water for cooling; do not use Class A/B/C foam."
     - **Safe-perimeter calculation**

       A recommended initial isolation and protective-action distance (e.g. establishing a 300 m safety perimeter) based on battery chemistry, load size, and current wind conditions.
     - **Real-time hazard warnings**

       Dynamic estimates of the probability and timing of secondary events, such as propagation to adjacent modules or container rupture.
     - **Evacuation and traffic control**

       Automated warnings for downstream traffic and suggested evacuation zones for nearby residences or businesses, informed by wind direction and dispersion modelling.

In this crisis scenario, the framework acts as an expert co-pilot for emergency responders, providing critical, context-specific intelligence that complements human expertise rather than replacing it [46–48]. By tightly coupling high-resolution sensing, Digital Twin visualisation, and Generative-AI decision support, the system improves the speed, consistency, and transparency of hazardous-materials incident management while still allowing incident commanders to retain ultimate authority over life-critical decisions.

5 Implementation Challenges and Strategic Considerations

While the proposed integrated framework offers a transformative vision for hazardous material management, its successful implementation depends on addressing a series of significant technical, computational, socio-technical, and ethical challenges. Acknowledging and proactively planning for these hurdles is crucial for any organisation seeking to deploy such a system at scale.

Technical and Computational Hurdles

While powerful, the framework's technical foundation introduces complexities in data security, system performance, and interoperability.

Data Security and Privacy

Deploying a vast network of IoT sensors across a distributed supply chain creates an expansive digital attack surface. Each sensor, gateway, and communication channel is a potential point of vulnerability. A malicious actor could intercept data streams, inject false information to trigger a spurious emergency response, or exploit a vulnerability to gain control of logistical assets [49, 50]. Therefore, a robust, multi-layered cybersecurity strategy is not an option but a core requirement. This must include end-to-end encryption of all data in transit and at rest, strong multi-factor authentication (MFA) for all system access, role-based access controls (RBAC), and the use of AI-driven security monitoring tools to detect anomalous network behaviour that could indicate a breach [49, 50].

Scalability and Performance

The computational demands of maintaining high-fidelity, real-time Digital Twins for a fleet that could eventually number in the millions of individual batteries are immense. The sheer volume of data processing, storage, and real-time synchronisation required can become a significant performance bottleneck and cost centre [51, 52]. Addressing this requires a sophisticated architectural approach. Scalable cloud-based infrastructure provides on-demand storage and processing power and supports elastic scaling during peak loads [51]. The strategic use of edge computing is critical: by processing data and running simpler analytical models locally on gateways near the assets, the system can reduce the volume of data transmitted to the cloud, lower latency for time-sensitive alerts, and maintain core functionality even with intermittent network connectivity [51, 52]. Furthermore, developing reduced-order models (ROMs) that approximate the behaviour of complex physical systems with far less computational overhead offers a pragmatic balance between simulation fidelity and real-time performance in Digital Twin environments [53].

Interoperability and Standardisation

The EoL battery ecosystem is inherently fragmented, comprising numerous battery manufacturers, vehicle OEMs, third-party logistics providers, and recyclers, each with proprietary systems and data formats. This lack of standardisation is a significant barrier to creating a cohesive, end-to-end data flow [51]. Achieving true interoperability requires industry-wide adoption of common data standards and communication protocols. Standards such as ISO 15118 for vehicle-to-grid communication, the Open Charge Point Protocol (OCPP) for charger-to-backend communication, and IEEE 2030.5 for smart energy profile communications provide foundational blueprints for how data can be structured and exchanged, enabling different systems to "speak the same language" and contribute to a unified data ecosystem [54–56].

Socio-Technical and Ethical Dimensions

Beyond the purely technical challenges, integrating autonomous decision-making into critical infrastructure management raises profound socio-technical and ethical questions that must be addressed to ensure trust, accountability, and fairness.

Explainable AI (XAI)

For a human operator to trust and effectively manage the system, the AI's decision-making process cannot be an opaque "black box". In high-stakes situations—such as recommending an emergency evacuation or flagging a specific company's shipments as high-risk—the system must provide a clear, understandable rationale for its conclusions. This necessitates integrating Explainable AI (XAI) techniques. Methods such as LIME (Local Interpretable Model-Agnostic Explanations) and SHAP (Shapley Additive exPlanations) can be used to highlight which specific data points (e.g., a sudden vibration spike followed by a temperature increase) most heavily influenced a particular prediction, making the AI's reasoning more transparent to human auditors [57–59].

Human-in-the-Loop (HITL) Frameworks

The proposed system is designed to augment human expertise rather than replace it. An essential architectural principle is the implementation of a Human-in-the-Loop (HITL) framework, whereby the most critical, high-consequence decisions generated by the AI require explicit human validation or override before execution [60–62]. This approach ensures human accountability, provides a crucial safeguard against catastrophic automated errors, and allows for the application of human judgment and ethical reasoning in complex, nuanced scenarios that may fall outside the AI's training data. The role of the human thus shifts from a manual operator constantly monitoring data streams to a strategic overseer tasked with validating the system's most critical outputs. This transition does not eliminate the need for human involvement but rather elevates it, demanding a new skill set focused on systems thinking, ethical oversight, and the ability to interpret and critically challenge complex AI recommendations [60–62].

Algorithmic Bias and Fairness

AI models are susceptible to learning and amplifying biases present in their training data. For example, suppose historical data show more incidents associated with a particular truck model or logistics provider (perhaps due to older equipment or more challenging routes). In that case, the AI may develop a bias that unfairly flags them as higher risk, with potentially significant economic and legal ramifications [63–65]. Mitigating this risk requires a concerted effort to ensure that training datasets are diverse and representative, coupled with regular algorithmic audits to detect and correct discriminatory outcomes [63, 64]. Establishing transparent risk-scoring criteria and providing a mechanism for partners to appeal or question the AI's assessments are crucial for maintaining fairness and trust within the supply-chain ecosystem [63–65].

Taken together, these technical, organisational, and ethical safeguards are not ancillary to the proposed framework, but constitutive of its legitimacy and long-term viability. A system that senses, simulates, and acts at scale on hazardous-materials risks must be not only accurate and efficient, but also secure, interoperable, explainable, and perceived as procedurally fair by all stakeholders across the reverse-logistics chain. Only under these conditions can the IoT–Digital Twin–Generative AI stack evolve from an experimental pilot into trusted critical infrastructure—and provide the socio-technical foundation for the data-driven circular economy described in Sect. 6.

6 Fostering a Secure and Sustainable Circular Economy

The integrated framework's primary function is to enhance safety and resilience in the management of hazardous materials associated with EoL EV batteries. However, its impact extends well beyond immediate risk mitigation. By creating an authoritative, data-rich, and continuous record for every battery throughout its lifecycle, the framework acts as a critical enabler of a secure and sustainable circular economy. In this paradigm, EoL batteries are no longer treated solely as high-risk waste streams; instead, they are transformed into traceable, value-bearing assets that can be safely reintegrated into secondary markets or optimised recycling pathways.

Figure 4 provides a high-level overview of how the proposed framework underpins a data-driven circular economy through the Digital Battery Passport concept. Acting as a data spine, the Digital Battery Passport connects all lifecycle stages—from in-vehicle use and EoL identification, through IoT- and AI-enabled risk and safety management, to second-life applications and chemistry-aware recycling. The resulting closed-loop system enables transparency, accountability, and value retention across the entire battery lifecycle, ultimately supporting both environmental sustainability and industrial resilience.

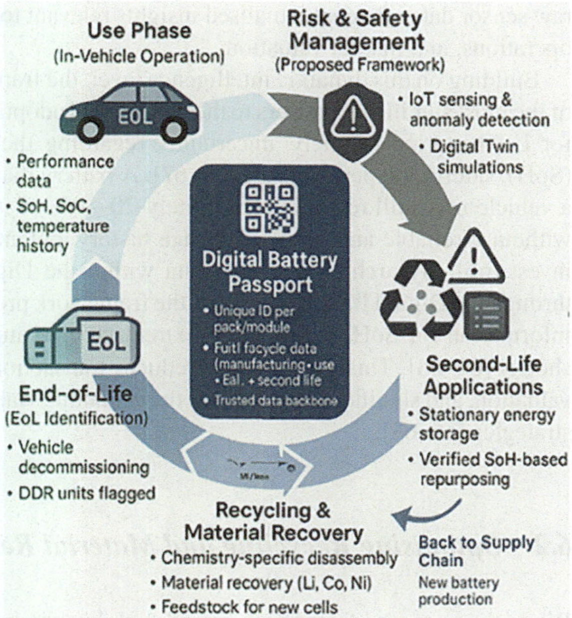

Fig. 4 Conceptual overview of the data-driven circular economy for EoL EV batteries enabled by the Digital Battery Passport

6.1 Selection of the Data Set Enabling Second-Life Applications

Current hazardous material management and battery tracking practices are predominantly designed to ensure regulatory compliance rather than to support value-preserving lifecycle decisions. Traditional hazmat management systems rely on periodic inspections, manual documentation, and incident-driven reporting, resulting in fragmented and time-delayed visibility into battery condition and risk exposure. While effective at meeting minimum safety requirements, these systems provide limited support for proactive decision-making throughout the battery lifecycle.

Similarly, static compliance-based approaches focus on fulfilling transport, storage, and disposal obligations, often treating EoL batteries as homogeneous waste units. Such approaches do not capture cumulative degradation patterns, thermal stress histories, or operational anomalies that are critical for assessing the potential for reuse. Even more advanced digital tracking solutions—such as barcode- or RFID-based systems—remain largely non-intelligent. Although they improve traceability, they typically function as passive data repositories, lacking analytical or predictive capabilities.

In contrast, the proposed framework introduces a fundamental shift from static traceability to real-time, intelligence-driven lifecycle management. Continuous data streams collected via IoT sensors are integrated into a Digital Twin that dynamically reflects the battery's evolving physical condition. The integration of Generative AI further enhances this capability by enabling adaptive decision support, transforming raw sensor data into contextualised insights relevant to safety management, logistics operations, and market valuation.

Building on this dynamic intelligence layer, the framework directly addresses one of the most significant barriers to the widespread adoption of second-life applications for EV batteries—namely, uncertainty regarding their remaining State of Health (SoH), safety, and performance [66, 67]. A battery that has completed its first life in a vehicle may still retain approximately 70–80% of its original capacity; however, without a reliable and verifiable usage history, potential buyers remain hesitant to invest [66]. By archiving sensor data within the Digital Twin and formalising it through the Digital Battery Passport, the framework provides precise and trustworthy information on SoH, exposure to extreme temperatures, and significant physical shocks [66–68]. This transparency reduces transaction risk, enables more accurate valuation, and significantly improves the economic viability of reuse and repurposing strategies [66–68].

6.2 *Optimising Recycling and Material Recovery*

While the proposed framework provides clear operational advantages, several limitations must be explicitly acknowledged. First, its effectiveness depends on the availability, completeness, and interoperability of lifecycle data across multiple stakeholders. In practice, the incoming stream of EoL batteries is highly heterogeneous—varying by pack design, size, chemistry, and damage status—and recyclers frequently receive units with incomplete or missing contextual information [69, 70]. Although the Digital Battery Passport concept is designed to address this fragmentation by providing pre-arrival chemistry and condition data [69–71], the framework's performance may still be constrained in environments where sensor coverage is inconsistent, historical data are unavailable, or data-sharing arrangements are limited.

Second, regulatory fragmentation across jurisdictions may complicate implementation. Battery and hazardous materials governance is evolving rapidly, yet requirements differ across regions and may not be fully harmonised, particularly in cross-border logistics chains. Even in policy-leading environments such as the European Union, compliance obligations related to traceability, performance history, and due diligence—together with mandatory Digital Battery Passport requirements—are being phased in and depend on standardised data structures and access mechanisms [71, 72]. This creates a risk that early implementations face uncertainty or rework as standards and secondary legislation mature.

Third, the framework introduces dependencies on IoT infrastructure, including sensor reliability, network connectivity, and long-term maintenance. Failures in data collection or interruptions in connectivity can degrade the accuracy of real-time risk monitoring and downstream decision support, especially for units with elevated safety risks, such as damaged battery packs entering recycling workflows [69, 70]. Closely related, cybersecurity and data governance represent critical constraints. Because the framework aggregates sensitive operational and safety information, robust safeguards are required to prevent unauthorised access, data manipulation, or systemic cyber risks—particularly if tamper-resistance is strengthened through blockchain integration to support a decentralised and auditable traceability record across stakeholders [72, 73].

Finally, while the integration of Generative AI enhances interpretation and adaptive decision support, its maturity in safety–critical environments remains uneven. Ensuring transparency, explainability, and effective human oversight is essential to avoid overreliance on automated outputs in contexts where incorrect recommendations could amplify hazards during transport, storage, or recycling operations [69–71]. These limitations do not undermine the framework's relevance; instead, they define the practical and governance conditions necessary for credible deployment at scale.

6.3 Alignment with Emerging Policy and Standards

Despite these limitations, the proposed framework is designed to be implementable within existing logistics and infrastructure ecosystems. In warehouse environments, real-time monitoring and Digital Twin–based insights can support safer storage practices, early detection of thermal risks, and informed segregation of high-risk units. At transport hubs and border crossings, the Digital Battery Passport enables rapid verification of battery status and regulatory compliance, reducing delays while enhancing safety oversight.

Within recycling facilities, advanced access to chemistry-specific and condition-related data allows recyclers to optimise sorting strategies, select appropriate processing routes (e.g., hydrometallurgical versus pyrometallurgical), and implement targeted safety protocols for damaged or high-risk batteries [69–71]. This data-driven approach improves both material recovery efficiency and worker safety [69, 70].

Finally, the framework supports enhanced coordination in emergency response scenarios. By providing responders with immediate access to battery-specific risk profiles, historical incident data, and predictive insights, emergency management teams can make faster, better-informed decisions, thereby reducing the likelihood of escalation and secondary damage.

Collectively, these applications demonstrate that the framework is not a purely theoretical construct, but a practically grounded system capable of supporting everyday logistics operations while advancing the broader objectives of a secure and sustainable circular economy.

7 Conclusion and Future Research Directions

The rapid global shift to electric mobility, while essential for a sustainable transportation future, presents an urgent and complex challenge to the resilience of critical infrastructure. The escalating volume of EoL EV batteries—classified as hazardous materials due to their significant fire, chemical, and environmental risks—cannot be safely managed using traditional, reactive methodologies. This paper presents a novel, integrated framework that leverages the synergistic capabilities of the Internet of Things, Digital Twins, and Generative Artificial Intelligence to directly address this challenge. By creating a system that can sense, simulate, predict, and dynamically respond to risks in real time, the framework enables a necessary paradigm shift from a reactive, compliance-focused posture to a proactive, predictive, and resilient model for hazardous material management.

The operational use cases demonstrate the framework's potential to prevent catastrophic events such as thermal runaway, optimise complex reverse logistics, and provide invaluable, context-aware intelligence to emergency responders during a crisis. Beyond immediate risk mitigation, the framework's ability to create an authoritative and transparent data record for each battery—a Digital Battery Passport—serves as a powerful enabler of a secure and sustainable circular economy, unlocking the value of second-life applications and enhancing the efficiency of critical material recovery. In doing so, the proposed system ties together safety, regulatory compliance, and circular-economy objectives within a single data-driven architecture for EoL EV batteries. As synthesised in Sect. 6, this lifecycle record is operationalised through the Digital Battery Passport as the central data spine linking second-life valuation, chemistry-aware recycling, and day-to-day logistics decision-making within a closed-loop circular economy.

Consistent with the limitations discussed in Sect. 6.2, the framework's benefits depend on reliable data capture, interoperability among stakeholders, and governance safeguards appropriate for safety–critical operations. Likewise, the implementation pathways outlined in Sect. 6.3 indicate that deployment can be phased through warehouses, transport nodes, recycling facilities, and emergency-response interfaces without requiring a wholesale redesign of existing logistics ecosystems. At the same time, several implementation limitations should be acknowledged to ensure a balanced and credible interpretation of the proposed approach. The framework's effectiveness depends on the availability, quality, and interoperability of lifecycle data across multiple stakeholders, as well as on reliable IoT infrastructure and connectivity. Practical deployment may also be affected by regulatory fragmentation across jurisdictions and by cybersecurity and data-governance requirements, particularly when sensitive operational and safety data are shared across organisational boundaries. Finally, while Generative AI strengthens interpretative and decision-support capabilities, its use in safety–critical settings requires transparency, explainability, and robust human oversight to prevent overreliance on automated outputs.

These constraints also translate into concrete recommendations. For industry stakeholders (manufacturers, logistics operators, storage providers, recyclers, and second-life market participants), a staged implementation approach is advisable—starting with standardised data capture for safety–critical variables, governance agreements for data sharing, and pilot deployments in controlled logistics corridors or facilities. For regulators and public institutions, the priority should be to support interoperable data standards, establish trusted mechanisms for secure access to Digital Battery Passport information, and enable regulatory "sandboxes" or pilot programs that accelerate adoption while ensuring accountability and auditability across borders.

While the proposed framework offers a comprehensive foundation, its implementation marks not an end point but a starting point for further research and innovation in intelligent infrastructure management. Several promising avenues for future work can build upon and extend the capabilities of this system:

- **Multi-Agent Systems**

  The current framework relies on a largely centralised AI core for decision-making. Future research could explore the use of Multi-Agent Systems to create a more decentralised and robust control architecture. In such a system, individual agents—representing autonomous entities such as trucks, battery containers, or storage facilities—could negotiate and coordinate actions directly with one another to optimise logistics and respond to local disruptions without relying on a single central controller. This could enhance system resilience, reduce single points of failure, and improve scalability [74–76].
- **Quantum Computing**

  The optimisation problems inherent in managing a global-scale reverse supply chain—such as multi-variable route planning and dynamic resource allocation—are often NP-hard, meaning their complexity grows exponentially and can become intractable for classical computers. The emerging field of quantum computing, particularly through quantum machine learning and quantum optimisation algorithms such as the Quantum Approximate Optimisation Algorithm (QAOA), holds the potential to solve these complex problems orders of magnitude faster than current systems [77, 78]. Future work should investigate integrating quantum algorithms to tackle the most computationally intensive optimization tasks within the framework, potentially unlocking unprecedented levels of efficiency in hazardous-material routing and network-wide risk minimization [79].
- **The Digital Thread**

  The Digital Twin concept focuses on mirroring the physical asset during its operational life. An even more ambitious vision is the creation of a comprehensive Digital Thread. This would represent an unbroken, authoritative data flow connecting every conceivable phase of the battery's existence—from the initial mining of raw materials and the model-based systems engineering (MBSE) of its design, through manufacturing, its first life in a vehicle, any second-life applications, and its final deconstruction and material recovery. Such a Digital Thread would provide end-to-end traceability, enabling a truly holistic and optimised circular economy and offering unparalleled insights into material provenance, performance degradation, and lifecycle environmental impact [80, 81].

By pursuing these advanced research directions, the foundational principles of the integrated framework proposed in this paper can be further expanded, paving the way for a future in which the lifecycle of critical and hazardous materials is managed with intelligence, safety, and sustainability commensurate with their societal importance.

References

1. Requia, W.J., Mohamed, M., Higgins, C.D., Arain, A., Ferguson, M.: How clean are electric vehicles? Evidence-based review of the effects of electric mobility on air pollutants, greenhouse gas emissions and human health. Atmos. Environ. **185**, 64–77 (2018)
2. Harper, G., Sommerville, R., Kendrick, E., Driscoll, L., Slater, P., Stolkin, R., Walton, A., Christensen, P., Heidrich, O., Lambert, S., Abbott, A., Ryder, K., Gaines, L., Anderson, P.: Recycling lithium-ion batteries from electric vehicles. Nature **575**(7781), 75–86 (2019)
3. Velázquez-Martínez, C., Valio, J., Santasalo-Aarnio, A., Reuter, M., Serna-Guerrero, R.: A critical review of lithium-ion battery recycling processes from a circular economy perspective. Batteries **5**(4), 68 (2019)
4. Feng, X., Ouyang, M., Liu, X., Lu, L., Xia, Y., He, X.: Thermal runaway mechanism of lithium ion battery for electric vehicles: a review. Energy Storage Mater. **10**, 246–267 (2018)
5. Heath, G.A., Ravikumar, D., Hansen, B., Kupets, E.: A critical review of the circular economy for lithium-ion batteries and photovoltaic modules: Status, challenges, and opportunities. J. Air Waste Manag. Assoc. **72**(6), 478–539 (2022)
6. U.S. Environmental Protection Agency (EPA): Defining Hazardous Waste: Listed, Characteristic and Mixed Radiological Wastes. 40 CFR Part 261 Subpart C—Hazardous Waste Characteristics. EPA, Washington, DC (2024)
7. 40 C.F.R. § 261.21: Characteristic of Ignitability. In: Code of Federal Regulations, Title 40—Protection of Environment. U.S. Government Publishing Office, Washington, DC (2023)
8. 40 C.F.R. § 261.23: Characteristic of Reactivity. In: Code of Federal Regulations, Title 40—Protection of Environment. U.S. Government Publishing Office, Washington, DC (2023)
9. U.S. Department of Transportation (DOT), Pipeline and Hazardous Materials Safety Administration (PHMSA): Hazardous Materials Regulations (HMR), 49 C.F.R. Parts 171–180. U.S. DOT, Washington, DC (2024)
10. 40 C.F.R. Part 273: Standards for Universal Waste Management. In: Code of Federal Regulations, Title 40—Protection of Environment. U.S. Government Publishing Office, Washington, DC (2024)
11. U.S. DOT PHMSA: Lithium-ion Batteries: Shipping & Emergency Response. Training presentation and guidance material, including updates to the Emergency Response Guidebook (ERG) 2024. PHMSA, Washington, DC (2024)
12. U.S. DOT PHMSA: Understanding the Risks of Damaged, Defective, or Recalled (DDR) Lithium Batteries. Training module and guidance document. PHMSA, Washington, DC (2024)
13. National Fire Protection Association (NFPA): NFPA 855—Standard for the Installation of Stationary Energy Storage Systems, 3rd edn. NFPA, Quincy, MA (2026)
14. United Nations Economic Commission for Europe (UNECE): European Agreement Concerning the International Carriage of Dangerous Goods by Road (ADR), vol. I–II. UNECE, Geneva (2025)
15. International Maritime Organization (IMO): International Maritime Dangerous Goods Code (IMDG Code), 2024 Edition. IMO, London (2024)
16. European Parliament and Council: Regulation (EU) 2023/1542 of 12 July 2023 concerning batteries and waste batteries. Off. J. Eur. Union L **191**, 1–192 (2023)

17. Nitta, N., Wu, F., Lee, J.T., Yushin, G.: Li-ion battery materials: present and future. Mater. Today **18**(5), 252–264 (2015)
18. Xu, K.: Electrolytes and interphases in Li-ion batteries and beyond. Chem. Rev. **114**(23), 11503–11618 (2014)
19. Mrozik, W., Rajaeifar, M.A., Heidrich, O., Christensen, P.: Environmental impacts, pollution sources and pathways of spent lithium-ion batteries from electric vehicles—a review. Energy Environ. Sci. **14**(12), 6099–6121 (2021)
20. Larsson, F., Andersson, P., Blomqvist, P., Mellander, B.-E.: Toxic fluoride gas emissions from lithium-ion battery fires. Sci. Rep. **7**, 10018 (2017)
21. Sun, J., Li, J., Zhou, T., Yang, K., Wei, S., Tang, N., Dang, N., Li, H., Qiu, X., Chen, L.: Toxicity, a serious concern of thermal runaway from commercial Li-ion battery. Nano Energy **27**, 313–319 (2016)
22. Bugryniec, P.J., Maynard, N., Marinescu, M., Offer, G.J.: Review of gas emissions from lithium-ion battery thermal runaway failure—considering toxic and flammable compounds. J. Energy Storage **87**, 111288 (2024)
23. Feng, X., Ouyang, M., He, X., Lu, L., Wang, L., Han, X., Li, J.: Thermal runaway mechanism of lithium-ion battery for electric vehicles: a review. Energy Storage Mater. **10**, 246–267 (2018)
24. Wang, Q., Mao, B., Stoliarov, S.I., Sun, J.: A review of lithium-ion battery failure mechanisms and fire prevention strategies. Prog. Energy Combust. Sci. **73**, 95–131 (2019)
25. Li, H., Wu, L., Wu, K., Zhang, L., Cheng, X., Xu, C.: Experimental study on cascading thermal runaway propagation of lithium-ion battery module under different states of charge. Energies **16**(13), 5172 (2023)
26. Essl, C., Golubkov, A.W., Fuchs, A.: Influence of aging on the failing behavior of automotive lithium-ion batteries. Batteries **7**(1), 23 (2021)
27. Li, W., Rentemeister, M., Badeda, J., Jöst, D., Schulte, D., Sauer, D.U.: Digital twin for battery systems: cloud battery management system with online state-of-charge and state-of-health estimation. J. Energy Storage **30**, 101557 (2020)
28. Naseri, F., Gil, S., Barbu, C., Cetkin, E., Mulder, G., Berecibar, M., Van Mierlo, J.: Digital twin of electric vehicle battery systems: comprehensive review of the use cases, requirements, and platforms. Renew. Sustain. Energy Rev. **179**, 113280 (2023)
29. Shi, W., Cao, J., Zhang, Q., Li, Y., Xu, L.: Edge computing: vision and challenges. IEEE Internet Things J. **3**(5), 637–646 (2016)
30. Mao, Y., You, C., Zhang, J., Huang, K., Letaief, K.B.: A survey on mobile edge computing: the communication perspective. IEEE Commun. Surv. Tutor. **19**(4), 2322–2358 (2017)
31. Wang, Y., Xu, R., Zhou, C., Kang, X., Chen, Z.: Digital twin and cloud–side–end collaboration for intelligent battery management system. J. Manuf. Syst. **62**, 124–134 (2022)
32. Kaleem, M.B., He, W., Li, H.: Machine learning driven digital twin model of Li-ion batteries in electric vehicles: a review. AI Auton. Syst. **1**, 0003 (2023)
33. Saba, I., Ullah, M., Tariq, M., Khan, I., Mehmood, A.: Advancing electric vehicle battery analysis with digital twins in intelligent transportation systems. IEEE Trans. Intell. Transp. Syst. **25**(9), 12141–12150 (2024)
34. Zhao, J., Feng, X., Pang, Q., Fowler, M., Li, W., Dahn, J.R., Burke, A.F.: Battery safety: machine learning-based prognostics. Prog. Energy Combust. Sci. **102**, 101142 (2024)
35. Xu, Y., Li, C., Wang, X., Zhang, H., Yang, F., Ma, L., Wang, Y.: Joint estimation method with multi-innovation unscented Kalman filter based on fractional-order model for state of charge and state of health estimation. Sustainability **14**(23), 15538 (2022)
36. Li, D., Nan, J., Burke, A.F., Zhao, J.: Battery prognostics and health management: AI and big data. World Electr. Veh. J. **16**(1), 10 (2024)
37. Tran, M.K., Fowler, M., Chen, B.: A review of lithium-ion battery thermal runaway modeling and diagnosis approaches. Processes **10**(6), 1192 (2022)
38. Yao, Q., Lu, D.D.C., Lei, G.: A surface temperature estimation method for lithium-ion battery using enhanced GRU-RNN. IEEE Trans. Transp. Electrif. **9**(1) (2023)
39. Chen, B., Shi, H., Wang, Q., et al.: Thermal safety of lithium-ion batteries: a critical review. Batteries **11**(3), 286 (2025)

40. Kolivand, H., Ahmad, R., Vanani, I., et al.: Artificial intelligence model applications in disaster governance: a scoping review. Processes **12**(4), 708 (2024)
41. Liu, L., Hu, J., Wen, X., Wu, Y.: Research on route optimization of hazardous materials road transportation considering risk equity. Sustainability **13**(16), 9427 (2021)
42. Erkut, E., Tjandra, S.A., Verter, V.: Hazardous materials transportation. In: Barnhart, C., Laporte, G. (eds.) Handbooks in Operations Research and Management Science, vol. 14, pp. 539–621. Elsevier, Amsterdam (2007)
43. Nozick, L.K., Turnquist, M.A.: Integrating inventory impacts into a fixed-charge model for locating routes for hazardous materials. Transp. Res. Rec. **1573**, 80–86 (1997)
44. Verma, M., Verter, V.: Railroad transportation of dangerous goods: population exposure to airborne toxins. Comput. Oper. Res. **34**(5), 1287–1303 (2007)
45. Ma, W., Li, J., Fan, B., Zhang, Y., Huang, Z.: A region-based multi-criteria decision-making approach for routing hazardous materials vehicles with time windows. Int. J. Environ. Res. Public Health **18**(2), 458 (2021)
46. Fang, Z., Zong, Z., Zhang, L., Wu, J.: A comprehensive survey of intelligent emergency management systems based on artificial intelligence techniques. IEEE Access **10**, 88460–88482 (2022)
47. Wang, R., Huang, D., Li, L., Zhang, J.: Digital-twin-driven intelligent emergency response for industrial accidents: a review and future perspectives. Process. Saf. Environ. Prot. **169**, 529–546 (2023)
48. Bajwa, A.S., Qadir, J., Al-Fuqaha, A.: Artificial intelligence for disaster management: a survey on the state of the art, challenges, and future directions. IEEE Access **9**, 116995–117029 (2021)
49. National Institute of Standards and Technology (NIST): Considerations for Managing Internet of Things (IoT) Cybersecurity and Privacy Risks. NISTIR 8228. NIST, Gaithersburg (2019)
50. Mrabet, H., Belguith, S., Alhomoud, A., Jemai, A.: A survey of IoT security based on a layered architecture of sensing and data analysis. Sensors **20**(13), 3625 (2020)
51. Qi, Q., Tao, F.: Digital twin and big data towards smart manufacturing and industry 4.0: 360 degree comparison. J. Manuf. Syst. **58**, 346–361 (2021)
52. Lu, Y., Liu, C., Wang, K., Huang, H., Xu, X.: Digital twin-driven smart manufacturing: connotation, reference model, applications and research issues. Robot. Comput. Integr. Manuf. **61**, 101837 (2020)
53. Benner, P., Gugercin, S., Willcox, K.: A survey of projection-based model reduction methods for parametric dynamical systems. SIAM Rev. **57**(4), 483–531 (2015)
54. International Organization for Standardization: ISO 15118–1:2019 Road Vehicles—Vehicle to Grid Communication Interface—Part 1: General Information and Use-Case Definition. ISO, Geneva (2019)
55. Open Charge Alliance: Open Charge Point Protocol (OCPP) 1.6 Specification. Open Charge Alliance, Arnhem (2015)
56. IEEE Standards Association: IEEE Std 2030.5–2018—IEEE Standard for Smart Energy Profile Application Protocol. IEEE, New York (2018)
57. Ribeiro, M.T., Singh, S., Guestrin, C.: "Why should I trust you?" Explaining the predictions of any classifier. In: Proc. 22nd ACM SIGKDD Int. Conf. Knowledge Discovery and Data Mining, pp. 1135–1144. ACM, New York (2016)
58. Lundberg, S.M., Lee, S.-I.: A unified approach to interpreting model predictions. In: Guyon, I., et al. (eds.) Advances in Neural Information Processing Systems, vol. 30, pp. 4765–4774. Curran Associates, Red Hook (2017)
59. Doshi-Velez, F., Kim, B.: Towards a rigorous science of interpretable machine learning. arXiv: 1702.08608 (2017)
60. Amershi, S., Weld, D.S., Vorvoreanu, M., et al.: Guidelines for human–AI interaction. In: Proc. 2019 CHI Conf. Human Factors in Computing Systems, Paper 3, pp. 1–13. ACM, New York (2019)
61. Raji, I.D., Smart, A., White, R.N., et al.: Closing the AI accountability gap: defining an end-to-end framework for internal algorithmic auditing. In: Proc. 2020 Conference on Fairness, Accountability, and Transparency (FAccT), pp. 33–44. ACM, New York (2020)

62. Rahwan, I., Cebrian, M., Obradovich, N., et al.: Machine behaviour. Nature **568**(7753), 477–486 (2019)
63. Mehrabi, N., Morstatter, F., Saxena, N., Lerman, K., Galstyan, A.: A survey on bias and fairness in machine learning. ACM Comput. Surv. **54**(6), 115 (2021)
64. Barocas, S., Hardt, M., Narayanan, A.: Fairness and Machine Learning. fairmlbook.org (2019)
65. Selbst, A.D., Barocas, S., Raghavan, M., Boyd, D.: Fairness and abstraction in sociotechnical systems. In: Proc. Conf. Fairness, Accountability, and Transparency (FAT*), pp. 59–68. ACM, New York (2019)
66. Gu, X., Bai, H., Cui, X., Zhu, J., Zhuang, W., Li, Z., Hu, X., Song, Z.: Challenges and opportunities for second-life batteries: a review of key technologies and economy. arXiv:2308.06786 (2023)
67. Neubauer, J., Pesaran, A.: The ability of battery second use strategies to impact plug-in electric vehicle prices and serve utility energy storage applications. J. Power. Sources **196**(23), 10351–10358 (2011)
68. Global Battery Alliance: A Vision for a Battery Passport for Sustainable Batteries. Global Battery Alliance, Geneva (2020)
69. Heelan, J., Gratz, E., Zheng, Z., et al.: Current and prospective Li-ion battery recycling and recovery processes. JOM **68**(10), 2632–2638 (2016)
70. Harper, G., Sommerville, R., Kendrick, E., et al.: Recycling lithium-ion batteries from electric vehicles. Nature **575**, 75–86 (2019)
71. Regulation (EU) 2023/1542 of the European Parliament and of the Council of 12 July 2023 concerning batteries and waste batteries, amending Regulation (EU) No 2019/1020 and Directive 2008/98/EC and repealing Directive 2006/66/EC. Off. J. Eur. Union L **191**, 1–128 (2023)
72. European Commission: Questions and Answers on Sustainable and Circular Batteries—New EU Regulatory Framework. European Commission, Brussels (2023)
73. Kouhizadeh, M., Zhu, Q., Sarkis, J.: Blockchain and the circular economy: potential tensions and synergies. Resour. Conserv. Recycl. **156**, 104437 (2020)
74. Wooldridge, M.: An Introduction to MultiAgent Systems, 2nd edn. Wiley, Chichester (2009)
75. Lang, N., Moonen, M., Srour, F., Zuidwijk, R.: Multi Agent Systems in Logistics: A Literature and State-of-the-Art Review. ERIM Report Series in Management. Erasmus University Rotterdam, Rotterdam (2009)
76. Davidsson, P., Henesey, L., Ramstedt, L., Törnquist, J., Wernstedt, F.: An analysis of agent-based approaches to transport logistics. Transp. Res. C **13**(4), 255–271 (2005)
77. Preskill, J.: Quantum computing in the NISQ era and beyond. Quantum **2**, 79 (2018)
78. Farhi, E., Goldstone, J., Gutmann, S.: A quantum approximate optimisation algorithm. arXiv: 1411.4028 (2014)
79. Jiang, Z., Balu, R., Chang, K., et al.: Quantum algorithms to speed up supply chain management. IEEE Trans. Quantum Eng. **3**, 1–16 (2022)
80. Hedberg, T., Jr., Lubell, J., Fischer, L., Rachuri, S.: Using graphs to link data across the product lifecycle. J. Manuf. Sci. Eng. **142**(2), 021004 (2020)
81. Madni, A.M., Madni, C.C., Lucero, S.D.: Leveraging digital twin and digital thread technologies in systems engineering. Systems **7**(1), 7 (2019)

Perceived and Modeled Safety of Road Tunnels: A Comparative Assessment Using Expert, Driver, and AI Evaluation

Aleksandar Trifunović, Aleksandar Senić, and Dragan Lazarević

Abstract Road tunnels represent one of the most complex and safety–critical segments of road infrastructure, as they involve restricted geometry, reduced visibility, and limited opportunities for evasive maneuvers. Inadequate lighting, poor ventilation, and insufficient signaling systems significantly increase the likelihood of traffic incidents, especially when drivers are not properly informed or when technical failures occur within confined spaces. In order to analyze risk perception and assess safety in road tunnels, a comprehensive study was conducted, covering three representative scenarios corresponding to different tunnel characteristics: short urban tunnels, long motorway tunnels, and bi-directional rural tunnels. The research included 108 drivers of various profiles (professional and non-professional drivers) and 8 experts from the fields of traffic and civil engineering. The results indicate a statistically significant difference in perceived safety between the artificial intelligence model, drivers, and experts. Notably, drivers often underestimate the hazards associated with tunnel environments, particularly those related to visibility, space limitation, and emergency response conditions. Based on the findings, an integrated AI-assisted safety assessment model is proposed, which can be implemented within intelligent transportation systems and navigation platforms to provide real-time alerts, support risk mitigation, and enhance safety in tunnel environments.

Keywords Road tunnels · Traffic safety · Risk assessment · Artificial intelligence · Driver perception · Expert evaluation

---

A. Trifunović (✉) · D. Lazarević
Faculty of Transport and Traffic Engineering—University of Belgrade, Belgrade, Serbia
e-mail: a.trifunovic@sf.bg.ac.rs

D. Lazarević
e-mail: d.lazarevic@sf.bg.ac.rs

A. Senić
Faculty of Civil Engineering—University of Belgrade, Belgrade, Serbia
e-mail: asenic@grf.bg.ac.rs

Z. Keković et al. (eds.), *Critical Infrastructure: Risk and Crisis Management in the New Reality*, Advanced Sciences and Technologies for Security Applications,
https://doi.org/10.1007/978-3-032-20948-1_6

1 Introduction

Road traffic safety remains one of the most significant public health and infrastructure challenges of the twenty-first century. According to the World Health Organization (WHO), more than one million people lose their lives each year as a result of road crashes, while several tens of millions sustain injuries of varying severity, many leading to permanent disability [1]. Despite improvements in vehicle design, road engineering, and emergency response systems, certain segments of the road network remain disproportionately hazardous. Among these, road tunnels represent one of the most critical environments in terms of safety management, evacuation efficiency, and driver behavior [2, 3].

Tunnels are characterized by restricted geometry, limited visibility, and complex interactions between drivers and infrastructure. Sudden changes in lighting conditions, narrow lanes, bidirectional traffic, and the absence of safe shoulders or escape zones contribute to heightened psychological stress and reduced reaction times among drivers [4, 5]. Empirical studies have shown that accident severity within tunnels is often higher than on open roads, due to secondary collisions, fire risks, and the confined nature of the space [6, 7]. Furthermore, inadequate ventilation, outdated lighting systems, and poor maintenance of emergency exits or communication systems may exacerbate the consequences of even minor incidents [7, 8].

Assessing and managing risk in tunnel environments requires a multidimensional approach, one that combines objective parameters such as geometry, traffic flow, and lighting intensity with subjective elements of driver perception and expert judgment [9, 10]. Traditional methods of safety evaluation, based primarily on statistical crash data or expert audits, often overlook the behavioral and perceptual aspects that influence how drivers experience risk. Drivers tend to underestimate hazards in familiar or monotonous tunnels, particularly when environmental cues are weak or when technological support systems are absent [11, 12].

In recent years, artificial intelligence (AI) has emerged as a valuable tool for traffic safety modeling and proactive risk prediction. AI-based systems can process large datasets, identify nonlinear patterns, and provide dynamic assessments of crash likelihood in complex environments (AI in traffic safety, predictive modeling). Machine learning and fuzzy logic techniques have been successfully applied to analyze road geometry, visibility, and vehicle behavior under varying environmental conditions [13]. However, the integration of AI-driven models with human perception and expert assessment in tunnel safety research remains limited.

To address this gap, the present study proposes a hybrid framework for tunnel safety assessment, combining three complementary perspectives: evaluations by traffic and civil engineering experts, subjective assessments from professional and non-professional drivers, and an artificial intelligence system used to estimate risk based on structured tunnel parameters. The study focuses on three representative tunnel types commonly found in Serbia: short urban tunnels, long motorway tunnels, and bi-directional rural tunnels. By comparing the differences in risk perception between human and algorithmic evaluators, the study aims to provide a deeper understanding of safety in confined traffic environments and identify opportunities for intelligent, context-aware risk management.

The structure of this paper follows a logical order. Section 1 introduces the background and rationale for studying tunnel safety and risk perception. Section 2 presents the applied methodology, including study design, participants, and statistical procedures. Section 3 reports the results of driver, expert, and AI evaluations. Section 4 discusses the findings in light of existing literature, and Sect. 5 concludes with key implications and recommendations for enhancing tunnel safety through human-centered and AI-supported approaches.

2 Methodology

The methodology applied in this study is structured into three segments: Study Design, Participants, and Statistical Analysis. This structure enables a comparative evaluation of safety perceptions in road tunnels among experts, drivers, and an artificial intelligence model.

2.1 Study Design

To assess the perceived and modeled safety of road tunnels, a comparative cross-sectional study was conducted. The research focused on three representative tunnel scenarios, each corresponding to different tunnel characteristics and traffic conditions commonly found in Serbia:

- a short urban tunnel (length <500 m),
- a long motorway tunnel (length >1000 m), and
- a two-way rural tunnel (bi-directional traffic flow).

Each scenario reflected real tunnel conditions observed on state roads (first and second category) and local roads/streets. Instead of using standardized images or video simulations, participants evaluated the safety of tunnels based on their own prior driving experience and memory of similar tunnel environments. The evaluation was grounded in how such tunnels typically appear in practice. The six predefined safety parameters were included in the evaluation: lighting quality, ventilation efficiency, visibility range, lane width, traffic flow density, and visibility of emergency exits. This approach aimed to capture authentic perception and expert judgment rooted in real-world exposure, rather than artificially prepared stimuli.

- Three groups of evaluators participated in the assessment process:
- Traffic and civil engineering experts,

- Professional and non-professional drivers, and
- An AI model trained to assess tunnel safety based on structured input variables.

Experts and drivers were asked to assign safety ratings for each tunnel type using the same predefined set of six parameters: lighting quality, ventilation efficiency, visibility range, lane width, traffic flow density, and visibility of emergency exits. These ratings reflected their subjective perception of safety based on real-life driving experience rather than responses to visual tests. In addition to the structured evaluation, participants could provide open-ended comments, and some mentioned additional aspects such as humidity or signage clarity. These qualitative remarks were used only for contextual interpretation and were not part of the statistical scoring procedure. In parallel, the AI model generated safety scores for the same tunnel types using the same predefined set of parameters evaluated by the human participants. It is important to note that the AI evaluation in this study was not based on a custom-trained model. Instead, an existing large-scale AI system was used, to which we provided a structured set of tunnel parameters. The AI system generated safety scores using its built-in reasoning and pattern-recognition capabilities. All scores were expressed on a 10-point Likert scale (1 = extremely unsafe, 10 = very safe), enabling direct comparison between AI-derived evaluations and human ratings.

2.2 *Participants*

The study involved a total of 116 participants, divided into two human subgroups and one AI system:

- Experts: 8 professionals (5 from the field of traffic engineering and 3 from civil engineering), each with at least 8 years of experience in tunnel design, road safety auditing, or infrastructure management.
- Drivers: 108 drivers, including 62 professional (freight and passenger transport) and 46 non-professional drivers. Participants were recruited through public announcements, driving schools, and transport companies. The sample was balanced by gender and age, ranging from 21 to 65 years, with an average driving experience of 13.6 years.
- AI Model: The AI system used in this study was an existing large-scale artificial intelligence tool. It received a structured set of tunnel parameters and generated safety scores using its built-in reasoning and pattern-recognition mechanisms. No custom training was performed for this study.

Participation was entirely voluntary, with informed consent obtained from all participants. No monetary compensation was offered, and respondents were informed that they could withdraw from the study at any time. The research followed ethical standards for human subject studies, in line with the principles of the Declaration of Helsinki. All data were anonymized, and no personally identifiable information was collected or stored.

2.3 *Statistical Analysis*

Collected data were processed using IBM SPSS Statistics 29. First, descriptive statistics (mean values and standard deviations) were calculated for all three evaluator groups and tunnel scenarios. To determine whether significant differences existed between the groups (experts, drivers, AI), a one-way analysis of variance (ANOVA) was conducted for each tunnel type. Where significant differences were found, Tukey's HSD post-hoc test was applied to identify pairwise differences between evaluator groups. Additionally, Pearson correlation coefficients were calculated to explore the degree of agreement between experts and AI, as well as between experts and drivers. The level of statistical significance was set at $p < 0.05$ for all analyses.

3 Results

The findings of this study are organized into five analytical sections that provide a comprehensive understanding of how safety in road tunnels was evaluated from different perspectives. These include: Profile of Study Participants, Expert-Based Safety Assessment, Driver Perception and Evaluation, AI-Derived Safety Indicators, and Cross-Group Statistical Comparison. This structure enables a detailed interpretation of the similarities and differences between expert judgments, driver perceptions, and AI model outputs, offering a clearer view of how each group interprets risk across various tunnel environments.

3.1 *Profile of Study Participants*

A total of 116 participants took part in the study. The sample included 8 experts (5 from traffic engineering and 3 from civil engineering) and 108 drivers, of whom 62 (57.4%) were professional drivers and 46 (42.6%) non-professional. The gender distribution included 71 male (61.2%) and 45 female (38.8%) respondents. The age range of participants was between 21 and 65 years, with an average age of 42.1 years ($SD = 9.8$). Professional drivers reported an average of 15.4 years of driving experience, while non-professional drivers averaged 11.2 years. All participants confirmed prior experience driving through tunnels in Serbia, including short urban tunnels, long motorway tunnels, and two-way rural tunnels, making their assessments relevant and experience-based.

Table 1 Mean safety ratings for road tunnels by expert group and tunnel type

Tunnel type	Mean	Standard deviation
Short urban tunnel	5.1	1.2
Long motorway tunnel	4.4	1.3
Two-way rural tunnel	3.9	1.1

3.2 Expert-Based Safety Assessment

Experts rated the safety of each tunnel type based on their professional experience and knowledge of real conditions in Serbian road infrastructure. The mean safety ratings (on a 10-point scale) are shown in Table 1.

Experts consistently identified two-way rural tunnels as the most critical in terms of safety, citing limited cross-sectional area, inadequate lighting, and the absence of emergency exits as the main risk factors. The long motorway tunnels were also rated relatively unsafe due to visibility and evacuation challenges, whereas short urban tunnels received slightly higher ratings owing to better monitoring and maintenance standards.

3.3 Driver Perception

Drivers evaluated each tunnel type based on their personal experience and perception of safety under typical driving conditions. The mean and standard deviation scores are shown in Table 2.

Compared to experts, drivers rated all tunnel types as significantly safer, particularly short urban tunnels. This difference may reflect limited awareness of hidden risks such as smoke accumulation, poor ventilation, or technical failures. Drivers with less experience tended to give slightly higher safety ratings, although this trend was not statistically analyzed in depth.

Table 2 Mean safety ratings for road tunnels by driver group and tunnel type

Tunnel type	Mean	Standard deviation
Short urban tunnel	6.7	1.5
Long motorway tunnel	6.0	1.7
Two-way rural tunnel	5.5	1.8

Table 3 Mean safety ratings for road tunnels by AI model and tunnel type

Tunnel type	Mean
Short urban tunnel	4.9
Long motorway tunnel	4.1
Two-way rural tunnel	3.8

3.4 AI-Derived Safety Indicators

The AI model provided safety ratings for each tunnel type based on structured input data and prior classifications of tunnel safety risk. The resulting mean scores are presented in Table 3.

The AI model's evaluations were most closely aligned with expert ratings. Both identified two-way rural tunnels as the most hazardous, primarily due to bidirectional traffic flow, poor lighting uniformity, and restricted visibility. The short urban tunnel scenario achieved the highest AI score but remained lower than the average driver evaluation, indicating that drivers may underestimate objective risk factors.

3.5 Statistical Comparison of Ratings

The average safety ratings across the three evaluator groups and tunnel types are presented in Table 4.

A one-way ANOVA was conducted to determine whether significant differences existed among the three evaluator groups (experts, drivers, and AI) for each tunnel type. For short urban tunnels, the difference was statistically significant ($F = 12.7$, $p < 0.001$), with drivers rating tunnels as safer than both experts and the AI model. For long motorway tunnels, the difference was also significant ($F = 10.8$, $p < 0.001$), again showing higher ratings by drivers. For two-way rural tunnels, the difference was highly significant ($F = 15.4$, $p < 0.001$), indicating the largest perception gap among the groups. Tukey's HSD post-hoc analysis confirmed that all pairwise comparisons between drivers, experts, and AI were statistically significant ($p < 0.05$) in the rural tunnel scenario. Pearson correlation analysis revealed a strong positive correlation between expert and AI ratings ($r = 0.87$), suggesting alignment in risk evaluation criteria. The correlation between drivers and experts was moderate ($r = 0.54$), while the correlation between drivers and the AI model was weaker ($r = 0.42$), indicating

Table 4 Average safety ratings by evaluator group and road category

Tunnel type	Experts (±SD)	Drivers (±SD)	AI
Short urban tunnel	5.1 ± 1.2	6.6 ± 1.4	4.9
Long motorway tunnel	4.4 ± 1.3	6.0 ± 1.6	4.1
Two-way rural tunnel	3.9 ± 1.1	5.3 ± 1.7	3.8

that drivers tend to underestimate tunnel-related risks, particularly those associated with limited visibility and bi-directional traffic.

4 Discussion

The findings of this study reveal noticeable differences in how drivers, experts, and the AI model perceive the safety of road tunnels. These discrepancies were consistent across all three tunnel types, short urban tunnels, long motorway tunnels, and two-way rural tunnels, indicating systematic variation in risk interpretation between human perception and algorithmic evaluation. The most prominent observation is that drivers consistently rated tunnels as safer than both experts and the AI model. This optimistic bias was particularly pronounced for short urban tunnels, which many drivers considered routine segments of their daily travel. Such results suggest that drivers tend to normalize tunnel environments and underestimate specific risks, such as poor lighting, abrupt luminance adaptation, or inadequate visibility of emergency exits. Previous studies have shown that drivers often experience a temporary loss of spatial orientation when entering or leaving a tunnel, yet few consciously recognize this as a safety risk.

Experts, on the other hand, provided more conservative ratings, especially for long motorway and rural tunnels. Their evaluations reflected a better awareness of structural and operational challenges, including ventilation limitations, emergency evacuation constraints, and the potential for secondary collisions in the event of breakdowns or fires. Experts emphasized that even minor technical faults, such as partial lighting failure or impaired communication systems, can dramatically increase risk levels within confined tunnel environments. The AI model's ratings closely mirrored those of the experts, indicating that data-driven approaches can effectively approximate expert judgment. The high correlation between expert and AI assessments ($r = 0.87$) suggests that AI-based evaluation systems can serve as a reliable supplementary tool in tunnel safety auditing. The model identified two-way rural tunnels as the most hazardous, primarily due to the combination of narrow cross-sections, opposing traffic, and limited escape routes, factors that are often overlooked in human perception.

The moderate correlation between experts and drivers ($r = 0.54$), along with the weaker alignment between drivers and AI ($r = 0.42$), highlights a persistent perceptual gap. Drivers' limited understanding of complex, invisible, or cumulative risks, such as smoke propagation, delayed braking response, or ventilation system failure, suggests that traditional signage and static warnings are insufficient. These findings underscore the need for adaptive and context-aware safety communication, potentially through in-vehicle navigation systems or variable message signs that provide real-time updates on tunnel conditions. Another important implication concerns driver training and awareness. Despite advancements in tunnel design and automation, safety outcomes remain highly dependent on driver behavior. Educational campaigns emphasizing appropriate speed adjustment, headlight use, and reaction to stopped vehicles could substantially reduce risk exposure. Finally, the study demonstrates that integrating AI-based risk models into intelligent transportation systems can enhance tunnel monitoring and decision support. Such systems could automatically detect patterns of elevated risk (e.g., reduced visibility or congestion) and trigger immediate alerts to both operators and drivers. Over time, these adaptive systems may contribute to a more harmonized understanding of tunnel safety among all stakeholders, bridging the gap between expert knowledge, driver experience, and automated intelligence.

5 Conclusion

This study examined the perceived and modeled safety of road tunnels through the comparative assessments of drivers, experts, and an artificial intelligence (AI) model. The results revealed significant differences between these groups, indicating that drivers consistently perceive tunnels as safer environments than experts or AI systems. This optimism bias suggests a limited awareness of hidden or technical risks, particularly in long and two-way tunnels where lighting, ventilation, and evacuation challenges are most critical.

Experts provided more conservative safety ratings, emphasizing structural and operational aspects such as inadequate emergency exits, insufficient monitoring, and poor illumination uniformity. Their assessments closely aligned with the AI model, confirming that algorithmic evaluations can effectively replicate expert reasoning when provided with relevant infrastructure and operational data. The strong correlation between expert and AI ratings highlights the potential of AI-based tools as reliable support instruments for tunnel safety audits and real-time monitoring.

The perceptual gap between drivers and data-driven evaluations underscores the importance of enhancing driver awareness and communication within tunnel environments. Educational initiatives focused on tunnel-specific hazards, such as speed control, adaptive lighting, and emergency procedures, could improve driver preparedness and response. Additionally, integrating AI-assisted risk assessment modules into intelligent transportation systems (ITS) and navigation platforms could enable real-time alerts based on changing tunnel conditions, improving both safety management and situational awareness.

While the study offers valuable insights, certain limitations should be acknowledged. The driver evaluations were based on subjective recall rather than controlled simulation, which may have introduced bias. Moreover, the AI model relied on a limited national dataset and should be validated using broader, international tunnel case studies. Future research should focus on incorporating real-time monitoring data, such as illumination levels, air quality, and traffic density, to dynamically update safety predictions. Expanding the model to include behavioral variables, such as stress, reaction time, and familiarity with tunnels, could further refine its predictive accuracy. In conclusion, by integrating human perception, expert judgment, and artificial intelligence, this study contributes to a more comprehensive and realistic understanding of tunnel safety. Bridging the perception gap among these stakeholders represents a critical step toward developing safer, smarter, and more resilient road tunnel systems.

References

1. World Health Organization: Global Status Report on Road Safety 2018. World Health Organization, Geneva, Switzerland (2019)
2. Bassan, S.: Overview of traffic safety aspects and design in road tunnels. IATSS Res. **40**(1), 35–46 (2016)
3. Tušer, I.,Hošková-Mayerová, Š.: Traffic safety sustainability and population protection in road tunnels. Qual. Quant. **57**(Suppl 3), 249–270 (2023)
4. Kirytopoulos, K., Konstandinidou, M., Nivolianitou, Z., Kazaras, K.: Embedding the human factor in road tunnel risk analysis. Process Saf. Environ. Prot. **92**(4), 329–337 (2014)
5. Bourdy, C., Chiron, A., Cottin, C., Monot, A.: Visibility at a tunnel entrance: effect of temporal adaptation. Light. Res. Technol. **19**(2), 35–44 (1987)
6. Pervez, A., Huang, H., Han, C., Wang, J., Li, Y.: Revisiting freeway single tunnel crash characteristics analysis: a six-zone analytic approach. Accid. Anal. Prev. **142**, 105542 (2020)
7. Łukasz, N., Natalia, S.P.: Verification of possible safe evacuation from road tunnels in the event of fire. Saf. Fire Technol. **47**(3), 96–110 (2017)
8. Zhang, Y., Huang, X.: A review of tunnel fire evacuation strategies and state-of-the-art research in China. Fire Technol. **60**(2), 859–892 (2024)
9. Ram, T., Chand, K.: Effect of drivers' risk perception and perception of driving tasks on road safety attitude. Transp. Res. Part F Traffic Psychol. Behav. **42**, 162–176 (2016)
10. de Oña, J., de Oña, R., Eboli, L., Forciniti, C., Mazzulla, G.: How to identify the key factors that affect driver perception of accident risk: a comparison between Italian and Spanish driver behavior. Accid. Anal. Prev. **73**, 225–235 (2014)
11. Trifunović, A., Antić, B., Pešić, D., Čičević, S.: Young drivers appraisement of road environment from the aspect of traffic safety. J. Road Traffic Eng. **63**(1), 43–47 (2017)
12. Wang, F., Ding, L., Love, P.E., Edwards, D.J.: Modeling tunnel construction risk dynamics: addressing the production versus protection problem. Saf. Sci. **87**, 101–115 (2016)
13. Zhou, Z.H.: Machine Learning. Springer Nature, Singapore (2021)

Beyond Anticipation: Composite Analytical Model of Organizational Resilience

Vladimir Ninković and Marina Dabetić

Abstract Organizational resilience has emerged as a multidimensional and contested concept, variously understood as an outcome, a process, or a strategy. This paper conceptualizes resilience as a strategic approach to managing systemic and non-routine risks and uncertainties, positioning it as a higher-order construct relative to crisis management and business continuity. The study operationalizes resilience through four interlinked capacities—anticipatory, absorptive, restorative, and adaptive—and distinguishes between first-order resilience (planned responses to known risks) and second-order resilience (coping with unforeseen disruptions through improvisation, flexibility, and learning). This study develops a composite analytical model that combines these four capacities with Gibson and Tarrant's "herringbone model," which differentiates between hard aspects (formal processes, systems, and procedures) and soft aspects (culture, leadership, learning, and adaptability). This framework provides both a conceptual and practical basis for evaluating organizational resilience and contributes to establishing resilience as a core concept within security studies. By integrating capacities and aspects, this study provides a novel approach for assessing resilience in organizations and highlights the importance of moving beyond anticipation toward more adaptive and holistic resilience strategies.

Keywords Resilience · Risk · Crisis · Critical infrastructure · Analytical model

---

V. Ninković (✉)
Center for Risk Analysis and Crisis Management, Belgrade, Serbia
e-mail: Vladamcc@gmail.com

M. Dabetić
Department of Physical Chemistry, Vinča Institute of Nuclear Sciences—National Institute of the Republic of Serbia, University of Belgrade, BelgradeBelgrade, Serbia
e-mail: fmarina@vin.bg.ac.rs

Z. Keković et al. (eds.), *Critical Infrastructure: Risk and Crisis Management in the New Reality*, Advanced Sciences and Technologies for Security Applications,
https://doi.org/10.1007/978-3-032-20948-1_7

1 Introduction

Organizations operating in complex and uncertain environments increasingly face non-routine risks—events that are low in probability but potentially catastrophic in their consequences. Such risks often emerge unexpectedly, cascade across systems, and overwhelm predefined response procedures. Traditional risk management approaches, grounded in anticipation, prediction, and prevention, remain important but are insufficient when dealing with disruptions whose timing, form, or scale cannot be reliably modeled. As Wildavsky [1] argued, anticipation alone cannot address the full spectrum of contemporary risks; organizations must also cultivate the ability to cope effectively when surprises occur.

This need becomes especially pronounced in sectors responsible for critical infrastructure, where disruptions may have far-reaching societal implications. The limitations of prediction in complex systems, highlighted by Taleb [2], reinforce the necessity of organizational strategies that extend beyond forecasting. Increasingly, scholars conceptualize organizational resilience as a strategic orientation that integrates anticipation with adaptability, enabling organizations to withstand, absorb, recover from, and learn through unexpected events. Sutcliffe and Vogus [3] describe resilience as the maintenance of positive functioning under strain, while Duchek [4] frames it as a process comprising anticipation, coping, and adaptation.

While the resilience concept has gained wide recognition across fields such as crisis management, organizational studies, and disaster research, it remains fragmented, with multiple definitions and divergent analytical approaches. Some scholars focus on resilience as a structural characteristic, others as a behavioural capacity, and still others as an emergent property of socio-technical systems. This conceptual pluralism, though intellectually rich, creates challenges for operationalizing and assessing resilience in real organizational contexts.

This paper addresses that challenge by proposing a Composite Analytical Model of Organizational Resilience. Combining the four widely recognized resilience capacities—anticipatory, absorptive, restorative, and adaptive—with the distinction between "hard" and "soft" organizational aspects developed by Gibson and Tarrant [5], the proposed analytical model offers a unified and practical framework for resilience assessment. The approach recognizes resilience as a composite and relational construct that emerges from the interaction of technical systems, organizational structures, human behaviour, and cultural dynamics.

The purpose of this article is therefore twofold:

1. To consolidate theoretical insights from resilience literature into a coherent analytical structure; and
2. To offer a model that can support future empirical research and practical assessments of resilience, particularly in high-risk domains such as critical infrastructure.

The following sections develop the theoretical basis for the proposed analytical model, outline its structure, and discuss its implications for understanding and assessing resilience in organizations exposed to non-routine risks.

2 Theoretical Background

2.1 Resilience as a Strategy

One of the central challenges in contemporary organizational and security practice is how to manage risks under conditions of deep uncertainty. Traditional risk management is built on a predictive logic: organizations identify potential threats, assess their likelihood, and estimate their consequences. This anticipatory strategy—reflected in classical probabilistic risk analysis—assumes that threats can be mapped in advance and that preventive measures can be designed to mitigate them. Kaplan and Garrick's [6] foundational model, based on the questions "What can go wrong? How likely is it? What are the consequences?" embodies this approach. Yet decades of research have shown that this logic becomes fragile in complex socio-technical systems.

Aven [7] and Renn [8] argue that probabilistic risk assessment is highly sensitive to uncertainty, vulnerable to cognitive biases, and structurally incapable of capturing unknown or emerging risks. Under real conditions, organizations often confront disruptive events not because they failed to anticipate them, but because uncertainty cannot always be reduced. In such contexts, threats may materialize despite preventive mechanisms, and decision-makers must act with incomplete information, high ambiguity, and time pressure. Taleb's work on fat-tailed risks reinforces this view: low-probability, high-impact events dominate real-world risk landscapes and cannot be forecast reliably [2]. Modern disruptions—pandemics, climate-driven extremes, technological accidents—often exceed planning assumptions and defy prediction [9, 10].

This insight has led to a significant shift in how risk is conceptualized. Instead of viewing risk solely as a calculable function of threat, probability, and consequence, scholars increasingly interpret it as a condition shaped by complexity, interdependence, and nonlinearity. Dunn Cavelty, Kaufmann, and Soeby Kristensen [11] observe that in complex systems, the full spectrum of risks cannot be known in advance. Unknown unknowns and emergent phenomena undermine the predictive assumptions of traditional risk assessment. As environmental complexity increases, Comfort [12] notes, system performance often degrades not because plans are insufficient, but because organizations cannot process the information required for centralised control.

Against this backdrop, resilience has emerged as a strategic approach to risk management, not merely a system property. Wildavsky [1] distinguished between two strategies: anticipation, which seeks to prevent harm by predicting threats; and resilience, which focuses on the ability to cope with harm when prevention fails. While anticipation aims at precision, control, and pre-emption, resilience emphasizes buffering, flexibility, improvisation, and adaptive learning. Renn [8] extends this argument by framing resilience both as an objective system attribute (the capacity to absorb, recover, and adapt) and a subjective one (the degree of confidence actors have in their ability to manage crises).

This reframing also aligns with contemporary complexity and resilience engineering literature. Hollnagel et al. [11] argue that high-reliability systems must be able to respond to the unexpected by adjusting performance, reallocating resources, and adapting on the fly. Longstaff [12] similarly highlights that organizations must be designed to maintain essential functions even when predictive models fail.

In this sense, resilience is not positioned in opposition to risk management—it is a form of risk management suited to environments where prediction is inherently limited. Rather than assuming that all threats can be prevented, resilience accepts uncertainty as a structural reality and focuses on strengthening the organization's capacity to anticipate, absorb, recover from, and adapt to disruptions.

Thus, besides observing resilience as a desired state of the system, or a process, resilience can be viewed as a comprehensive strategy for managing non-routine risks and uncertainties. It complements traditional approaches but also corrects their blind spots, shifting attention from forecasting individual threats to enhancing the organization's overall ability to function under uncertainty.

2.2 Non-Routine Risks

The concept of non-routine risks has been present in academic and professional literature for nearly five decades and broadly refers to risks that are low in probability or predictability but high in potential consequences [13]. However, similar formulations appeared even earlier in U.S. Nuclear Regulatory Commission reports describing "non-routine events" in the 1970s [14]. The terminology itself draws inspiration from non-routine mathematical problems, which require creative reasoning, multiple solution paths, and the ability to operate without clear procedural guidance [15].

A major contemporary contribution to this field comes from Tarrant [16], who argues that non-routine risks force organizations into non-routine modes of operation, requiring departures from standard procedures. These risks differ from routine risks not simply because they are rare or severe, but because they push organizations beyond their normal functional boundaries. At the extreme end of this continuum lie catastrophic events, whose dynamics correspond to Taleb's fat-tail distributions [2], rare but disproportionately destructive events whose impacts are vastly greater than anticipated by normal statistical models.

Crucially, non-routine risks can be known and included in risk registers; they are not synonymous with epistemically uncertain "Black Swan" events. What differentiates them from routine risks is the degree of organizational disruption they produce. A large system with extensive buffers may absorb an event routinely, whereas the same event may push a smaller organization into non-routine functioning.

Bier et al. [17] highlight several analytical and practical challenges posed by non-routine risks. First, their consequences may exceed available coping capacities. Second, their rarity limits experiential learning and the accumulation of institutional knowledge. Third, human, organizational, and technical behaviour can deviate significantly under unusual conditions, making modelling difficult. Fourth, even when not statistically uncertain, extreme events may resist interpretation because decision-makers lack relevant first-hand experience or contextual cues.

These challenges explain why non-routine risks have become relevant to contemporary thinking on organizational resilience, business continuity, crisis management, and crisis communication. Tarrant [16] argues that these approaches provide essential preparation for operating in a non-routine mode when standard processes fail. As Crichton, Ramsay, and Kelly [18] show, crises often generate radically altered tasks, objectives, and working conditions, increasing coordination demands and underscoring the importance of pre-established interorganizational relationships. Comfort [12] similarly argues that adaptive capacity is essential in dynamic conditions where formal procedures quickly become outdated.

Taken together, the literature demonstrates that non-routine risks cannot be managed through anticipation alone. They require organizations to cultivate flexibility, improvisation, shared sensemaking, adaptive coordination, and ability for learning—the defining elements of resilience-based approaches to risk management.

2.3 *First-Order and Second-Order Resilience*

A significant number of researchers distinguish between an organization's ability to cope with known threats using existing, pre-planned capabilities, and its ability to respond to unexpected problems by learning, adapting, and developing new capabilities in real time. These abilities are often referred to as first-order and second-order resilience, terms used by Hollnagel et al. [19], Kendra and Wachtendorf [20], and Lee et al. [21]:

First-Order Resilience refers to an organization's ability to respond effectively to known risks using predefined procedures, technologies, and control mechanisms. It is closely associated with risk management systems, crisis plans, business continuity procedures, and standard operating processes.

Second-Order Resilience refers to the ability to cope with unexpected, unanticipated, or novel threats through improvisation, experimentation, decentralized decision-making, and organizational learning. Sutcliffe and Vogus [3] and Weick and Roberts [22] have shown that second-order resilience often determines performance under true uncertainty.

To summarize, first-order resilience is grounded in stability and control; second-order resilience is grounded in flexibility and adaptation.

2.4 *The Capacities of Organizational Resilience*

ISO Standard 22,316:2016 *Organizational Resilience – Guidelines* defines organizational resilience as *the outcome of an organization's ability to anticipate and respond to disruptions caused by risks, as well as its capacity to adapt to complex and changing circumstances under conditions of uncertainty* [23]. The standard further defines *capacity* as the combination of all available strengths and resources within an organization, community, or society that can reduce either the level of risk or the impact of a crisis [23]. This framing positions resilience not merely as a static attribute but as a dynamic expression of organizational capability.

Across literature, scholars commonly identify four core resilience capacities—anticipatory, absorptive, restorative, and adaptive. This formulation is reflected in the works of authors such as Bruneau et al. [24], Madni and Jackson [25], Francis and Bekera [26], and Keković et al. [27]. These capacities capture the essential processes through which organizations prepare for, respond to, recover from, and learn through disruptive events.

- **Anticipatory capacity** refers to the ability to foresee and prepare for potential disruptions through monitoring, early warning systems, and scenario-based planning.
- **Absorptive capacity** denotes the ability to withstand shocks with minimal loss of function, relying on robustness, redundancies, and protective mechanisms.
- **Restorative capacity** encompasses the organization's ability to recover and restore essential functions after disruption, aligning closely with business continuity and crisis management efforts.
- **Adaptive capacity** reflects the ability to modify practices, structures, or behaviours in response to changing conditions, enabling the organization to learn, innovate, and transform.

These four capacities form the conceptual foundation of the analytical model developed in this study. Capacities are only loosely temporally defined, as they will often overlap and intertwine. Thus, rather than treating resilience as a linear sequence of phases, this approach emphasizes the interdependence and simultaneity of capacities that may be activated before, during, or after a disruptive event. The model therefore provides a multidimensional view of resilience that aligns with contemporary standards and leading academic perspectives.

2.5 Hard and Soft Aspects of Organizational Resilience

Gibson and Tarrant [5] proposed a so called "herringbone model" for organizational resilience. This model recognizes that an organization possesses a broad range of capabilities and undertakes a variety of activities—collectively, what the organization does—that contribute to its resilience. In this study, we refer to these "capabilities" and "activities" as the hard aspect of resilience. These include:

- risk management frameworks,
- crisis management plans,
- business continuity arrangements,

- redundancies, buffers, and technical safeguards,
- defined decision-making protocols and communication structures.

Hard aspects enable predictable and coordinated responses, reinforcing first-order resilience.

On the other hand, organizations also display numerous characteristics—the *how* of organizational functioning—that influence the effectiveness of their capabilities and activities and therefore contribute to strengthening resilience. In this study, we refer to these “characteristics” as the soft aspect of resilience.

These include:

- leadership style,
- organizational culture and values,
- trust and social cohesion,
- improvisation and creativity,
- learning orientation,
- sensemaking and communication quality.

While most hard and soft aspects are critical for functioning in routine environments, it is the way in which they can be adjusted to non-routine conditions that generates resilience in non-routine situations. Soft aspects underpin second-order resilience and are often decisive when organizations face non-routine or rapidly evolving disruptions. Braes and Brooks [28], as well as Tveiten et al. [29], emphasize that soft dimensions significantly influence resilience outcomes, sometimes more than formalized planning. The following section introduces the composite analytical model that integrates these capacities and aspects into a single cohesive resilience framework.

3 Composite Analytical Model of Organizational Resilience

Existing approaches to organizational resilience frequently emphasize either formal technical structures and systems (the hard aspects – activities and capabilities) or behavioural, relational, and cultural characteristics (the soft aspects) [12, 19]. Yet extensive research shows that resilience emerges from the interaction between these domains rather than from one alone. Robust crisis plans and redundant systems can fail if teams lack coordination, trust, or sensemaking capacity, while strong improvisational culture cannot compensate for missing redundancies or weak continuity structures [22, 30].

To overcome the conceptual fragmentation that characterizes much of the resilience literature, where diverse definitions, dimensions, and capacities coexist without integration, this paper introduces and Composite Analytical Model of Organizational Resilience. By synthesizing the four dominant resilience capacities model [24–27] with Gibson and Tarrant’s distinction between hard and soft organizational aspects [5], our proposed model offers a holistic, multidimensional, and relational understanding of resilience.

The logic for integrating the two models becomes clearer when examining the specific indicators associated with each domain. The hard indicators in Table 1 represent the structural and procedural foundations through which organizations prepare for and manage disruptions. These are the systems, tools, and formalized arrangements that enable organizations to anticipate risks, absorb shocks, restore services, and support adaptive reconfiguration.

In contrast, the soft indicators capture the behavioural and cultural mechanisms that shape how these systems are enacted in practice. These characteristics influence interpretation of warning signals, coordination during emergencies, the speed and quality of decision-making, and the capacity to innovate when established procedures fail.

The model is structured as a 4×2 matrix, where each of the four resilience capacities—anticipatory, absorptive, restorative, and adaptive—contains:

- Hard indicators → formal systems, tools, processes, technologies.
- Soft indicators → behavioural, relational, cognitive, and cultural elements.

This structure ensures that both planned and adaptive elements of resilience are captured systematically.

What the table reveals is that each resilience capacity is inherently dual in nature:

- Anticipatory capacity depends on both formal processes (e.g., risk assessments, early warning systems) and behavioural elements (e.g., foresight culture, participatory planning).
- Absorptive capacity requires engineered redundancies and protective systems, but also team cohesion, psychological readiness, and tolerance for ambiguity.
- Restorative capacity is supported by continuity plans and recovery protocols, but its effectiveness hinges on leadership stability, trust, and clarity of roles.

Table 1 Resilience capacities and hard/soft indicators

Capacity	Hard Indicators (Systems and Processes)	Soft Indicators (Behavioural and Cultural)
Anticipatory	Risk assessments; early warning systems; monitoring tools; scenario planning;	Acuity; foresight culture; participatory approach to planning; risk communication
Absorptive	Technical redundancies; buffers; protective systems; emergency and incident response systems;	Team cohesion; sensemaking; psychological readiness; ambiguity tolerance;
Restorative	Business continuity planning; recovery protocols; resource prioritization	Leadership stability; inter-team trust; role clarity; interdependency; coping with stress
Adaptive	Flexible systems; dynamic adaptive planning; modular processes; alternative resource pathways;	Improvisation; creativity; non-linear learning orientation; adaptive leadership

- Adaptive capacity involves flexible systems, modular processes, and dynamic planning, yet ultimately relies on improvisation, creativity, learning orientation, and adaptive leadership.

Each proposed indicator represents a specific organizational practice or characteristic associated with resilience. The indicators are drawn from established literature that emphasizes both formal capacities and human/behavioural factors [3, 5, 22, 24, 25, 29]:

By linking indicators to capacities, organizations can determine where resilience strengths and weaknesses lie. For example:

- Strong anticipatory capacity with weak adaptive capacity indicates over-reliance on planning.
- Strong absorptive capacity with weak restorative capacity suggests that organizations can withstand shocks but struggle to recover quickly.
- Strong hard aspects but weak soft aspects indicate rigid resilience: high compliance, low adaptability.

We believe that this diagnostic insight may be significant for all organizational systems, particularly for critical infrastructure operators.

Thus, the integration of the capacity model with the "herringbone model" is not merely conceptual, but it reflects the actual structure of resilience as observed in organizations. The indicators demonstrate that no capacity can be adequately understood, assessed, or strengthened without considering both domains simultaneously.

3.1 Analytical Advantages of the Model

The proposed composite, integrative model offers several advantages for both theory and practice. First, it enables a holistic assessment of organizational resilience by synthesizing technical, behavioural, structural, and cultural determinants into a single analytical framework. This directly addresses long-standing critiques that resilience scholarship is fragmented across risk governance, crisis management, business continuity, and organizational behaviour, each emphasizing different facets of resilience without adequate integration [12, 30]. Second, the model facilitates a balanced evaluation by jointly examining hard and soft indicators within each resilience capacity. This helps identify when organizations rely too heavily on formal, compliance-oriented systems at the expense of adaptive behaviours and sensemaking processes.

The alignment of the model with established organizational processes and international standards such as ISO 22316 ensures that the framework can be embedded within existing risk and continuity systems without major structural change [23]. Moreover, by combining resilience capacities with the distinction between hard and soft organizational aspects, the model provides enhanced diagnostic power for identifying capability gaps and prioritizing targeted interventions. As each resilience capacity is assessed through both hard and soft indicators, it allows organizations to determine whether their resilience is primarily structural/procedural or behavioural/cultural—and how these components interact.

Finally, its clearly defined indicators offer a strong foundation for empirical work, supporting validation studies, cross-sector benchmarking, and resilience maturity assessments, an area widely recognized as underdeveloped in current resilience research [26].

4 Discussion

This section interprets the value and practical relevance of the composite model by examining how it addresses persistent conceptual and operational gaps in resilience research. While the concept of organizational resilience has grown in prominence, its practical application has been hindered by conceptual fragmentation, inconsistent terminology, and the absence of comprehensive analytical tools. This discussion outlines how the proposed model contributes to closing these gaps and advancing the strategic assessment of resilience in complex organizational environments.

4.1 Benefits of a Capacity-Based Approach

The capacity-based framework adopted in this model follows a growing body of literature emphasizing resilience as a multidimensional system rather than a single attribute [24, 25], with resilient organizations developing capacities that are activated across different phases of a disruption or crisis [31]. In particular, the model's emphasis on adaptive capacity avoids the common organizational trap of over-investing in planning while neglecting adaptability, a shortcoming observed repeatedly in crisis inquiries [32].

Moreover, the model reinforces principles embedded in ISO 22316, which conceptualizes resilience as a combination of anticipation, adaptation, and continuity under uncertainty [23].

By distinguishing anticipatory, absorptive, restorative, and adaptive capacities, the model provides:

1. Clarity—avoiding the conceptual ambiguity associated with "phases," which may not occur sequentially.
2. Flexibility—allowing capacities to overlap temporally, particularly adaptive capacity, which may be activated before, during, or after a crisis.
3. Analytical precision—enabling organizations to diagnose which capabilities require strengthening.

This structure reinforces the idea that resilience emerges through the dynamic interplay of multiple capabilities rather than through static attributes.

4.2 Hard and Soft Aspects: The Missing Link in Resilience Modelling

Hard aspects are essential for enabling first-order resilience, whereby organizations respond effectively to known threats using predefined mechanisms. They provide stability, predictability, and coordination—qualities necessary for managing routine disruptions. However, exclusive reliance on hard aspects can lead to rigid, compliance-driven structures that perform poorly when disruptions deviate from planned scenarios.

On the other hand, soft aspects are essential for second-order resilience. As Sutcliffe and Vogus [3] and Tveiten et al. [29] show, these behavioural and cognitive factors often determine whether organizations can adapt during high-uncertainty events. Namely, empirical studies show that failures in crisis response rarely result from the absence of plans; rather, they stem from deficits in coordination, trust, adaptability, or decision-making under uncertainty [12, 30].

Our proposed model addresses this gap by pairing hard and soft indicators under each capacity, enabling organizations to diagnose structural–behavioural imbalances that constrain performance during non-routine events.

4.3 Implications for Critical Infrastructure Operators

Critical infrastructure systems operate within tightly coupled socio-technical environments in which disruptions propagate rapidly across sectors. Research consistently demonstrates that interdependence among energy, transport, water, ICT, and supply-chain networks creates conditions where even minor disturbances can escalate into cascading failures with cross-sector and cross-border consequences [9]. These interdependencies are intensified by shared dependencies on digital platforms, SCADA/OT systems, and real-time communications, increasing the potential for systemic disruption. At the same time, the growing frequency and intensity of climate-driven hazards, technological accidents, cyber incidents, and pandemics challenge the operational reliability and continuity of infrastructure services [33]. Under such circumstances, traditional probabilistic risk management—focused on threat identification, likelihood estimation, and protective measures—cannot adequately capture the complex, dynamic, and uncertain nature of infrastructure risk [32, 34].

The analysis presented in this paper reinforces the conclusion that critical infrastructures cannot rely solely on preventive or anticipatory strategies. Uncertainty, non-linear escalation, and interdependency mean that even robust protective systems may be overwhelmed or neutralized during major disruptions [35, 36]. Crisis research shows that critical infrastructure operators often face conditions that exceed design assumptions: power systems entering protective shutdown modes, transport corridors becoming inaccessible, telecommunications nodes becoming saturated, or control-room staff having to operate with incomplete situational awareness [32]. In such contexts, the ability to coordinate, improvise, and adapt becomes as essential as the technical functionality of assets and systems [12, 30].

The composite model proposed in this paper offers particular value for critical infrastructure operators as it allows operators to assess not only the robustness of engineered systems, redundancies, and continuity plans (hard aspects), but also the behavioural, relational, and cultural factors that determine how those systems perform under stress (soft aspects) [24, 25]. This dual perspective is crucial during high-pressure, non-routine events such as cyber-physical incidents, large-scale blackouts, major floods, or supply-chain disruptions, when rapid sensemaking, flexible decision-making, and cross-sector collaboration become essential for maintaining or restoring critical functions. Accordingly, the model shifts analytical attention from individual asset protection toward system-wide resilience, offering an operationally relevant framework for preparedness, planning, and real-time response [23].

Furthermore, the model provides operators with a diagnostic tool for identifying imbalances between formal management systems and adaptive capacities. This is especially important in sectors governed by strict regulatory and compliance regimes, where emphasis on procedures, audits, and technical standards may unintentionally obscure deficits in leadership, communication, improvisation skills, or inter-organizational trust [36, 37]. By revealing gaps—whether in anticipatory monitoring and forecasting, absorptive redundancy and surge capacity, restorative capability and continuity arrangements, or adaptive learning and after-action processes—the composite model helps organizations prioritize targeted interventions and allocate resources more effectively [26]. In doing so, it supports not only organizational resilience but also societal expectations that essential services must be restored, stabilized, or adapted rapidly following large-scale disruptions.

5 Conclusion

The increasing complexity, systemic interdependence, and escalating volatility characterizing contemporary risk environment place unprecedented demands on organizational systems. In such an environment, traditional, probability-based approaches to risk management, however sophisticated, cannot fully address the wide spectrum of disruptions that organizations may encounter [2, 34]. This gap has elevated the importance of resilience as a strategic and operational paradigm, yet the concept itself has remained fragmented across disciplinary silos, variably defined as an outcome, a process, a set of capacities, or a property of socio-technical systems [12, 19].

This paper sought to address these conceptual limitations by developing a composite analytical model of organizational resilience that combines the four resilience capacities, anticipatory, absorptive, restorative, and adaptive [24, 25], with the distinction between hard and soft organizational aspects conceptualised in the herringbone model [5]. The rationale for this integration rests on a growing body of empirical and theoretical work demonstrating that resilience is neither exclusively a technical achievement nor solely a behavioural phenomenon. In addition, in practice, critical failures rarely result solely from missing redundancies or inadequate plans [35]. Equally often, they arise from breakdowns in coordination, cognitive overload, poor communication, or an inability to improvise when disruptions deviate from expected patterns [12, 22]. The hard–soft integration embedded in the composite model therefore helps illuminate why certain organizations cope effectively with severe disruptions while others, equally resourced and formally structured, experience catastrophic failures. By mapping each resilience capacity against both hard and soft indicators, the model exposes organizational imbalances that may remain hidden when assessments focus exclusively on compliance, engineering robustness, or procedural adequacy [36, 37].

This integrative logic has relevance for critical infrastructure operators. As infrastructures grow more interconnected, a single-point failure in one sector can propagate rapidly across networks [9], magnifying societal consequences and compressing the time window available for effective intervention. In domains where cross-sector collaboration, shared situational awareness, and contextually informed decision-making are essential, the soft dimensions of resilience become indispensable complements to formal systems. Accordingly, the proposed model provides a foundation for resilience planning that is more aligned with the realities of complex, interdependent systems than approaches grounded solely in prediction and control.

While the model presented here offers a conceptually robust and operationally meaningful framework, it also opens several avenues for future research. Empirical validation is the next logical step, as the main limitation of this study is that the model has only been piloted, and not empirically validated at scale. Testing the model across sectors, organizational sizes, and national contexts would enable assessment of its diagnostic utility and the relative weight of hard and soft indicators in different operational environments [26]. Longitudinal studies could further illuminate how resilience capacities evolve over time and how investments in specific indicators contribute to overall system performance. The indicators proposed in this model should be regarded as an initial approximation and will require further scrutiny, empirical testing, and refinement to ensure their applicability across diverse organizational and sectoral settings. Finally, integration with resilience maturity models or simulation-based assessment tools may support more systematic benchmarking across the organizational, sectoral, and critical infrastructure landscape.

Acknowledgements This work was carried out within the scientific research activities of the'Vinča' Institute of Nuclear Sciences—National Institute of the Republic of Serbia, funded by the Ministry of Science, Technology, and Innovation, grant number 451-03-136/2025-03/ 200017.

References

1. Wildavsky, A.: Searching for Safety. Transaction Publishers, New Brunswick, NJ (1988)
2. Taleb, N.N.: The Black Swan: The Impact of the Highly Improbable. Random House, New York (2007)
3. Sutcliffe, K.M., Vogus, T.J.: Organizing for resilience. In: Cameron, K., Dutton, J., Quinn, R. (eds.) Positive Organizational Scholarship, pp. 94–110. San Francisco: Berrett-Koehler (2003)
4. Duchek, S.: Organizational resilience: a capability-based conceptualization. Bus. Res. **13**, 215–246 (2020)
5. Gibson, C., Tarrant, R.: A conceptual model of organizational resilience. Int. J. Emergency Manage. **7**(1), 37–63 (2010)
6. Kaplan, S., Garrick, B.J.: On the quantitative definition of risk. Risk Anal. **1**, 11–27 (1981)
7. Aven, T.: On some foundational issues concerning the relationship between risk and resilience. Risk Anal. **42**(9), 2062–2074 (2022)
8. Renn, O.: Risk governance: coping with uncertainty in a complex world. Lond.: Earthscan 456 (2008)
9. Jin, A.S., Trump, B.D., Golan, M., Hynes, W., Young, M., Linkov, I.: Building resilience will require compromise on efficiency. Nat. Energy **6**, 997–999 (2021)
10. Horton, R., Kiker, G.A., Trump, B.D., Linkov, I.: International airports as agents of resilience. J. Contingencies Cris. Management. **30**, 217–220 (2022)
11. Dunn Cavelty, M., Kaufmann, M., Soeby Kristensen, K.: Resilience and (in)security: practices, subjects, temporalities. Secur. Dialogue **46**(1), 3–14 (2015)
12. Comfort, L.K.: Shared Risk: Complex Systems in Seismic Response. Pergamon, New York (1999)
13. Wilkin, L., Sutton, A. (eds.): The Management of Uncertainty: Approaches, Methods and Applications. NATO ASI Series (D: Behavioural and Social Sciences), vol. 32. Springer, Dordrecht (1986)
14. U.S. Nuclear Regulatory Commission: Reactor Safety Study. An assessment of Accident Risk in U.S. Commercial Nuclear Power Plants. Washington, NUREG-75/014. (1975)
15. Polya, G.: How to Solve It—A New Aspect of Mathematical Method. Double Day Anchor Books, New York, USA, p. 253 (1957)
16. Tarrant, M.: The organisation: risk, resilience and governance. Aust. J. Emerg. Management. **25**(2), 15–19 (2010)
17. Bier, V., Haimes, Y.Y., Lambert, J., Mathalas, N., Zimmerman, R.: A survey of approaches for assessing and managing the risk of extremes. Risk Anal. **19**(1), 83–94 (1999)
18. Crichton, M., Ramsay, C., Kelly, T.: Enhancing organisational resilience through emergency planning. J. Contingencies Cris. Manag. **17**(1), 24–37 (2009)
19. Hollnagel, E., Woods, D.D., Leveson, N.: Resilience Engineering: Concepts and Precepts. Ashgate, Aldershot (2008)
20. Kendra, J., Wachtendorf, T.: Elements of resilience after the world trade center disaster: reconstituting New York City's emergency operations centre. Disasters **27**(1), 37–53 (2003)
21. Lee, A.V., Vargo, J., Seville, E.: Developing a tool to measure and compare organizations' resilience. Nat. Hazard. Rev. **14**(1), 29–41 (2013)
22. Weick, K.E., Roberts, K.H.: Collective mind in organizations: heedful interrelating on flight decks. Adm. Sci. Q. **38**(3), 357–381 (1993)
23. International Standards Organization: ISO 22316:2016, Organizational Resilience (2016)
24. Bruneau, M., Chang, S., Eguchi, R., Lee, G., O'Rourke, T., Reinhorn, A., Shinozuka, M., Tierney, K., Wallace, W., von Winterfeldt, D.: A framework to quantitatively assess and enhance the seismic resilience of communities. Earthq. Spectra **19**(4), 733–752 (2003)
25. Madni, A.M., Jackson, S.: Towards a conceptual framework for resilience engineering. IEEE Syst. J. **3**(2), 181–191 (2009)
26. Francis, R., Bekera, B.: A metric and frameworks for resilience analysis of engineered and infrastructure systems. Reliab. Eng. Syst. Saf. **121**, 90–103 (2014)

27. Keković, Z., Dragišić, Z., Ninković, V.: Towards Resilient Critical Infrastructure against Terrorism Risk. In: Čaleta, D., Radović, V. (eds.) Comprehensive approach as "sine qua non" for critical infrastructure protection, pp. 45–59. IOS Press, NATO ARW, Amsterdam (2014)
28. Braes, B., Brooks, D.: Organisational resilience: a propositional study to understand and identify the essential concepts. J. Bus. Contin. Emer. Plan. **5**(3), 262–273 (2010)
29. Tveiten, C.K., Albrechtsen, E., Wærø, I.: Building resilience in emergency management: leadership and learning. Int. J. Emergency Manage. **8**(2), 114–131 (2012)
30. Dalgaard-Nielsen, A.: Organizational resilience in national security bureaucracies: realities and opportunities. J. Contingencies Cris. Manag. **25**(4), 341–350 (2017)
31. Paton, D., Johnston, D.: Disasters and communities: vulnerability, resilience and preparedness. Disaster Prev. Manag. **10**(4), 270–277 (2001)
32. Curt, C., Tacnet, J.M.: Resilience of critical infrastructures: review and analysis of current approaches: resilience of critical infrastructures. Risk Anal. **38**, 1–18 (2018)
33. Boin, A., Mcconnell, A.: Preparing for critical infrastructure breakdowns: the limits of crisis management and the need for resilience. J. Contingencies Cris. Manag. **15**, 50–59 (2007)
34. Fritzon, A., Ljungkvist, K., Boin, A., Rhinard, M.: Protecting Europe's critical infrastructures: problems and prospects. J. Contingencies Cris. Management. **15**(1), 30–41 (2007)
35. Cools, M., Pashley, V.: Pieces of a larger puzzle: the almost impossible task of protecting critical infrastructures, Cahiers inlichtingenstudies—Cahiers d'études du renseignement: BISC 3 Maklu **3**, 51–63 (2013)
36. Khakzad, N., Khan, F., Amyotte, P., Cozzani, V.: Risk management of domino effects considering dynamic consequence analysis. Risk Anal. **34**(6), 1128–1138 (2014)
37. Landucci, G., Argenti, F., Tugnoli, A., Cozzani, V.: Quantitative assessment of safety barrier performance in the prevention of domino scenarios triggered by fire. Reliab. Eng. Syst. Saf. **143**(C), 30–43 (2015)
38. Longstaff, P.H.: Security, Resilience, and Communication in Unpredictable Environments Such as Terrorism, Natural Disasters, and Complex Technology. Center for Information Policy Research, Harvard University, Boston (2005)

Risk Assessment and Crisis Management in Wildfire Protection: The Case of Montenegro and the Role of the Union Civil Protection Mechanism

Sanja Grbović and Ivan Jeknić

Abstract This paper examines the European Union Civil Protection Mechanism (UCPM) with a particular focus on its legal framework, operational structure and potential future reforms aimed at strengthening disaster-response capacities. It further assesses Montenegro's preparedness for natural hazards, especially wildfires, within the context of its EU accession process and in light of the major wildfire outbreak of August 2025 in central Montenegro. Through an analysis of relevant EU and national legislation, policy documents, and strategic frameworks, the paper identifies key capacities, structural challenges, and areas where Montenegro's civil protection system needs to be aligned with the EU standards. The study contributes to a better understanding of the effectiveness of the UCPM and the readiness of candidate countries such as Montenegro to integrate into the EU's broader crisis-management architecture.

Keywords Wildfire risk management · Critical infrastructure protection · Crisis response · Union civil protection mechanism (UCPM) · International cooperation · Resilience · Montenegro

1 Introduction: The Growing Importance of Civil Protection

Modern societies are increasingly exposed to complex and recurrent threats, ranging from extreme weather events and natural hazards to technological accidents and crises with cross-border impacts. Climate change, through rising temperatures, shifting

---

S. Grbović (✉)
Faculty of Law, University of Montenegro, Podgorica, Montenegro
e-mail: sanjag@ucg.ac.me

I. Jeknić
Faculty of Political Science, University of Montenegro, Podgorica, Montenegro

Z. Keković et al. (eds.), *Critical Infrastructure: Risk and Crisis Management in the New Reality*, Advanced Sciences and Technologies for Security Applications,
https://doi.org/10.1007/978-3-032-20948-1_8

precipitation patterns and sea-level rise, continues to amplify the intensity and unpredictability of these events [1]. Consequently, phenomena such as droughts, wildfires, floods and cyclones are occurring more frequently and generating growing risks for people, infrastructure and the environment These challenges are further compounded by rapid urbanization and the increasing interdependence of critical infrastructures, underscoring the need for integrated, resilient and forward-looking disaster-management systems [2]. Within this context, civil protection has emerged as a central component of security governance and societal resilience, encompassing measures aimed at prevention, mitigation, preparedness and response.

Civil protection comprises a set of activities designed to safeguard human life, material assets, and the environment from natural and human-induced hazards [3]. In addition to emergency response operations, it encompasses systematic planning, risk assessment, and the development of organizational and operational procedures necessary for coordinated action [3]. As such, civil protection constitutes a fundamental element of institutional resilience and provides a framework for cooperation among the various actors involved in crisis management.

The aim of this paper is to provide a comprehensive understanding of both the functioning of the EU Civil Protection Mechanism and the readiness of Montenegrin authorities to respond to natural hazards, in order to determine measures and recommendations for their improvement. The first part of this paper examines the EU Civil Protection Mechanism, its legal and institutional framework and modalities of cooperation with non-EU states, such as Montenegro. The second part of the paper analyzes the legal and strategic framework for disaster response and risk protection in Montenegro given the country's size, population, geographical characteristics, and status as a candidate for EU accession. The paper focuses primarily on wildfires, as the most frequent and severe natural hazards in Montenegro during the summer months. The third part of this paper assesses the country's preparedness to respond to natural hazards, in the case of a major wildfire outbreak in August 2025 in Montenegro. Finally, the paper provides conclusions and recommendations for improving local and national disaster-response capacities of Montenegro, based on previous insights.

2 Methodology

The analysis of this paper is based primarily on a review of relevant legal and policy documents, strategic frameworks, and reports issued by EU institutions, news outlets and national authorities. Particular attention is devoted to the regulatory and operational foundations of the EU Civil Protection Mechanism, including ongoing discussions and proposals concerning its potential future legal reforms.

On the other hand, to assess Montenegro's preparedness for natural disasters, especially wildfires, the paper examines national legislation, civil protection strategies, and publicly available data. This approach enables the identification of existing capacities, structural challenges, and areas requiring further alignment with EU standards. In order to evaluate Montenegro's disaster readiness in practice, the paper also analyzes the country's response to the major wildfire outbreak in August 2025 in central Montenegro, providing insights into the limitations of the country's disaster protection services, and the operational value of the international support mechanisms.

Accordingly, the research is guided by the following questions:

1. How does the EU Civil Protection Mechanism operate in legal and operational terms, and what future reforms are being considered?
2. To what extent is Montenegro prepared to effectively respond to natural disasters, especially wildfires?

Together, these questions frame a comparative assessment that highlights both the evolution of EU civil protection governance and Montenegro's progress, and remaining gaps, in strengthening its disaster-management system.

3 Civil Protection in EU—Solidarity in the Face of Disaster

At the level of the European Union, the legal foundation for civil protection is established in Article 196 of the Treaty on the Functioning of the European Union (TFEU) [4]. This provision delineates the Union's role in supporting and complementing the efforts of Member States at national, regional and local levels. The principal objectives include strengthening risk-prevention policies, enhancing the capacities of civil protection services, improving preparedness and facilitating rapid and coordinated operational cooperation among Member States. Article 222 TFEU, the solidarity clause, further obliges the Union, upon the request of a Member State affected by a disaster or attack, to mobilize all available instruments of assistance, with coordination led by the Council of the European Union [4].

The operational implementation of EU civil protection objectives is ensured through the Union Civil Protection Mechanism (UCPM), established by Decision No. 1313/2013/EU [5]. The Mechanism enables joint action and more efficient coordination in emergency situations. Subsequent amendments, specifically Decision 2019/420 [6] and Regulation 2021/836 [7], have further reinforced the Union's capacities in disaster prevention, preparedness and response.

A central element of the UCPM is the establishment of shared capacities and coordination structures that enable effective assistance to affected states. Particular importance is attributed to RescEU, an additional European reserve of disaster-response capacities created to supplement national resources. RescEU is designed as a safety net that is activated when Member States' own capacities and the existing European Civil Protection Pool prove insufficient. It consists of strategic reserves, including material assets, transport and logistics resources, medical stockpiles, and specialized equipment that can be rapidly mobilized and deployed. Their establishment, management and operation require substantial financial resources at EU level, a need explicitly recognized in the Mechanism's legal framework.

The operational hub of the Mechanism is the Emergency Response Coordination Centre (ERCC), which functions as a 24/7 monitoring and coordination center. The ERCC tracks developments in real time both inside and outside the EU, supports needs assessments, directs offers of assistance and ensures that resources are allocated efficiently and complementarily. Scientific and technical expertise, particularly from the Joint Research Centre of the European Commission, plays an essential role in supporting ERCC operations [8]. The integration of EU space infrastructure further enhances situational awareness and response capacity. Systems such as Copernicus [9], Galileo and GOVSATCOM provide critical tools for disaster monitoring, early warning, satellite communication and navigation support.

Although civil protection cooperation belongs formally to the EU's internal policies, the Mechanism possesses a significant external dimension. UCPM assistance may be deployed to non-EU countries in duly justified circumstances and in line with humanitarian principles. Currently, six non-EU European states, Iceland, Norway, Montenegro, North Macedonia, Serbia and Turkey, participate in the Mechanism, enhancing its regional reach and positioning civil protection as a wider instrument of European solidarity.

3.1 *How Does the UCPM Work?*

Civil protection in the EU is based on the principle of multi-level governance. While the EU provides a strategic framework and shared capacities, responsibility for prevention, preparedness and initial response rests predominantly with Member States. Within Member States, regional and local authorities often constitute the first line of defense. The UCPM framework emphasizes the importance of supporting local communities through technical assistance and capacity building, as well as the need for early public communication on protective measures. Particular attention is given to vulnerable populations and the gender dimension of disasters [10]. The EU commits to a gender-sensitive approach, recognizing specific vulnerabilities and promoting best practices, including support for victims of gender-based violence during and after emergencies.

Once a request for assistance is submitted through the Mechanism, the Emergency Response Coordination Centre (ERCC) mobilizes the necessary support, including specialized teams and technical experts [8]. Operating around the clock, the ERCC monitors global events 24/7 and ensures the rapid deployment of emergency assistance through direct communication with national civil protection authorities. Specialized assets, such as firefighting aircraft, search-and-rescue teams and medical teams, can be deployed at very short notice, both within Europe and beyond its borders. Satellite-based mapping produced by the Copernicus Emergency Management Service further supports civil protection operations by providing timely and accurate geospatial data essential for identifying affected areas and planning relief interventions [9].

In the world's most vulnerable regions, civil protection assistance is frequently delivered alongside EU humanitarian aid. Experts from both fields closely coordinate their assessments and responses, particularly in complex emergencies where humanitarian and civil protection needs overlap [10]. Any country worldwide, as well as the United Nations, its agencies and other relevant international organizations, may request support through the EU Civil Protection Mechanism.

Since its establishment, the EU Civil Protection Mechanism has been activated more than 820 times to respond to emergencies, as of 2025 [11]. In 2024, the Mechanism was activated 58 times in response to a wide range of emergencies, including the war in Ukraine, the conflict in the Middle East, severe flooding in France, the Czech Republic, Poland, Bosnia and Herzegovina and Spain, major wildfires across Europe and Latin America and tropical cyclones affecting Mayotte [11]. In addition, EU Member States and participating countries may activate the Mechanism to provide consular assistance to their citizens, for instance during evacuation operations. In 2024, the Mechanism supported the safe evacuation of nearly 1,400 European citizens from countries such as Haiti, the Middle East, and Vanuatu [11].

The effective functioning of the Mechanism requires a stable and flexible financial framework. Decision 1313/2013/EU initially established a budgetary envelope for 2014–2020, later updated as part of the Multiannual Financial Framework and the EU Recovery Instrument [5]. Additional funding has been directed toward crisis preparedness, strategic stockpiling and rapid-response infrastructure.

Given the inherent unpredictability of disasters, financial flexibility is crucial. The UCPM therefore allows for the carry-over of unused appropriations to the following year and employs multiple funding modalities in line with EU financial rules. Safeguards concerning fraud prevention, monitoring and accountability are implemented in cooperation with OLAF and the European Public Prosecutor's Office.

3.2 Recent Reform Initiatives

On 16 July 2025, the European Commission presented a proposal to reform the UCPM as part of the broader EU Preparedness Union Strategy. The Finnish Government communicated its position to Parliament on 13 November [12].

The proposal envisions integrating funding for health emergency preparedness and response into the existing Mechanism and foresees the creation of a cross-sectoral crisis coordination hub operating alongside the ERCC [12]. Additional elements include enhanced civil–military cooperation and initiatives aimed at strengthening preparedness at the population level. The overarching goals of the reform are to reinforce cooperation between the EU and its Member States in preventing, preparing for and managing disasters both within and outside the Union, and to improve the Mechanism's capacity to address cross-sectoral crises such as major cross-border health threats. This initiative responds to increasingly complex crisis scenarios, illustrated by the COVID-19 pandemic and the Russia-Ukraine conflict. The reform proposal is also linked to the Commission's suggested Multiannual Financial Framework

(MFF) for 2028–2034, which foresees a substantial budgetary increase for the UCPM compared with the 2021–2027 period, including expanded funding for health security and broader preparedness activities [12].

4 Civil Protection Capacities in Montenegro—Managing Forest Vulnerabilities and Wildfire Risks

As a candidate for EU accession, Montenegro must maintain and develop civil protection services in accordance with EU standards and requirements, adapting to the natural hazards native to the region. Given the climate and geography of Montenegro, this necessitates special considerations for protecting forests and low vegetation (shrubland) from wildfire during dry summer months.

Forests are among the most important natural ecosystems and form a key foundation for Montenegro's sustainable development. They provide numerous essential functions, productive, ecological and social. Forests cover 59.5%, and forest land an additional 9.9%, amounting to 69.4% of Montenegro's total territory. This forest coverage is significantly above the European average (46%) and the global average (30%). Furthermore, 13.41% of Montenegro's territory is under formal protection [13, 14]. Preserving forests is of crucial importance for national development, and investment in forest protection, conservation, and regeneration has no alternative. Montenegro has committed to a development model based on the principles of an ecological state, which places special emphasis on the sustainable management of forest resources, ecosystems, and biodiversity.

Forest fires represent a major threat to forest ecosystems and are a significant global environmental and economic challenge. Due to Montenegro's geographic position in the Mediterranean, general terrain characteristics, and the increasingly severe impacts of climate change, Montenegrin forests are particularly vulnerable. Forest fires and open-space fires are becoming more frequent every year, typically associated with summer drought periods and directly driven by extreme meteorological conditions, as well as human activities.

When several meteorological parameters occur simultaneously, they create conditions that trigger fires and enable their rapid, uncontrolled spread. These fires typically affect low and medium vegetation and occur mostly in inaccessible terrain [13]. Almost all such fires escalate into forest fires, threatening commercial forests, olive groves, orchards, agricultural crops, parks, and green areas.

Beyond forest complexes and unique landscape units, fires also endanger numerous tourist, transport and industrial facilities, cultural and historical sites, valuable artifacts and archives, authentic natural settings, as well as critical infrastructure, many of whose facilities fall under the category of having high fire-load potential. This leads to potentially severe consequences for the population, economy and environment.

The disaster protection and rescue system in Montenegro is based on a combination of professional and volunteer capacities, a well-defined legal and institutional framework, and extensive international cooperation. Operational capacities currently include approximately 700 professional firefighters, around 200 volunteer firefighters, and about 170 vehicles, which is considered insufficient to ensure an effective response to the full spectrum of natural and man-made hazards [13].

Through the Disaster Risk Management Capacity Assessment of Montenegro, the most significant challenges and priorities for reducing fire risk were identified [15]. As part of the Project for Building and Strengthening Disaster Risk Reduction Capacities, Montenegro established the Montenegrin Forest Fire Information System (MEFFIS) and implemented ecosystem-based disaster risk reduction (Eco-DRR) activities (16). Montenegro received specialized MEFFIS software that can significantly improve the national early wildfire detection system [16].

Montenegro has been a member of the Union Civil Protection Mechanism (UCPM) since 2014, which enables the country to participate in a wide range of activities offered through this mechanism, including professional training and capacity building of officials through participation in workshops, seminars, courses, exercises, expert exchanges, and other similar programs. Montenegro also takes part in meetings of the Committee and expert working groups, as well as various projects and initiatives.

Permanent communication has been established with the Emergency Response Coordination Centre (ERCC), and Montenegro has the ability to use the Common Emergency Communication and Information System (CECIS). Through the Union Civil Protection Mechanism, Montenegro has both received assistance and provided support to other countries, primarily those in the region, during disaster events.

The legal basis of Montenegro's disaster risk management system consists of several legal documents, including: Law on Protection and Rescue ("Official Gazette of Montenegro", nos. 13/07, 32/11, 54/16, and 146/21), Law on the Determination and Protection of Critical Infrastructure ("Official Gazette of Montenegro", no. 72/2019), Law on Local Self-Government ("Official Gazette of Montenegro", nos. 2/18, 34/19, and 38/20), Law on Transport of Dangerous Goods, Law on Explosive Substances, Law on Flammable Liquids and Gases, and others [17, 18].

The current strategic framework is based on The Disaster Risk Reduction Strategy for the period 2025–2030 [19], continuing and building on the National Platform for Disaster Risk Reduction (established in 2014) and the National Fire Protection and Rescue Plan of 2018 [20], along with two major analytical assessments: the *Disaster Risk Assessment of Montenegro* (2021) and the *Disaster Risk Management Capability Assessment* (2023).

Regarding the international cooperation, Montenegro has a well-developed network of international partnerships in disaster management. Bilateral agreements on cooperation in the field of disaster protection have been signed with Greece, Slovakia, Ukraine, Slovenia, Serbia, North Macedonia, Croatia, Bosnia and Herzegovina, Albania, Turkey, Azerbaijan, and Bulgaria. Memoranda of cooperation have also been concluded with Italy, the Russian Federation, and Armenia.

A major milestone in Montenegro's international involvement is its membership in the EU Civil Protection Mechanism. This membership enables continuous cooperation with DG ECHO, access to the Emergency Response Coordination Centre (ERCC), participation in trainings, exercises, expert exchanges, and other mechanisms, provision and receipt of international assistance.

In the past, Montenegro had repeatedly provided assistance to other states. Some notable examples include support to Albania following the 2019 earthquake, as well as assistance during floods in 2014 (Serbia), 2015 and 2017 (Albania).

4.1 Institutional Structure

The central institution responsible for the disaster risk management system is the Directorate for Protection and Rescue, established in 2007 within the Ministry of the Interior. Protection and rescue activities are planned and implemented in keeping with the provisions of the Rulebook on the Content and Methodology for Preparing Disaster Risk Assessments ("Official Gazette of Montenegro", 31/17) and the Rulebook on the Content and Methodology for Preparing Protection and Rescue Plans ("Official Gazette of Montenegro", 34/17). Disaster risk assessments form the basis for three types of protection and rescue plans: national and municipal plans, as well as plans of business entities, other legal persons, and entrepreneurs [17].

To ensure effective management and coordination during emergencies, Montenegro has established special operational bodies, such as the National Coordination Team for Protection and Rescue, and municipal protection and rescue teams. The Coordination Team is headed by the Prime Minister, with the Minister responsible for protection and rescue serving as deputy head. Members include ministers from key sectors, a representative of the General Secretariat of the Government, and the chair of the government body responsible for damage assessment. Additional officials, institutional representatives, and experts may be invited as needed. The team has a secretary responsible for ensuring implementation of orders and decisions [17].

4.2 Montenegro's EU Accession Program

Montenegro's EU Accession Program outlines the scope and timeline of reforms required for alignment with EU law. Under Negotiation Chapter 27, Environment and Climate Change, the adoption of the Strategy for the period from 2025 to 2030 is explicitly required. With its adoption, the document becomes fully aligned with the EU Accession Program. Within Chapter 27, the Government, as mentioned before, adopted the Action Plan for Meeting the Final Benchmarks on 18 February 2021, which remains the most important document in this field, as its implementation will fulfill all EU requirements defined by the final benchmarks and the EU's Common Position for this chapter [21].

In September 2023, Montenegro successfully completed the Disaster Risk Management Capacity Assessment, also funded by the European Commission. This document provides a detailed analysis of Montenegro's administrative, technical and financial capacities relative to the risks identified in the National Disaster Risk Assessment.

The most recent strategic document, *The Disaster Risk Reduction Strategy with an Action Plan for the Implementation of the Strategy for the period 2025–2030* (hereinafter: the Strategy), aims to reduce and prevent the emergence of new risks, while strengthening the capacities of society and of state and local institutions to respond to various types of natural and other disasters. The Strategy provides a systematic approach to identifying, assessing and reducing disaster risks, with the overarching goal of promoting sustainable development by enhancing community resilience to all types of disasters. It represents a key public policy instrument that defines objectives and tasks across different timeframes, accompanied by specific targets and indicators, thereby ensuring an effective disaster risk management framework [21].

Reducing disaster risk requires strengthening capacities for disaster risk management and improving cooperation and connectivity among all relevant institutions, sectors, and stakeholders at all levels. The purpose of this document is to ensure an adequate level of disaster risk reduction through proper disaster-risk mapping and the assessment of community vulnerabilities, thereby decreasing overall vulnerability, mitigating disaster impacts, and supporting sustainable development. The Strategy continues the work of the *Disaster Risk Reduction Strategy with the Dynamic Action Plan for the period 2018–2023*, with a primary focus on implementing previously unrealized activities and analyzing those already completed. This enables targeted and effective planning for the upcoming period and improves the protection and rescue system at both the local and national levels.

The European Commission identified five disaster resilience goals that all Member States of the Union Civil Protection Mechanism should address [22]:

1. Anticipation – improving risk assessment, forecasting, and disaster risk management planning;
2. Preparedness – increasing public awareness and readiness with regard to risks;
3. Alerting – strengthening early warning systems;
4. Response – enhancing response capacities;
5. Safety – ensuring a robust civil protection system.

As a candidate country for EU membership and a participating state in the Union Civil Protection Mechanism, Montenegro is making significant efforts to improve the effectiveness and efficiency of its protection and rescue system, that is, its civil protection system. Since the Disaster Risk Reduction Strategy defines a clear pathway for further development, the EU Disaster Resilience Goals have been incorporated as operational solutions for strengthening disaster resilience. These goals will serve to measure Montenegro's progress in the short, medium, and long term.

5 Disaster Response Capacities in Practice- the August 2025 Wildfires in Montenegro

Local and international disaster response services were put to the test in Montenegro in August 2025. The Montenegrin wildfire incident in the summer of 2025 was a major, severe and widespread event, primarily affecting areas around Podgorica (the Capital) and the central region of Montenegro. Around August 10th a series of wildfires erupted and raged for about eight days. The Montenegrin Army, firefighters, volunteers, and international aid were mobilized to fight the fires [23]. Persistent winds made containment difficult, several homes were destroyed, multiple evacuations were ordered, and one Montenegrin soldier lost his life [20]. Every day for the duration of the event, the authorities conducted extensive firefighting efforts with local personnel and international support [23].

International aid played a crucial role in combating the Montenegrin wildfires of August 2025, with Montenegro activating the EU Civil Protection Mechanism and NATO's EADRCC (including Copernicus mapping support) for urgent assistance after local services were overwhelmed by the wildfires. Countries from Europe and NATO deployed aircraft, helicopters, ground teams, and equipment to battle flames threatening Podgorica and other regions [24, 25]. The EU mechanism was triggered early in the crisis (August 11th), mobilizing rescEU assets from Czechia, Croatia, and Italy, including two firefighting airplanes and four helicopters [25]. Additional aircraft were provided and deployed by Serbia, Hungary and Bosnia and Herzegovina, as part of bilateral support. Switzerland, Slovenia and Austria also provided ground firefighting teams, alongside technical equipment [23, 26].

Protection and rescue services in Montenegro number around 780 employees across 23 municipalities, with 636 employees in operational roles [25]. Yet, this was not enough to effectively contain the wildfires, even with army involvement and international support. After fighting the wildfires for over a week, rain began to fall over the affected areas, helping to bring the fires under control. The containment and suppression efforts were concluded soon after that. In the end, ground operations and international assistance proved insufficient alone, as this precipitation played a key part in extinguishing the flames [23]. The wildfires affected around 7 thousand hectares of land [27]. It is estimated that around 58 000 citizens were exposed to the wildfires [24].

6 Conclusion

The Union Civil Protection Mechanism has evolved into a cornerstone of European crisis-management architecture. Through its legal foundations, shared capacities, operational hubs and financial instruments, it embodies the principles of solidarity, collective responsibility and multi-level governance. While this Mechanism provides great support in disaster relief, it alone is not enough to effectively protect individual countries from major incidents. This was evident in the case of Montenegro, during the August 2025 wildfires, where local and national disaster response services were quickly overwhelmed.

To enhance the effectiveness of disaster protection and risk reduction, the administrative capacities of Montenegro must be significantly strengthened. This includes increasing the number of staff dedicated exclusively to disaster-related responsibilities and clearly defining their mandates within the internal acts of state institutions, local governments, and other relevant bodies. In parallel, inspection oversight should be reinforced, supported by improved technical resources and enhanced professional competencies to ensure consistent and effective supervision.

The protection and rescue system should be further consolidated through the establishment of functional mass-support structures, civil protection units and teams, along with enhanced training for specialist and volunteer services in accordance with existing regulations. Modernizing and expanding the range of software tools and technical equipment, alongside training personnel who use them, represents an important step in building overall system resilience.

Given the central role of local governments in disaster risk reduction, all municipalities should prepare risk assessments, protection and rescue plans, and local disaster risk reduction strategies. Strengthening cooperation with the private sector is also necessary, as its human and technical capacities can significantly contribute to reducing overall community vulnerability.

Considering that financial resources remain a limiting factor, it is recommended to establish a clearly defined budget program dedicated to planning and implementing disaster risk reduction measures, including the material reserves required for an effective protection and rescue system.

All proposed recommendations should be aligned with the provisions of the Law on Protection and Rescue, as well as with the results of the National Disaster Risk Assessment and the Disaster Risk Management Capacity Assessment. This will lay the foundation for a stable, comprehensive, and sustainable disaster risk reduction system that strengthens community resilience and enhances the safety of citizens.

As crises grow more complex under the combined pressures of climate change, geopolitical instability and technological interdependence, the UCPM will continue to play a pivotal role in strengthening the EU's resilience and ensuring a coordinated response to disasters both within and beyond its borders.

References

1. Furtak, K., Wolińska, A.: The impact of extreme weather events as a consequence of climate change on the soil moisture and on the quality of the soil environment and agriculture—a review. CATENA **231**, 107378 (2023). ISSN 0341-8162. https://doi.org/10.1016/j.catena.2023.107378
2. Rezvani, S.M., Falcão, M.J., Komljenovic, D., de Almeida, N.M.: A systematic literature review on urban resilience enabled with asset and disaster risk management approaches and GIS-based decision support tools. Appl. Sci. **13**(4), 2223 (2023). https://doi.org/10.3390/app13042223

3. United Nations International Strategy for Disaster Reduction: The Structure, Role and Mandate of Civil Protection in Europe (ISDR Document No. 9346) (n.d.). https://www.unisdr.org/files/9346_Europe.pdf
4. European Union: Treaty on the Functioning of the European Union: Summary (2017).https://eur-lex.europa.eu/EN/legal-content/summary/treaty-on-the-functioning-of-the-european-union.html
5. European Parliament & European Council of the European Union: Decision No. 1313/2013/EU of the European Parliament and of the Council on a Union Civil Protection Mechanism (OJ L 347) (2013). https://eur-lex.europa.eu/eli/dec/2013/1313/oj/eng
6. European Parliament & Council of the European Union: Decision (EU) 2019/420 of the European Parliament and of the Council amending Decision No 1313/2013/EU on a Union Civil Protection Mechanism (OJ L 70) (2019). https://eur-lex.europa.eu/eli/dec/2019/420/oj/eng
7. European Parliament & Council of the European Union: Regulation (EU) 2021/836 of the European Parliament and of the Council on a Union civil protection mechanism (OJ L 189) (2021). https://eur-lex.europa.eu/eli/reg/2021/836/oj/eng
8. European Commission: Emergency Response Coordination Centre (ERCC). Directorate-General for European Civil Protection and Humanitarian Aid Operations (2025). https://civil-protection-humanitarian-aid.ec.europa.eu/what/civil-protection/emergency-response-coordination-centre-ercc_en
9. Organization for Security and Co-operation in Europe: Gender-Responsive Disaster Risk Reduction Practices in Europe (2023). https://www.osce.org/files/f/documents/a/b/536858.pdf
10. European Commission: European Consensus on Humanitarian Aid. Directorate-General for European Civil Protection and Humanitarian Aid Operations (n.d.). https://civil-protection-humanitarian-aid.ec.europa.eu/who/european-consensus_en
11. European Commission: EU Civil Protection Mechanism. Directorate-General for European Civil Protection and Humanitarian Aid Operations (2025). https://civil-protection-humanitarian-aid.ec.europa.eu/what/civil-protection/eu-civil-protection-mechanism_en
12. Government of Finland, Ministry of the Interior: Reform of the EU Civil Protection Mechanism would strengthen crisis preparedness [Press release] (2025). https://valtioneuvosto.fi/en/-/1410869/reform-of-the-eu-civil-protection-mechanism-would-strengthen-crisis-preparedness
13. Ministry of Sustainable Development and Tourism: National Biodiversity Strategy with Action Plan for the Period 2016–2020 (Montenegro) [PDF]. Convention on Biological Diversity (2015). https://www.cbd.int/doc/world/me/me-nbsap-v2-en.pdf
14. Food and Agriculture Organization of the United Nations: Report: Montenegro. FAO (2020). https://openknowledge.fao.org/server/api/core/bitstreams/d8a96853-f21e-4795-bfc3-9ef44ab9614c/content
15. United Nations Office for Disaster Risk Reduction: Midterm review of the Sendai Framework for Disaster Risk Reduction 2015–2030 [PDF] (2023). https://sendaiframework-mtr.undrr.org/media/84438/download?startDownload=20251121
16. Japan International Cooperation Agency: NFFIS & Eco-DRR Newsletter: Project on Capacity Building for Disaster Risk Reduction through National Forest Fire Information System (NFFIS) and Eco-DRR (Vol. 9) (2025). https://www.jica.go.jp/oda/project/1900443/en_news/__icsFiles/afieldfile/2025/09/18/newsletter_09_en.pd
17. Parliament of Montenegro: Zakon o zaštiti i spašavanju (Law on Protection and Rescue), "Official Gazette of Montenegro", 13/2007, 5/2008, 86/2009, 32/2011, 54/2016, 146/2021, 3/2023 and 82/2025 (2007)
18. Parliament of Montenegro: Zakon o određivanju i zaštiti kritične infrastrukture (Law on the Determination and Protection of Critical Infrastructure). "Official Gazette of Montenegro", 72/2019 (2019)
19. Ministarstvo unutrašnjih poslova: Strategija za smanjenje rizika od katastrofa za period 2025–2030, sa Akcionim planom za 2025–2026 (The Disaster Risk Reduction Strategy for the period 2025–2030, with the Action Plan for 2025–2026) (2024). https://www.gov.me/dokumenta/ffa2c895-bcc8-47bf-a1e4-dd797a8db7ab

20. Government of Montenegro: Nacionalni plan zaštite i spašavanja od požara (National Fire Protection and Rescue Plan) (2018). https://www.gov.me/dokumenta/8ae4a008-3ea8-45a7-b695-c9b0d475b211
21. UCP Knowledge Network: Disaster Risk Reduction Strategy for the period 2025–2030 with the Action Plan for 2025/2026 (2024). https://civil-protection-knowledge-network.europa.eu/system/files/2024-12/drrs-mne-executive-summary.pdf
22. Salomonsson, F.: Guideline for Duty Officers: Activation of and Requests for Assistance via the EU Civil Protection Mechanism (PPRD East 3). UCP Knowledge Network (2024). https://civil-protection-knowledge-network.europa.eu/system/files/2025-01/guideline-for-duty-officers-activation-of-and-requests-for-assistance-via-the-eu-civil-protection-mechanism.pdf
23. EU solidarity knows no borders: WeBalkans.eu, News Portal (2025). https://webalkans.eu/en/stories/eu-solidarity-knows-no-borders/
24. Copernicus Emergency Management Service: Wildfire in Montenegro, EMSR836- Situational reporting (2025). https://storymaps.arcgis.com/stories/e3bb1d811f3541be92ba11f5297b7b6a
25. Montenegrin soldier killed, another injured as wildfires rage across country: BNE Intellinews, News Portal (2025). https://www.intellinews.com/montenegrin-soldier-killed-another-injured-as-wildfires-rage-across-country-395800/
26. European Civil Protection and Humanitarian Aid Operations: EU Civil Protection Mechanism responds to multiple wildfires across the continent (2025). EU Commission Website. https://civil-protection-humanitarian-aid.ec.europa.eu/news-stories/news/eu-civil-protection-mechanism-responds-multiple-wildfires-across-continent-2025-08-14_en
27. Global Disaster Awareness and Coordination System (GDACS): Forest Fire (7408 ha) in Montenegro 11 Aug 2025. https://www.gdacs.org/report.aspx?eventid=1024575&episodeid=5&eventtype=WF

Risk Management, Cyber Security and Business Continuity from the Aspect of Standardization in the Protection of Critical Infrastructure

Tatjana Bojanić and Violeta Nešković-Popović

Abstract Modern society increasingly depends on complex and interconnected systems, whose availability, reliability, and security form the foundation of everyday life. The growing complexity and interdependence of these systems lead to greater vulnerability, where even minor disruptions can cause serious consequences for society and the economy. These systems, including energy networks, transport infrastructure, water supply, healthcare and financial systems, information and communication technologies, public administration, and environmental protection, are referred to as critical infrastructure. Their protection against disruptions has become a strategic priority for every state and society as a whole. In this context, the integration of risk management, cybersecurity, and business continuity, based on international and European standards, emerges as the only sustainable framework for preserving social stability and plays a key role in protection strategies. Standards in the fields of risk management, cybersecurity, and business continuity provide a common language, methodological framework, and best practices that can be applied across different sectors. Over the years, information and communication systems have become an integral part of many activities that constitute elements of critical infrastructure across all sectors. Failures and security incidents in these systems directly threaten business continuity, as most critical processes depend on information and communication systems. According to ISO/IEC 27031:2025, readiness of information and communication systems/technologies (ICT) for business continuity means that ICT and its operational capabilities demonstrate the ability to achieve desired business continuity objectives in case of a disruption affecting ICT.

Keywords Standardization · Risk management · Business continuity · Cyber security

T. Bojanić · V. Nešković-Popović (✉)
Institute for Standardization of Serbia, Belgrade, Serbia
e-mail: violeta.neskovic@iss.rs

T. Bojanić
e-mail: tatjana.bojanic@iss.rs

Z. Keković et al. (eds.), *Critical Infrastructure: Risk and Crisis Management in the New Reality*, Advanced Sciences and Technologies for Security Applications,
https://doi.org/10.1007/978-3-032-20948-1_9

1 Introduction

The evolution of modern societies has been characterized by a gradual shift away from reactive and intuitive security approaches toward systematic risk management grounded in analytical methods and internationally recognized standards. Within this framework, the concept of risk management has become one of the key pillars of the contemporary security paradigm: a risk-based approach enables the identification, analysis, and controlled mitigation of threats that may compromise the stability and functionality of complex societal systems. This is particularly relevant in the field of critical infrastructure protection, where interdependencies and interconnectedness among subsystems are more prominent than in any other domain. The development of mature risk assessment and risk management models is therefore a necessary prerequisite for safeguarding vital societal functions. Nevertheless, both theoretically and operationally, risk management alone is not sufficient to ensure the long-term resilience of critical systems. Its purpose and full scope become evident only when it is integrated with a business continuity plan – an instrument that translates risk analysis into the system's operational capability to maintain essential functions under disruptive conditions. While risk management answers the question "What is the worst that could happen?", the business continuity plan addresses the question "How can essential functions be sustained even if the worst occurs?". Thus, business continuity represents the natural and indispensable extension of risk management, as it operationalizes protective and mitigation measures through structured procedures, resources, and organizational arrangements, embedding them into a broader resilience framework.

From this perspective, a business continuity plan performs a dual conceptual function: first, it materializes the principles of risk management by transforming them into concrete capacities for response, recovery, and the continuation of operations; and second, it provides mechanisms through which critical infrastructure can fulfil its societal role – ensuring uninterrupted delivery of services and meeting stakeholder expectations even in conditions of hazards, crises, or operational disruptions. In this sense, business continuity is not merely a technical or procedural component of security management but a strategic instrument that supports societal stability, economic sustainability, and public safety. Consequently, contemporary literature increasingly recognizes the business continuity plan as a key link connecting risk management, cybersecurity, resilience, and sustainable development, thereby giving full meaning to the overall risk management framework applied in the protection of critical infrastructure.

The purpose of this paper is to theoretically and normatively substantiate an integrated framework for risk management, cybersecurity, and business continuity in the protection of critical infrastructure, grounded in contemporary concepts of societal and human security and operationalized through international and European standards. The paper responds to the growing gap between sector-specific and technically fragmented approaches to critical infrastructure protection and the need for systemic, standardized, and interdisciplinary models capable of addressing complex, cascading, and cross-sectoral risks.

The contribution of the paper lies in its integrative approach, which connects security theory, the concept of complex adaptive systems (CAS), and international standardization into a coherent model of resilience governance. Unlike studies that treat risk management, cybersecurity, and business continuity as separate or parallel domains, this paper demonstrates how international standards – particularly ISO 31000, ISO/IEC 27001, ISO 22301, ISO/IEC 27031, and the emerging ISO 22371–22396 series – function as socio-technical instruments for translating abstract security concepts into operational, measurable, and verifiable management practices. By situating this standards-based framework within the contemporary European regulatory environment and illustrating its relevance through comparative and regional examples, the paper contributes both to the theoretical development of critical infrastructure studies and to the practical advancement of resilience-oriented security governance.

2 Critical Infrastructure, Societal Security, and the Preservation of Vital Societal Functions

The term critical infrastructure did not originate in the academic community but within the security-policy discourse of the United States during the Cold War. Its development can be traced through four phases, the first of which is essential for the emergence of the concept itself. The term initially appears in U.S. military and defense documents, used to describe facilities essential to national defense – primarily electricity generation and transmission assets, telecommunications systems, and industrial installations. At that stage, the term was not formally defined, but it was applied in practice. A formal, modern definition was first introduced in Presidential Decision Directive 63 (1998), which also established the initial multisector classification, including energy, transportation, telecommunications, banking and finance, water supply, emergency services, and government operations [1]. After the attacks of 11 September 2001, the term became central to U.S. national security policy. The USA PATRIOT Act (2001) provided an expanded definition: "Critical infrastructure means systems and assets vital to national security, national economic security, national public health, or any combination thereof." Thus, the concept of critical infrastructure emerged as a defense-oriented category but gradually expanded to encompass societal, economic, and public health dimensions. The European Union later adopted and further broadened the concept, introducing the notion of infrastructure resilience as a legally binding standard.

The concept of societal security has developed along two parallel traditions: the Anglo-American tradition, based on human security and social risks, and the Scandinavian/EU tradition, which conceptualizes societal security within the framework of the Copenhagen School. Both strands have subsequently influenced the standardization of resilience and shaped the understanding of critical infrastructure. In this sense, societal security evolved as a broader systemic framework for understanding community safety and societal stability, whereas critical infrastructure protection represents a technical–operational subsystem within that framework. While critical infrastructure protection aims to safeguard the functionality of essential systems (energy, water, transport, ICT), societal security seeks to protect social cohesion, institutional integrity, and a society's overall ability to withstand and adapt to disruptions.

As Keković and Kešetović note in their analysis of societal security: "If state security is no longer threatened exclusively by other states, but also by challenges and threats that originate from within, then its security must also be analyzed from an internal perspective, which significantly expands the scope of security and requires inclusion of additional levels: individual and societal" [2]. They further emphasize: "Recent decades have been marked by dramatic events resulting in mass civilian casualties. Contemporary societies are confronted with numerous threats linked to socio-demographic vulnerabilities. Although some may be interpreted as threats to the 'identity' of affected social, ethnic, or religious ethnicities, their destructive potential usually stems from the combined effects of social and economic factors, along with the driving forces inherent in nature and society. Despite the diversity of causes and consequences, several common characteristics describe most hazards arising from natural disasters, terrorist attacks, technical or technological accidents, and other emergencies affecting the modern world. Their psychological impact and public visibility, complexity, interdependence, and cascading effects make crisis management a top priority in global and national security policy" [2]. This understanding of the societal dimension of risk confirms that the stability of critical infrastructure protection, as a subsystem of societal security, depends on the resilience of the community as a whole. Therefore, an integrated management model must simultaneously encompass technical, institutional, and societal factors, because only through their interdependence can the true sources of vulnerability and the pathways for effective response be fully understood.

The academic "birth" of the term societal security is associated with the book Barry Buzan, Ole Wæver, and Jaap de Wilde, Security: A New Framework for Analysis [3]. In this context, societal security refers to the ability of a community to "maintain its identity, cohesion, and essential functions under conditions of external and internal threats", and its key elements include protection of collective identity, institutional stability, an interdisciplinary approach to risks, and a strong emphasis on social and cultural threats.

This school of thought has had a significant impact on EU security policy, including policies governing critical infrastructure. Scandinavian literature later introduced a resilience-oriented interpretation of societal security, focusing on the ability of institutions, organizations, and local communities to absorb and adapt to disruptions within CAS. Accordingly, societal security can be understood as the capacity of societal systems – institutional, cultural, economic, and infrastructural – to preserve essential values, cohesion, and functionality under conditions of complex risks and disruptions. This approach integrates both intellectual roots: UNDP (human security) and Buzan/Wæver (societal security).

When comparing the concepts of societal security and critical infrastructure protection, the following distinctions emerge (Table 1):

From these distinctions, it follows that societal security is built upon:

- securitization theory,
- the concept of human security,
- resilience theories within CAS systems [4].

Critical infrastructure protection, by contrast, is grounded in:

- national security and defense,
- risk and risk management (ISO 31000),
- cybersecurity (ISO/IEC 27001),
- business continuity (ISO 22301).

By the end of the twentieth century, the two domains – societal security and critical infrastructure protection – began to converge conceptually:

- The CER Directive (2022/2557) introduces resilience governance, directly derived from the societal security model.
- The ISO 22371–22396 series integrates societal and infrastructural dimensions of resilience.
- CAS models confirm that infrastructure cannot be resilient without societal mechanisms such as trust, institutional capacity, and communication.

Thus, critical infrastructure protection becomes the technical subsystem of the broader concept of societal security. Societal security provides the systemic, community-oriented framework, while critical infrastructure protection represents its operational-technical component. Contemporary European regulation (CER 2022/2557) and ISO standards (22371–22396) demonstrate that these concepts are no longer separate, but jointly form a model of resilience governance – a socio-technical system of managing resilience as an integrated whole.

In contemporary concepts of societal and human security – as articulated by Keković and Kešetović [2] on the one hand, and by Đorđević and Keković [5] on

Table 1 Distinctions between the concepts of societal security and critical infrastructure

	Societal security	Critical infrastructure protection
Origin	Copenhagen School (1990) and UNDP (1994)	USA – PDD-63 (1998), Patriot Act (2001)
Object of protection	Society, community, identity, institutions	Physical and digital systems of vital importance
Types of risk	Social, cultural, political, hybrid	Technical, cyber, infrastructural, cascading
Model	Cohesion + Resilience	Functionality + Continuity

the other – it is evident that security can no longer be understood merely as the protection of state territory and institutions, but rather as the protection of societal functions and the conditions of life of citizens within an increasingly interdependent and complex system. In this sense, the protection of critical infrastructure emerges as the "backbone" of societal and human security: its functioning enables the continuity of essential societal functions, while its vulnerability becomes a central subject of security analysis.

Keković and Kešetović demonstrate that the development of the concept of societal security arose as a response to the limitations of the traditional, state-centric understanding of security. Drawing on the Copenhagen School, the authors highlight that society becomes the referent object of security, whereby identity, cohesion, and the functioning of key societal systems are considered equally important as the territorial integrity of the state. The development of globalization, transnational crime, migration, ecological and technological risks has made the boundary between internal and external security – and between state and societal security – increasingly difficult to maintain. Within this framework, societal security encompasses not only protection from traditional threats, but also crisis management, civil protection, the safeguarding of critical societal functions, and the development of standards that support such practices [2].

At the same time, the concept of human security, as interpreted by Đorđević and Keković, provides an analytical framework based on quality of life and seven thematic dimensions (economic, food, health, environmental, personal, community, and political security). Drawing on the UNDP Human Development Report [6], the authors demonstrate that insecurity most often stems from institutional dysfunction, unequal distribution of risks and benefits, and regulatory gaps that allow the interests of large capital actors to prevail over the public interest. In this sense, the concept of human security offers a matrix of indicators through which insecurity is measured by its effects on people – not merely on the state or the market – enabling normative and technical instruments, including standards, to be interpreted as public policy tools aimed at improving quality of life [5].

Both approaches – societal security and human security – converge around the notion of critical societal functions, that is, the infrastructures and systems without which basic living conditions, social stability, and political legitimacy cannot be sustained. For Keković and Kešetović, societal security requires that all key actors – state institutions, local communities, the private sector, and civil society – be incorporated into risk management through a holistic "top-down" and "bottom-up" process. Đorđević and Keković, operating within the same normative horizon, show that the international standards system can act as a mechanism for translating the concept of human security into operational requirements for organizations, infrastructure operators, and public institutions.

Within this context, international standards – primarily the ISO 31000, ISO 22300, and ISO 27000 families and related documents – become a key element of the theoretical framework for integrated critical infrastructure management. Đorđević and

Keković show that the standards system gradually forms a National Quality Infrastructure (NQI), encompassing metrology, certification, accreditation, and standardization, and that this infrastructure is essential to ensuring that the effects of globalization do not manifest in asymmetrical disadvantages for less-developed societies. Standards addressing risk management, business continuity, information security, environmental protection, and organizational social responsibility are interpreted as instruments through which the seven dimensions of human security and the pillars of societal security are operationalized.

It is particularly important that both works emphasize the cascading, complex, and cross-sectoral nature of contemporary security threats. Keković and Kešetović refer to the "mutual vulnerability" of threatened infrastructures and the necessity of standardized procedures in crisis management, while Đorđević and Keković argue that international standards, when applied consistently, can reduce structural inequalities and limit the "export" of risks to weaker states. Theoretically, this brings the analysis of critical infrastructure closer to the concept of CAS: infrastructure is viewed as an interconnected, evolving system of systems in which local disruptions can produce global effects. Managing such systems cannot rely on classical, sector-based regulation, but instead requires integrated, standardized, and interoperable frameworks.

On this basis, critical infrastructure is interpreted as the nexus of societal and human security, and the system of international standards as the primary mechanism for its institutionalization and governance. The concept of societal security provides the referent object ("society" and its essential functions), the concept of human security provides the analytical matrix (quality of life across multiple dimensions), while ISO standards provide the normative and technical framework for managing risks, cyber threats, business continuity, and societal resilience. Integrated critical infrastructure management is therefore theoretically positioned as a synthesis of three levels: conceptual (societal and human security as frameworks for understanding what is "at risk"), systemic (critical infrastructure as a complex, interdependent system of systems), and normative-instrumental (ISO standards as the operationalization of these concepts into organizational management systems and public policies).

Critical infrastructure is thus not protected merely as a set of technical assets, but as the material foundation of societal and human security – whose governance must be rooted in standardized principles of risk management, business continuity, and resilience.

3 Critical Infrastructure as an Interdependent and Systemically Vulnerable Complex: Normative Framework, Structural Complexity and Security Implications

Critical infrastructure represents the foundation of the functioning of modern society, as it comprises energy systems, transport networks, healthcare institutions, water supply, financial institutions, and ICT systems. Moreover, the category of critical infrastructure is so broad that some authors, half-jokingly and half-seriously, argue that it is difficult to identify sectors that are not critical [7]. These systems do not exist independently of one another; instead, they form a highly integrated, mutually conditioned technical-organizational system in which disruptions can produce consequences far beyond the initial incident point.

It is precisely this interdependence – which increases efficiency and integration – that simultaneously generates systemic vulnerabilities. Modern infrastructures operate under a high level of technical, organizational, and process complexity, meaning that even localized disruptions can trigger cascading effects across other sectors – commonly described as domino disruptions. The European Union Agency for Cybersecurity (ENISA) reports that more than 53% of incidents in critical sectors have had multisectoral, regional, or even transnational consequences [8]. These data empirically confirm that modern infrastructure can no longer be approached through a strictly sectoral model, but exclusively through an integrated, systemic, and interdisciplinary model of risk management and business continuity.

In addition to technical factors, heightened vulnerabilities arise from broader structural processes: digitalization, excessive reliance on ICT systems, automation, global supply chains, and increasing exposure to cyber threats. The combination of these factors results in critical infrastructure functioning as a complex system in which the failure of a single component can generate far-reaching effects. This perspective corresponds to the concept of CAS, which has become a fundamental explanatory model in security studies and crisis management for understanding why traditional protection models are no longer sufficient. The concept, most clearly articulated by John H. Holland in his work about CAS [9], defines these systems as "composed of numerous interacting components that self-organize, learn, and evolve in dynamic environments".

In this context, multisectoral analysis, cross-sector risk management, interdependency assessment, and business continuity planning represent key instruments for strengthening resilience. Contemporary approaches, such as those based on ISO 31000 (Risk Management), ISO 22301 (Business Continuity Management Systems), and related normative documents, require states and critical infrastructure operators to adopt a dynamic, holistic protection model – one in which continuity of operations becomes as important as physical or cyber protection.

Thus, critical infrastructure becomes a socio-economic resource whose resilience directly affects state stability and societal well-being. For this reason, the field requires an advanced, integrated, and scientifically grounded approach centered on understanding vulnerabilities, interdependencies, and the need for operational continuity amid increasingly complex risks. Modern regulation governing the resilience of critical entities rests on normative convergence between the most developed risk management systems and infrastructure resilience frameworks. The European Union, the United States, the United Kingdom, Australia, Canada, and New Zealand all develop comprehensive regulatory regimes based on an integrated all-hazards approach, cyber-physical security, and continuity of operations as a central value.

Within the European Union, the Critical Entities Resilience Directive – CER (EU) 2022/2557 [10] represents the most significant and comprehensive normative document. For the first time, it systematizes the obligations of Member States and critical entities across all phases of resilience: prevention, protection, response, robustness, mitigation, absorption, adaptation, and recovery. CER also corrects the limitations of the former Directive 2008/114/EC, which covered only the physical protection of energy and transport systems, without integrating cross-sector interdependencies – today one of the primary sources of systemic risk. Complementing CER, the NIS2 Directive (EU) 2022/2555 [11] forms the second pillar of the European framework by regulating cybersecurity and technical-organizational measures. Together, these two acts constitute a normative "dual system" governing the physical, technical, and operational resilience of critical infrastructure.

The United States applies a heterogeneous model structured around Presidential Policy Directive-21 (PPD-21) [12] and the National Infrastructure Protection Plan (NIPP) [13]. These documents define 16 critical sectors and articulate a national partnership architecture between public and private stakeholders, with strong reliance on the operational capacities of the Cybersecurity and Infrastructure Security Agency (CISA) in the cyber domain. Although not codified in a single law, the U.S. system is considered one of the most advanced due to the maturity of its incident management and cross-sector dependency mechanisms.

In the United Kingdom, the fundamental document remains the Civil Contingencies Act (CCA) 2004 [14], the first legal act to introduce the obligation of Business Continuity Management (BCM) into national legislation. This act structurally divides organizations into "Category 1" and "Category 2 responders", specifying their roles in planning, coordination, and response, and continues to be regarded as one of the strongest legal frameworks in the field of resilience.

Australia's Security of Critical Infrastructure Act (SOCI), adopted in 2018 and significantly expanded in 2021 [15], introduces one of the strictest oversight regimes in the world. SOCI mandates risk management programs, mandatory reporting, cyber and physical security measures, interdependency management, and even allows direct governmental intervention in critical infrastructure systems when national security is at stake.

Canada employs a multi-level model grounded in the National Strategy for Critical Infrastructure (2009) [16], which is not centralized but built on cooperative governance between the federal government, provinces, and the private sector. Despite its decentralized nature, it remains one of the most stable models of integrated protection.

New Zealand, through the Civil Defense Emergency Management Act (CDEM) [17] and its "lifeline utilities" regime, establishes a legally binding framework for organizations whose service disruption may compromise state stability. The New Zealand system is regarded as one of the most advanced, as it treats infrastructure as the "backbone of society" and requires all vital services to maintain minimum service levels during crises, supported by strong regional coordination.

A comparative analysis of these regulatory systems shows that the world's leading states converge around three key principles: (1) Critical infrastructure is a systemic category essential to the survival of the state and society; (2) Risks are interconnected, transboundary, and cascading in nature; (3) Business continuity is the logical outcome of risk management and the central mechanism for achieving resilience.

For this reason, modern regulation tends toward a universal paradigm in which risk management, cyber-physical security, and business continuity become the cornerstone of contemporary security policy.

4 Historical and Conceptual Development of Management Systems in the Protection of Critical Infrastructure

The development of the management systems that today form the foundation of risk management, information and cybersecurity, and business continuity is not the result of a single unified process but rather of decades of accumulated knowledge across various scientific disciplines, societal shifts driven by technological change, and institutional experience. The historical currents that shaped these systems show that they emerged as responses to the growing complexity of modern organizations and their increasing dependence on technological, digital, and infrastructure networks. Critical infrastructure, as a network of "essential societal functions", requires precisely such integrated, mutually aligned standards, because its subsystems – energy, healthcare, water supply, transportation, finance, and public administration – are deeply interconnected through numerous technical and organizational interactions.

For these reasons, international standards have been developed gradually, in line with technological and societal changes. Each of them reflects not only professional consensus but also collective institutional experience regarding how organizations behave, react, fail, and evolve under conditions of uncertainty, crises, and operational disruption.

4.1 From the Evolution of the Concept of Risk to Standardized Managerial Practice – ISO 31000:2018 – Risk Management – Guidelines

The development of ISO 31000 marks a major turning point in the contemporary understanding of risk. While throughout most of the twentieth century risks were viewed as isolated threats to be eliminated, the gradual evolution of scientific approaches – particularly in the financial sector – led to a new perspective: risk is not merely a danger but an inevitable, manageable element of decision-making. Financial institutions were among the first to develop formalized risk management methods in the 1970s and 1980s [18], and these methods were later adopted by the environmental, quality, safety, and corporate governance sectors. Globalization and the growing interdependence of systems created the need for these diverse practices to be harmonized into a unified conceptual framework.

ISO 31000:2009 [19] emerged from this need as the first international synthesis of risk management principles and practices. The revised 2018 edition emphasizes the integration of risk management into organizational culture, highlights leadership, context, communication, and continual improvement, and, for the first time, positions risk management as a managerial function rather than a technical procedure [20]. The focus shifts from managing threats to proactive decision-making under uncertainty. Research conducted by ISO/TC 262 and the OECD [21] shows that organizations implementing ISO 31000 achieve more mature management practices, more rational resource allocation, and greater resilience in crisis situations. This standard thus becomes the methodological backbone of the entire management architecture in critical infrastructure: the connecting vertical that enables all other standardized systems – from information security to business continuity – to operate within a consistent and coherent framework.

4.2 *Institutionalizing Information and Cybersecurity – ISO/IEC 27001:2022 – Information Security, Cybersecurity and Privacy Protection – ISMS – Requirements*

In parallel with the development of risk management, the rise of digital technologies, computer networks, global electronic communication, and the explosion of data volumes created the need for a standardized approach to information protection. Early models emerged in military and telecommunications contexts during the 1970s and 1980s, but the civilian foundation for information security management was established only with the British standard BS 7799 from 1995 [22], which was the first to introduce a systemic, process-based approach to protecting information. These foundations later produced the first international edition, ISO/IEC 27001:2005, which formally integrated information security into the broader architecture of management systems. This edition introduced the PDCA cycle, systematic risk assessment, and the implementation of controls outlined in ISO/IEC 27002. The 2013 revision aligned the standard with the unified Annex SL structure [23], enabling strong integration with other management systems, while the 2022 edition introduced concepts tailored to the contemporary cyber environment, including cloud security, zero-trust architecture, and digital resilience [24].

The significance of ISO/IEC 27001 for critical infrastructure stems from the fact that information systems form the nervous system of all other infrastructure subsystems. Cyber incidents are no longer limited technical disruptions but events capable of affecting healthcare, energy, transport, and finance. ISO/IEC 27001 is therefore adopted in more than 180 countries as the key standard for protecting digital assets [25], and its implementation is one of the fundamental mechanisms for enforcing the EU NIS2 Directive. Studies by the OECD, ENISA, and ISO/IEC JTC 1/SC 27 consistently show that implementing ISO/IEC 27001 significantly reduces security incidents, increases operational resilience, and strengthens public trust [26]. Thus, much like ISO 31000, this standard functions not only as a technical document but also as an instrument for institutional and societal preparedness for cyber risks.

4.3 Business Continuity as an Element of Infrastructure Stability – ISO 22301:2019 – Security and Resilience – Business Continuity Management Systems – Requirements

The concept of business continuity emerged as a response to accelerated technological transformation and the growing dependence on digital and logistical systems. Financial institutions and telecommunications operators were among the first to develop recovery plans during the 1970s, motivated by the vulnerabilities of centralized computer systems [27]. By the 1990s, the need arose to integrate these practices into the broader framework of corporate risk management, leading to the development of the British standard BS 25999.

Building on these foundations, ISO developed ISO 22301:2012, the first global standard for business continuity. The 2019 revision fully aligned the standard with the unified structure of other management systems and strengthened requirements relating to business impact analysis, the determination of maximum acceptable outage, recovery planning, and leadership in ensuring institutional stability [28]. In the domain of critical infrastructure – where disruptions have direct implications for public safety – business continuity acquires the status of a core element of resilience. The COVID-19 pandemic confirmed this in practice: the functioning of hospitals, energy systems, public services, and state administration depended heavily on well-established BCM systems [29].

4.4 *The ICT Bridge Between Information Security and Business Continuity – ISO/IEC 27031 – Guidelines for ICT Readiness for Business Continuity (IRBC)*

While ISO/IEC 27001 addresses information security and ISO 22301 focuses on business continuity more broadly, ISO/IEC 27031 is the first standard to formalize the concept of ICT readiness to support continuity. First published in 2011 and revised in 2025, this standard covers methodological elements such as ICT-oriented business impact analysis, recovery parameter determination, loss-of-infrastructure scenario testing, and integration of information security with recovery planning. In the context of critical infrastructure, its role is particularly significant because it bridges two domains that historically operated in parallel: information security and the operational feasibility of recovery [30]. The standard connects the theoretical requirements of ISO 22301 with the technical mechanisms of ISO/IEC 27001, ensuring that digital systems can withstand disruptions without losing essential functions.

The historical development of these standards reveals a profound relationship between technological change, institutional capacity, and societal learning. As risks evolved through global digitalization, cyber threats, and hybrid crises, the standards evolved as well. Each standard represents a lesson drawn from past experience – a synthesis of practices and knowledge gained as organizations and societies faced crises. Thus, standards are not merely technical documents but dynamic mechanisms of institutional evolution. Viewed through the lens of critical infrastructure development, they illustrate that the stability of modern society depends not only on establishing systems but also on the ability to protect them, govern them, adapt them, and restore them. Risk management, cybersecurity, business continuity, and ICT readiness constitute the four fundamental pillars of this resilience. Each developed gradually in response to technological shifts and societal experience, and together they form the integrated system that enables societies to function under conditions of complexity and uncertainty. As the complexity of risks increased, the standards correspondingly became more sophisticated, comprehensive, and conceptually mature.

5 Development of the Standards Series in the Field of Societal Security and Resilience (ISO 22371–22396)

The recent generation of standards dedicated to societal security and resilience represents a significant evolutionary step within the ISO 22300 family. While earlier documents focused primarily on internal organizational capabilities and on a technical-organizational approach to business continuity, the standards of the ISO 22371–22396 series introduce a much broader conceptual horizon. They shift the focus from organizational self-sufficiency toward collaborative, multi-stakeholder governance of risks and resilience, in which urban, organizational, and social systems are understood as interdependent and deeply interconnected.

In this sense, the Community Resilience series constitutes a normative response to the realization that modern society is not only a technical and infrastructural system but a complex structure of relationships, trust, information flows, and shared decision-making. For this reason, these standards introduce the concept of resilience governance – the governance of resilience as an integrated process encompassing not only institutional but also social and economic dimensions of functions of public interest. This marks a departure from a purely technical understanding of resilience and an acknowledgement that systemic stability can be achieved only when all stakeholders are involved, from public services and local authorities to communities and individuals.

5.1 Standards as a Governance Framework for Societal Security and Resilience

Within this group of standards, ISO 22371:2024 – *Security and resilience – Community resilience – Principles*, framework and guidelines on urban resilience establishes a framework for assessing urban resilience and introduces a systematic methodology for analyzing interdependencies among key sectors – energy supply, water services, transportation, waste management, and ICT infrastructure. It emphasizes that urban resilience cannot be understood in isolation but only in the context of the interconnections and potential domino effects among infrastructural subsystems. Within this standard, the Resilience Index for Cities has been developed as an indicative tool enabling cities to systematically measure their readiness and steer investments toward strengthening critical functions.

ISO 22372:2025 – *Security and resilience – Community resilience – Guidelines for infrastructure resilience* provides a comprehensive framework for measuring, managing, and continuously improving infrastructure resilience. It builds upon the concepts of redundancy, modularity, and flexibility, which are recognized as the most effective adaptation models under conditions of unexpected disruptions. By introducing this standard, countries are enabled to develop their own resilience metrics for infrastructure systems, representing a critical step toward establishing predictable, measurable, and comparable infrastructure security policies.

The standard ISO 22393:2023 – *Security and resilience – Community resilience – Guidelines for planning recovery and renewal* complements the previous one by focusing on the recovery and renewal phase after a crisis. It elaborates the build back better concept, which implies that the reconstruction of infrastructure, public services, and institutions should not be limited to restoring pre-disruption conditions, but should transform systems so they become functionally, structurally, and socially more resilient. This approach is especially relevant for critical infrastructure, where recovery after natural hazards or technological incidents must be accompanied by long-term planning and the integration of future risks.

A specific dimension within the series is introduced by ISO 22395:2018 – *Security and resilience – Community resilience – Guidelines for supporting vulnerable persons in an emergency*, which addresses the protection of vulnerable population groups in emergency situations. This document demonstrates that system resilience is measured not only by technical and operational parameters but also by the ability to preserve the rights, dignity, and safety of the most vulnerable groups during crises. It is applicable in sectors such as healthcare, social services, education, and public transportation, as well as in evacuation planning and communication protocols for critical infrastructure systems.

Finally, ISO 22396:2020 – *Security and resilience – Community resilience – Guidelines for information exchange between organizations* establishes principles for inter-organizational information exchange – an essential prerequisite for interoperability under crisis conditions. The standard formulates principles of trust, transparency, and data protection, stressing that effective coordination among energy operators, telecommunications providers, transport systems, and public administration depends on harmonized protocols and shared information-exchange platforms. In practice, it provides a normative foundation for national CERT/CSIRT mechanisms and multisectoral early-warning systems.

Taken together, these standards demonstrate that resilience is neither a static nor merely technical concept, but a dynamic process requiring continuous learning, institutional connectivity, and active involvement of the wider community. For this reason, the ISO 22371–22396 series represents a logical extension of existing standards by introducing the societal and institutional dimension into the protection of critical infrastructure. In doing so, it builds an integrated model of societal resilience aligned with contemporary security challenges.

5.2 *Standards Under Development: ISO/DIS 22354 (Local Resilience Capability) and ISO/FDIS 22366 (Framework and Principles for Energy Resilience)*

The development of the new generation of resilience standards represents one of the most significant advancements in the contemporary approach to risk management. While previous decades focused primarily on operational stability and crisis response, the current drafts of international standards demonstrate a clear shift toward a holistic, integrated, and preventive model of resilience – one that connects technical, institutional, and societal factors.

ISO/DIS 22354 – *Local Resilience Capability* establishes a foundation for understanding and assessing local resilience as a key subsystem of national security. It develops a methodology enabling local communities to identify their vulnerabilities, capacities, and response capabilities under conditions of disruptions, crises, or prolonged disturbances. Of particular importance is its intention to integrate different dimensions of local functioning – from institutional coordination and infrastructure resources to social capital and community self-organization. In this sense, the standard serves as a necessary mechanism for contextualizing and operationalizing the EU CER Directive at the local level, thereby contributing to harmonized resilience practices within states.

As energy systems grow increasingly complex, the need for a standardized approach to their resilience becomes imperative. ISO/FDIS 22366 – *Framework and Principles for Energy Resilience* is the first document to systematically define the concept of energy resilience – from long-term disruption analysis, risk and dependency assessment involving ICT systems, to guidance on planning and recovery. Its application is intended for a broad range of actors: from national energy operators and electricity distribution system managers to entities active in the field of renewable energy sources. This standard lays the foundation for future alignment of energy policy with the principles of integrated resilience.

Taken together, these standards mark a strategic turning point in the approach to critical infrastructure resilience. Instead of fragmented, sector-specific measures, they introduce the concept of resilience governance – the governance of resilience as a shared, integrated function of national authorities, local communities, infrastructure operators, and society as a whole. At the methodological level, this series of standards supports the transition from reactive policies to preventive, adaptive, and coordinated models of crisis management.

6 European and Regional Application of Standards and Examples of Serbia in the Field of Critical Infrastructure

The development of critical infrastructure protection policy in Europe is based on a long-term institutional and legal process that accelerated after the adoption of the European Directive on the identification and designation of European Critical Infrastructure (2008/114/EC), and is today being fully transferred into a new regulatory framework through the Directive on the resilience of critical entities (Directive (EU) 2022/2557 – CER) and the Directive on network and information security (Directive (EU) 2022/2555 – NIS2). Their essence lies in the requirement that physical, cyber, and operational security are no longer viewed separately, but as interdependent parts of the same system of resilience.

In addition to the listed international standards, the European Committee for Standardization (CEN) has published the technical specification CEN/TS 14383–2, *Crime prevention through environmental design, urban planning and facility management – Part 2: Principles and processes.* This document addresses Crime Prevention Through Environmental Design (CPTED) in urban and infrastructure contexts and is significant because physical security and crime prevention are inseparable from the resilience of facilities and infrastructure.

This standard has multiple parts (up to 8) – which indicates a comprehensive series covering various aspects of design, urban planning, and maintenance.

6.1 Critical Infrastructure Protection Systems in Europe

In the European Union, the critical infrastructure protection system is institutionalized through three pillars:

- Regulatory Framework (NIS2 and CER) – defines the obligations of states and operators to enhance the resilience and security of essential services (energy, healthcare, transport, water, finance, public administration).
- Standards (ISO, IEC, CEN, ETSI) – ensure a harmonized methodological and operational framework for implementation.
- Institutional Coordination – national and European resilience hubs are established, such as ENISA, the European Union Agency for Civil Protection and Humanitarian Aid, JRC (Joint Research Centre), and CERT-EU.

The application of ISO standards in the European context is visible through resilience management frameworks in several countries:

- Finland has integrated ISO 22301 and ISO 27001 into its national system for the protection of energy and telecommunications (Finnish National Emergency Supply Agency);
- Germany (Bundesamt für Sicherheit in der Informationstechnik – BSI) applies ISO/IEC 27001 as the basis for the national certification scheme for critical infrastructure operators (KRITIS);
- The Netherlands uses ISO 31000 as the normative basis for all public risk management strategies in urban systems (Dutch Safety Regions Act);
- France and Spain have developed national platforms based on ISO 22301 for business continuity in the public and private sectors.

The European Commission and ENISA emphasize in their reports (e.g. ENISA 2024 Report on the State of Cybersecurity in the Union) that the application of these standards has increased transparency, consistency in risk management, and interoperability between sectors, which is a fundamental prerequisite for a rapid and coordinated response to incidents.

6.2 Regional Application of Standards and Integration of the Western Balkans into European Critical Infrastructure Protection Systems

The countries of the region, as candidates and potential candidates for EU membership, are gradually adopting European models. The biggest challenges relate to the varying levels of institutional maturity, limited resources, and the lack of an integrated approach that unifies cyber, physical, and organizational security.

There are good examples of the application of standards in critical infrastructure protection systems in the region, for instance:

- Slovenia has adopted the National Critical Infrastructure Act harmonized with the CER Directive and is already applying ISO 22301 in public services within the energy and utility sectors.
- Croatia has a developed risk management system in public enterprises that is structured according to ISO 31000 and ISO/IEC 27001 systems.
- Montenegro has, for the first time, introduced the ISO/IEC 27001 standard as a mandatory framework for public administration bodies in its Cybersecurity Strategy (2022–2026).
- Bosnia and Herzegovina, through the work of the State Regulatory Commission for Electricity (DERK), is implementing elements of the ISO 22301 standard for business continuity in the electricity supply sector.
- The Republic of North Macedonia initiated the integration of the ISO 31000 system into the national Administration for Protection and Rescue system (2023).

6.3 Examples of Standard Application in the Critical Infrastructure Sector in the Republic of Serbia

The Republic of Serbia has made significant strides in the last decade toward the institutional and normative improvement of critical infrastructure protection.

According to the Law on Critical Infrastructure („Official Gazette of the Republic of Serbia", No. 87/2018) and relevant by-laws, a framework has been established for the identification, categorization, and protection of systems classified as critical infrastructure.

Key areas where standards are applied or can be directly applied include:

Energy

- Public companies such as Elektroprivreda Srbije (EPS) and Elektromreža Srbije (EMS) have introduced elements of ISO 31000 and ISO 22301 for risk management and business continuity.
- The gas sector (Srbijagas) applies ISO/IEC 27001 in managing SCADA (Supervisory Control and Data Acquisition) systems and data protection.

Telecommunications and ICT

- The National CERT, within the Regulatory Agency for Electronic Communications and Postal Services (RATEL), uses ISO/IEC 27001 and ISO/IEC 27035 as the basis for incident response.
- Private operators (Telekom Srbija, Yettel, A1) certify their systems according to ISO/IEC 27001:2022 and ISO 22301:2019.

Water Supply and Wastewater Management

- The public utility company "Beogradski vodovod i kanalizacija" (Belgrade Waterworks and Sewerage) is developing a risk management system based on the ISO 31000 framework.

Healthcare System

- Institutes and clinical centers are gradually introducing ISO/IEC 27001 to protect medical information and ensure the continuity of electronic services (e-Zdravlje, eRecept).

Public Administration and State Systems

- The Office for IT and eGovernment uses ISO 22301 as a framework for the continuity of state digital services and backup data centers (e.g., the Disaster Recovery Data Center in Kragujevac).

In addition, since 2020, the Institute for Standardization of Serbia (ISS) has adopted and published all the mentioned standards in the field of Security and Resilience as SRPS ISO (IEC) versions (including ISO 22395, ISO 22396, ISO 22301, ISO/IEC 27001, and ISO 31000). This has established the institutional foundation for widespread application in public and private organizations.

7 Standards for the Field of Artificial Intelligence and Their Application in Critical Infrastructure

The development of Artificial Intelligence (AI) represents one of the most significant factors shaping the future of critical infrastructure. On one hand, AI enables faster data analysis, risk detection, and the automation of incident response; on the other hand, it introduces new forms of vulnerability and ethical risks that require systematic regulation.

In this context, the International Organization for Standardization (ISO) has established a framework – standards in the field of Artificial Intelligence – which includes over 30 documents in various stages of development. The following are the most relevant for critical infrastructure (Table 2):

AI technologies are increasingly being used for:

- Predictive maintenance and fault detection (energy, transport, water supply);

Table 2 List of AI standards relevant for Critical Infrastructure

Standard	Name and application
ISO/IEC 22989:2022	*Artificial Intelligence – Concepts and terminology*. Establishes the conceptual and terminological foundations for the application of AI
ISO/IEC 23053:2022	*Framework for AI systems using machine learning*. Describes the architecture and life cycle of AI systems, including data governance and model control. It is applied in systems for monitoring and predictive maintenance of infrastructure
ISO/IEC 23894:2023	*Artificial Intelligence – Guidance on risk management*. Defines an approach to managing risks specific to AI, which is based on ISO 31000
ISO/IEC 42001:2023	*Artificial Intelligence Management System (AIMS) – Requirements*. This is the first standard for an AI Management System, based on the ISO Annex SL structure. It enables the certification of organizations that apply AI, including ethical, security, and technical controls
ISO/IEC 38507:2022	*Governance implications of AI for organizations*. Applied at the strategic level, to ensure that the governance structures of public and private operators have clear mechanisms for accountability, transparency, and monitoring of AI-based decisions

- Real-time analysis of cyber incidents;
- Vulnerability assessment and risk simulations (digital twins);
- Optimizing the flow of energy, water, or traffic based on adaptive models.

For example, ENISA, in its report *AI and Cybersecurity in the Energy Sector* (2024), highlights that machine learning can reduce incident detection time by 30–40%, but simultaneously warns that algorithms can be exposed to manipulation or „data poisoning" attacks. This is why there is insistence on the introduction of ISO/IEC 23,894 and ISO/IEC 42,001 standards into national regulatory mechanisms for critical entities.

In the European Union, the AI Act [31] establishes the legal framework for the application of AI systems, particularly in high-risk domains such as critical infrastructure, healthcare, transport, and energy. The direct link between the AI Act and ISO standards is reflected in the fact that ISO/IEC 42,001 serves as the technical standard for demonstrating compliance with European requirements for the security and transparency of AI systems.

8 Advantages of Applying ISO 31000 – ISO/IEC 27001 – ISO 22301 Standards in Critical Infrastructure

The development of modern societies is rapidly increasing dependence on technological systems, which simultaneously brings progress but also increases exposure to risks. In such an environment, critical infrastructure is the basis of social stability and economic development, but also the most vulnerable point of any state.

The application of international standards represents the most reliable way to ensure a systemic, transparent, and verifiable approach to risk and resilience management. The standards ISO 31000, ISO/IEC 27001, and ISO 22301 do not operate in isolation: they form a system of integrated security and business continuity management, which allows prevention, response, and recovery to be viewed as parts of a single unified process. In the long term, the integrated application of these standards is not just a matter of compliance, but also an indicator of societal maturity – its ability to anticipate risks, manage them, and maintain functioning despite disruptions. In other words, resilience becomes the new form of security, and standardization – its most reliable instrument.

All the mentioned standards together form the conceptual and operational foundation of the critical infrastructure protection system, particularly those in risk management, information security, and business continuity, in such a way that ISO 31000 establishes the strategy (prevention), ISO/IEC 27001 protects the digital infrastructure (response), and ISO 22301 enables recovery (restoration). This model ensures a holistic approach: predict – protect – recover, and transforms critical infrastructure into a dynamic system that not only withstands disruptions but is also improved by them (resilience by learning).

References

1. Presidential Decision Directive 63, The White House (1998). https://irp.fas.org/offdocs/pdd/pdd-63.htm, last accessed 2025/12/05
2. Keković, Z., Kešetović, Ž.: Društvena bezbednost u Srbiji. Srpska politička misao 1–2 (2008)
3. Buzan, B., Wæver, O., De Wilde, J.: Security: A New Framework for Analysis, 1st edn. Lynne Rienner Publishers, Colorado, Boulder, CO (1998)
4. Gunderson, L.H., Holling, C.S.: Panarchy: Understanding Transformations in Human and Natural Systems. Island Press, Washington, DC (2002)
5. Djordjević, I., Keković, Z.: The impact of international standards on the quality of people's life. Int. J. Qual. Res. **4** (2023)
6. UNDP Human Development Report 1994, New York, Oxford University Press (1994). https://hdr.undp.org/system/files/documents/hdr1994encompletenostats.pdf, last accesed 2025/11/22
7. Keković, Z., Ninković V.: Zaštita Kritične Infrastrukture: Sistemski Pristup, 2nd edn. CARUK, Beograd (2025)
8. ENISA Threat Landscape Report 2025. https://www.enisa.europa.eu/publications/enisa-threat-landscape-2025, last accessed 2025/12/19
9. Holland, J.H.: Complex adaptive systems. Daedalus **121**(1), 17–30 (1992)
10. Directive (EU) 2022/2557 of the European Parliament and the Council of 14 December 2022 on the resilience of critical entities and repealing Council Directive 2008/114/EC (2022). https://eur-lex.europa.eu/legal-content/EN/TXT/?uri=CELEX:32022L2557, last accessed 2025/11/22
11. Directive (EU) 2022/2555 of the European Parliament and of the Council of 14 December 2022 on measures for a high common level of cybersecurity across the Union (2022). https://eur-lex.europa.eu/legal-content/EN/TXT/?uri=CELEX:32022L2555, last accessed 2025/11/22
12. Presidential Policy Directive 21 (PPD-21): Critical Infrastructure Security and Resilience, The White House (2013). https://obamawhitehouse.archives.gov/the-press-office/2013/02/12/presidential-policy-directive-critical-infrastructure-security-and-resil, last accessed 2025/11/30

13. National Infrastructure Protection Plan (NIPP): Partnering for Critical Infrastructure Security and Resilience, U.S. Department of Homeland Security (2013). https://www.dhs.gov/xlibrary/assets/NIPP_Plan_noApps.pdf, last accessed 2025/11/30
14. Civil Contingencies Act, United Kingdom Government (2004). https://www.legislation.gov.uk/ukpga/2004/36/contents, last accessed 2025/11/30
15. Security of Critical Infrastructure Act 2018 (SOCI), Consolidated Act., Government of Australia (2018). https://www.legislation.gov.au/Series/C2018A00029, last accessed 2025/11/30
16. National Strategy for Critical Infrastructure, Public Safety Canada (2009). https://www.publicsafety.gc.ca/cnt/rsrcs/pblctns/srtg-crtcl-nfrstrctr/index-en.aspx, last accessed 2025/30/11
17. Civil Defense Emergency Management Act, Government of New Zealand (2002). https://www.legislation.govt.nz/act/public/2002/0033/latest/DLM149789.html, last accessed 2025/30/11
18. Jorion P.: Value at risk: the new benchmark for managing financial risk. McGraw-Hill (2007)
19. ISO 31000:2018—Risk management—Guidelines. ISO, Geneva (2018)
20. Aven, T.: An emerging new risk analysis science: foundations and implications. Risk Anal. **38**(5), 876–888 (2018)
21. Governance of Critical Risks. OECD, Paris (2014)
22. BS 7799-1:1995—Code of Practice for Information Security Management. BSI, London (1995)
23. Annex SL—High-Level Structure. ISO, Geneva (2012)
24. ISO/IEC 27001:2022—Information security, cybersecurity and privacy protection—ISMS—Requirements. ISO/IEC, Geneva (2022)
25. ISO Survey 2023: ISO Management System Certifications. https://www.iafcertsearch.org/services/iso-survey, last accessed 2025/12/01
26. ENISA Threat Landscape Report 2022. https://www.enisa.europa.eu/publications/enisa-threat-landscape-2022, last accessed 2025/12/19
27. Herbane, B.: The evolution of business continuity management: a historical review of practices and drivers. Bus. Hist. **52**(6), 978–1002 (2010)
28. ISO 22301:2019—Security and resilience—BCMS—Requirements. ISO, Geneva (2019)
29. COVID-19 Lessons for Resilience. WHO/UNDRR (2022)
30. ISO/IEC 27031:2011—Guidelines for ICT Readiness for Business Continuity. ISO/IEC, Geneva (2011)
31. Regulation (EU) 2024/1689 of the European Parliament and of the Council of 13 June 2024 laying down harmonised rules on artificial intelligence and amending Regulations (EC) No 300/2008, (EU) No 167/2013, (EU) No 168/2013, (EU) 2018/858, (EU) 2018/1139 and (EU) 2019/2144 and Directives 2014/90/EU, (EU) 2016/797 and (EU) 2020/1828 (Artificial Intelligence Act), 2024. https://eur-lex.europa.eu/legal-content/EN/TXT/?uri=OJ:L_202401689, last accessed 2025/12/19

Resilience of Critical Infrastructure to Cyber Attacks in the Western Balkan States

Aleksandar Bogićević, Sanela Veljković, and Marina Dabetić

Abstract The emergence of cyberspace and its growing militarization have transformed the contemporary security environment, leading to the appearance of entirely new forms of state activity within the international context. Parallel to the development of the digital domain, the international order has begun to undergo significant changes, triggering tensions between hegemonic powers and challengers of the order. Within such a security context, cyber attacks have become an instrument of hybrid warfare, with the targeting of critical infrastructure representing one form of malicious activity. The increasing digitalization of state institutions and economic systems, combined with the complexity of attack attribution and the emergence of sophisticated state and non-state cyber actors, further accentuates the vulnerability of critical infrastructure, whose functioning constitutes the foundation of state functioning. Given that disabling critical infrastructure may generate significant socio-economic consequences, strengthening its resilience has emerged as a priority of modern state security policies. The Western Balkan states, characterized by limited material and human resources, face pronounced vulnerability in an increasingly anarchic cyber environment. The geopolitical position of the region, as well as the differing foreign policy alignments of individual states toward ongoing international crises, have made them targets of coordinated cyber operations. The aim of this paper is to analyze the resilience of critical infrastructure in the Western Balkan states through case studies of cyber attacks on Montenegro, Albania, and Serbia, which illustrate the structural deficits that small states face in the domain of cyber security. The research employs a qualitative methodology based on a series of case studies. The findings indicate a pronounced vulnerability of Western Balkan

A. Bogićević
Institute for Strategic Research, University of Defence, Belgrade, Serbia
e-mail: aleksandar.bogicevic@mod.gov.rs

S. Veljković · M. Dabetić (✉)
Vinča Institute of Nuclear Sciences—National Institute of the Republic of Serbia, University of Belgrade, Department of Physical Chemistry, Belgrade, Serbia
e-mail: fmarina@vin.bg.ac.rs

S. Veljković
e-mail: sanela.veljkovic@vin.bg.ac.rs

Z. Keković et al. (eds.), *Critical Infrastructure: Risk and Crisis Management in the New Reality*, Advanced Sciences and Technologies for Security Applications,
https://doi.org/10.1007/978-3-032-20948-1_10

states, as well as the necessity of a comprehensive approach that integrates the development of human resources, the establishment of public–private partnerships, and the deepening of regional and international cooperation as essential prerequisites for building resilient critical infrastructure amidst growing cyber threats.

Keywords Critical infrastructure · Resilience · Cyber attacks · Western Balkans · Hybrid threats

1 Introduction

The emergence of cyberspace has radically influenced the transformation of the contemporary security environment in which states operate, generating entirely new types of threats, incomparable to those they had faced throughout their history. This view is supported by the fact that in contemporary military doctrines cyberspace is considered a new, fifth domain in which conflict can unfold, alongside land, sea, air, and space [1]. The shift of conflict dynamics from the physical realm to the abstract cyber domain has resulted in the appearance of new actors, new types of threats, and the need for states to adjust their policies to these changed circumstances. In addition to the state, as the traditional actor of security and foreign policy, the emergence of the cyber sphere has enabled the expansion of activity by non-state actors, whose capabilities often surpass those possessed by states. Due to the complexity of attack attribution, cyber operations are perceived as an exceptionally potent means of waging asymmetric conflict, primarily through the targeting of elements of another state's critical infrastructure in order to achieve strategic advantage and, ultimately, a different distribution of power in the international order.

The increasing digitalization of various functions of state administration, as well as of the economy, in the pursuit of improved efficiency and productivity, has made states even more vulnerable to malicious activities by various cyber actors. The constant expansion of digitalization across different spheres of life has inevitably led to the growth of the digital space that states must defend, which, conversely, has increased the risks of "successful" cyber attacks. The strain on national cyber resources, caused by the growing number of targets that may be exposed to foreign actors, has had a negative impact on the security of critical infrastructure, which can now be disabled for shorter or longer periods even without the use of kinetic force. Potential disruption of the functioning of institutions, industrial facilities, or telecommunications systems may generate serious challenges to the normal functioning of the state, which is why their protection has become a focus of both national and international security [2]. For this reason, strengthening the resilience of critical infrastructure ranks among the highest security priorities of contemporary states.

The erosion of the existing international system, marked by the dominance of the United States and the rise of powers challenging this order such as the People's Republic of China, Russia, Iran, and others has led to the increasing militarization of cyberspace, within which the targeting of critical infrastructure has become a key component. This trend has not bypassed the Western Balkan states, which have found themselves exposed to attacks by various state and non-state actors. However, their highly limited resources, both material and human, prevent them from responding adequately to such threats, placing them in an extremely unfavorable position.

The aim of this paper is to present and analyze the resilience of critical infrastructure in Western Balkan states through case studies of cyber attacks on Montenegro, Albania, and Serbia as illustrative examples of the challenges faced by small states in an increasingly anarchic and militarized cyber environment. This research employs a qualitative analysis based on case studies, treating the attacks on critical infrastructure in Montenegro, Albania, and Serbia as empirical examples that fundamentally illustrate the challenges faced by small states under conditions of escalating cyber threats. The research was conducted through systematic analysis of secondary sources, including relevant academic literature, reports of specialized organizations, and media dana, combined with a comparative approach aimed at identifying shared characteristics of vulnerability and structural factors that affect the cyber resilience of the observed states. Such a methodological framework enables the identification of patterns, trends, and implications that transcend the context of individual cases, thus contributing to a broader understanding of the regional context of cyber security.

2 Conceptual Definition of Critical Infrastructure and Resilience

The protection of critical infrastructure emerged as a security issue after the end of the Cold War in the 1990s, and the term itself originated in the United States during the administration of Bill Clinton [3]. The literature offers numerous definitions of this concept, and Lehto states that "critical infrastructure encompasses physical and cyber systems and resources that are so vital to a state that their disruption or destruction would have a serious negative impact on physical or economic security, public health, or public safety" [4]. On the other hand, Keković and Ninković argue that what should first and foremost be protected is the continuity of critical infrastructure operations. For them, "the importance of critical infrastructure is not reflected in infrastructural objects (whether referring to facilities, physical and logical networks, etc.), but rather in the services and products that are of critical importance for the functioning of society as a whole" [2]. When discussing critical infrastructure, it is necessary to note that there is no single list of sectors and systems that must be protected. Moreover, even if such a list existed, it would not be universal or equally applicable in every state. "No division of critical infrastructure is absolute, but is largely based on assessments by experts and/or political decision makers" [2]. What is protected and what is classified as critical infrastructure varies from state to state, although certain systems and sectors are commonly categorized as critical, such as energy, transportation, telecommunications, healthcare, and the financial sector.

Threats to a given infrastructure may manifest as internal and external failures. While the former relate to irregular changes occurring within the system, the latter originate outside the system and may arise in the form of natural phenomena, malicious activities, or accidents [5]. The literature also distinguishes between dependence and interdependence that may exist among different infrastructures. Dependence refers to a relationship between two infrastructures through which the state of one affects the state of the other, whereas interdependence implies a two-way relationship between infrastructures, through which the state of each affects or correlates with the state of the other [6]. Interdependence may be physical, cyber, geographic, or logical, whereby "cyber interdependence links infrastructures through electronic and informational connections: the outputs of informational infrastructure represent the inputs of other infrastructural systems, while the 'goods' exchanged between them are—information" [6]. Owing to the development of information technologies, a cyber attack against one infrastructure may produce not only direct consequences for that system, but also indirect consequences for other infrastructures closely linked to it. Therefore, "national security policies have always listed the protection of national infrastructures, infrastructures necessary for the normal functioning of key systems (sectors) within the state, and thus its security, among their priority tasks" [1].

Closely connected to the concept of critical infrastructure is the concept of resilience, which has its roots in the natural sciences and has recently been increasingly used in the social sciences as well. In security studies, resilience is associated with the protection of critical infrastructure, and this term is used to describe "the ability of a system, community, or society exposed to hazards to resist, absorb, accommodate, adapt to, transform, and recover from their effects in a timely and efficient manner, including through the preservation and restoration of its essential basic structures and functions through risk management" [7]. Attempts have been made to determine the dimensions of systems considered resilient; however, "the dimensions, or constitutive elements of a system, will differ depending on whether we are observing an organization, a state, a community, or another system, whereas the resilience capacities are equal for each type of system" [2]. When it comes to the resilience of critical infrastructure, there are internal and external dimensions. Technical, organizational, and economic resilience constitute parts of both dimensions, while the external dimension additionally includes social resilience [8]. Each segment of resilience must be adequately addressed, especially since the resilience of critical infrastructure is gaining importance in an environment where the possibility of attacks from the digital domain increasingly exists. Given that key systems are largely dependent on information technologies, their ability to quickly recover and continue functioning even after a cyber attack becomes crucial. Therefore, the next section of the paper will analyze the concept of cyber attack, its specific characteristics, and possible consequences for critical infrastructure.

3 Cyber Attacks as a Threat to Critical Infrastructure

In order to prevent the further development of Iran's nuclear program and to deter Israel from potentially employing kinetic strikes against targets in Iran, the administration of George W. Bush in 2006 opted to use cyber instruments against the Iranian nuclear facility in Natanz [9]. The Israeli-American operation *Olympic Games*,[1] which resulted in disabling part of Iran's uranium enrichment capabilities, became synonymous with cyber attacks on critical infrastructure [10] and radically influenced the way cyber threats are perceived. Subsequent cases, such as the Shamoon attack on the Saudi oil giant *Aramco* in 2012, Russian cyber operations directed at the Ukrainian energy system in 2015 and 2017, and the attack on the U.S. *Colonial Pipeline* in 2021, further confirmed the high level of risk and the potential for serious, socio-economically destabilizing consequences arising from the endangerment of critical infrastructure, bringing cyber conflicts closer to the destructive character of traditional, armed conflict [11].

Although Operation Olympic Games demonstrated the ability of certain states to inflict significant damage on adversaries through the use of cyber instruments, numerous other attacks that have occurred in the intervening period indicate a worrying trend of increasing participation of non-state actors in the execution of complex operations. Evidence of the growing concern among state actors is provided by the 2014 statement of the former head of the *U.S. Cyber Command* and the *National Security Agency*, Admiral *Michael S. Rogers*, who pointed to the development of cyber-criminal groups that not only possess increasingly advanced technical capabilities for conducting attacks, but also cooperate more frequently with states attempting to conceal their malicious activity in cyberspace [12]. The increasing availability of information concerning the execution of cyber operations [13] and various forms of cooperation between states and such non-state actors directly weaken the ability of targeted victims to accurately attribute responsibility for an attack—a phenomenon that will be particularly evident in the subsequent examples from the Western Balkan states.

The ability of states to conceal their activities behind various cyber-criminal groups (non-state actors) has become a key factor in the digital domain, especially in the context of rising instability within the international order. States such as China, Russia, North Korea, and Iran increasingly use the digital domain to carry out cyber operations within what Western authors define as hybrid warfare [14]. In the context of competition between the hegemon—embodied by the United States and its allies—and the challenger powers, the Western Balkan states have found themselves caught in between, as actors with considerably limited resources, both human and material, for an adequate response to this new form of hybrid threat. Although the discourse on cyber threats in the Western Balkans had been primarily focused on the Russo-Ukrainian conflict that began in 2014, rather than on domestic cases, the escalation of this conflict into open war in February 2022 marked a qualitatively new phase in the intensity and presence of cyber threats in the region. The use of cyber proxies—in

[1] In the literature, this incident is often referred to by the name of the virus that was used - StuxNet.

other words, the employment of cyber-criminal groups as a façade for attacks on the critical infrastructure of states—has become a recognizable *modus operandi* of states that have developed the concept of hybrid warfare.

3.1 Cyber Attacks on the Critical Infrastructure of Western Balkan States

The convergence of cyber crime, the digital domain, and the deterioration of relations between key actors on the international stage has made the foreign policy component of states an important element of cyber defence. Given that a significant number of capable cyber-criminal groups are under the partial or full control of intelligence agencies of various states, the selection of potential targets of cyber attacks is largely conditioned by the patron state's perception of allies and adversaries in international relations. The Western Balkan states illustrate particularly well how the adoption of different positions on certain foreign policy issues, such as support for Ukraine in the current conflict, can influence the degree to which they are exposed to the activities of various state and non-state cyber actors.

In the context of analyzing sophisticated cyber attacks on the critical infrastructure of Western Balkan states and the difficulties these states face in attributing responsibility for such attacks, the case of Montenegro from August 2022 stands out as especially indicative. The cyber attack carried out by the hacker group *Cuba Ransomware* resulted in significant disruption of Montenegro's state apparatus, with degradation of its functionality occurring in two ways: through the encryption of compromised data by means of ransomware, and through the actions of competent authorities aimed at localizing the damage and preventing the further spread of malicious software across the network [15]. This cyber attack disrupted the work of courts, prosecutors' offices, and other institutions, leading to temporary suspension of certain administrative processes [16] which in turn caused delays in social welfare payments and months-long postponements of public procurement procedures [15]. NATO allies, primarily the United States and France [17] played a key role in mitigating the wider consequences and addressing the resulting damage, providing the Montenegrin government with significant technical and forensic support in incident remediation and investigative processes. Following the initial attack and the incapacitation of the Montenegrin administration, the question of responsibility remained unresolved in the public sphere. Despite the fact that the attack bore all the characteristics of a ransomware operation whose primary objective was extortion of money from the victim in exchange for data restitution, representatives of Montenegrin authorities, above all the Minister of Defence and the National Security Agency, publicly qualified the threat as hybrid activity by the Russian Federation [18]. However, the National Security Agency did not substantiate its claims with credible evidence, and the investigation, carried out with the support of American (FBI) and French cyber units, failed to prove a direct link between the hacker group and Moscow [19].

This case illustrates the complexity of attributing malicious activity in cyberspace, and thus the ability of the victim to respond adequately. *Cuba Ransomware* is a hacker group from the Russian-speaking area that U.S. cyber security agencies nominally classify as an organization with exclusively financial motives [20, 21]. However, due to its activity in the Ukrainian cyber domain and its continuous attacks on the critical infrastructure of Western states, primarily the United States [21]some experts categorize this group among cyber proxies with a certain level of association with Russian intelligence agencies [22]. Jens Stoltenberg, the Secretary General of NATO, also stated that Montenegro was the target of "Russian espionage and malign activities" [23]. The Montenegrin government's open support for Ukraine and condemnation of Russian aggression, as well as the perception of Montenegro as the weakest link in NATO's cyber defence, may have served as reasons for the attack on Montenegro's critical infrastructure. The ambivalence of the status of the *Cuba Ransomware* group positioned between an independent criminal entity and a potential state proxy illustrates a fundamental challenge of contemporary cyber security: the impossibility of reliable attribution as a precondition for a legitimate and proportional response.

Only a month earlier, in July 2022, Albania also faced cyber incidents involving critical infrastructure. The trigger for this cyber attack was Albania's acceptance and support for several thousand members of the largest Iranian opposition group, *Mujahedin-e-Khalq*, in 2016 [24], as well as the organization of its annual conference [25]. The hackers, who identified themselves as the group *Homeland Justice* directly linked to state structures in Tehran [26, 27] infiltrated the computer network of Albanian state institutions with the aim of destroying data [28] and publishing sensitive information [25]. Over the following months, Albanian state institutions continued to suffer sophisticated cyber attacks, even after the government of Edi Rama decided in early September 2022 to sever diplomatic relations with Iran as a direct response to the attacks [29], making Albania the first state to take such a step as a response to another state's activity in the digital domain [30].

Despite attempts to punish Iran, Tehran's activity through cyber proxies against various elements of Albania's critical infrastructure did not cease. The banking sector was among the targets, where the online services of one of the largest banks were disabled [31] causing panic among citizens already unsettled by a disinformation campaign distributed via SMS messages [32]. Particular public attention was drawn to cases of leaked confidential information such as the identities of informants and agents of the Albanian intelligence service, email correspondence of the former head of the *State Intelligence Service* and the chief of police [32]. The border control system was also disabled, leading to collapse at border crossings, and the data it collected on entries and exits from the state were completely erased [24, 33].

Although government officials attempted to limit the scope of damage by prohibiting the publication of leaked information in the media [34] and in some cases by outright denying the incidents [35] this did not prevent further attacks on critical infrastructure or the leakage of citizens' personal data. In one of the attacks on the Ministry of Interior's computer network, personal data of around 100,000 Albanian citizens suspected by police of various illegal acts were stolen and published [35].

However, the attack with the most drastic consequences for personal data confidentiality occurred in July 2022, when biometric data of 910,000 Albanian citizens were stolen during an attack on the e-Albania portal [36]. The acquisition of such highly sensitive biometric data by foreign cyber actors represents a particularly serious security incident, as such information cannot be altered or replaced, which increases the risk of misuse in the long term and may further undermine citizens' trust in state institutions.

In addition to the aforementioned Iranian hacker group, the activity of Russian hacker groups was also identified in the Albanian cyber domain during the observed period. These groups targeted the critical infrastructure of Montenegro and North Macedonia during the same period, which may indicate coordinated action by cyber actors from these two states in the digital space. An additional indicator of such cooperation is the fact that stolen data from Albanian servers primarily appeared on Russian websites and Telegram channels, which suggests mutual sharing of information between actors on both sides [25]. The coordinated activity of cyber actors from multiple states, especially those that present themselves as non-state and independent, may prove to be the future of offensive operations in the digital domain. Through information exchange and joint action against the critical infrastructure of states—as demonstrated in the case of the Western Balkans—attackers may achieve certain advantages. First, they can synchronize and coordinate activities, which may positively affect the outcome of an attack, particularly when targeting states that are dependent on external assistance, such as the Western Balkan states. Second, through the exchange of experience and knowledge regarding malicious programs, knowledge of the target's infrastructure, and tactics, techniques, and procedures (TTP), hacker organizations may enhance their own capabilities and, equally importantly, further mask their identity in the digital space.

The issue of resilience of critical infrastructure to cyber threats has also emerged in other Western Balkan states, primarily in Serbia, where various state bodies and the energy sector have been exposed to intensive attacks by hacker groups. However, unlike the previously mentioned cases of Montenegro and Albania both NATO members that can rely on support from other states and their expertise in cyber defence—Serbia, as a state that does not belong to any military-political alliance, cannot do so and must rely largely on its own capacities. Such a foreign policy orientation, in the context of contemporary non-conventional threats, may produce serious negative consequences at moments of confrontation between opposing actors on the international stage. Providing support to one side or the other—such as arms sales to Ukraine, condemnation of the Russian invasion at the UN General Assembly, or refusal to impose sanctions on Moscow—may serve as a pretext for initiating cyber attacks as a means of signalling dissatisfaction with the policy pursued by the targeted state.

The potential consequences of cyber attacks on critical infrastructure may also be observed through the example of Elektroprivreda Srbije (EPS), which became the focus of the hacker group *Qilin* or a related group. Operating through a *Ransomware-as-a-Service* (RaaS) model,[2] *Qilin* has become globally known for conducting or assisting in the execution of highly sophisticated operations most often targeting Western companies and institutions [37, 38]. In addition to the disruption of EPS operations, cyber criminals may have gained access to personal data of citizens held by the company, as well as data on the electricity supply system itself, according to experts' assessments [39]. A particularly sensitive case was the Bajina Bašta hydropower plant, whose computer network was fully compromised by hackers, resulting in the theft of business documentation, employee identity cards, and even architectural drawings and plans of the entire complex [40, 41].

Soon it was discovered that other institutions had been compromised as well, and credentials for employee accounts at Post of Serbia, Srbijagas, and the National Employment Service appeared on black markets, creating opportunities for further attacks and greater damage [31]. The takeover of institutional email accounts would allow hackers to access correspondence and initiate phishing attacks while hiding behind official accounts. Although these attacks did not cause major direct material damage, as was the case with cyber operations in Ukraine prior to the start of the war in 2022, such actions aim not only at financial gain, but also at generating secondary costs in the form of endangering the security of citizen and institutional data, as well as undermining public trust in the ability of state institutions to prevent such attacks. However, unlike other Western Balkan states, where the question of the perpetrators was raised immediately by the political leadership after the attacks, no culprit was identified in the case of attacks against Serbian institutions. Apart from assurances from EPS leadership that the consequences were negligible and that the security of user data was not threatened [42] no additional information about the incident was released to the public, and responsible institutions did not issue official statements on the matter [40].

The governments of Western Balkan states face a complex security challenge posed by the activities of various state-backed hacker groups and associated cyber criminals, predominantly originating from Russia and Iran. The objectives of attacks on the critical infrastructure of Western Balkan states include unauthorized acquisition of sensitive personal data, disabling of institutions, potential financial profit, and creating platforms for further cyber operations directed at Western states [43]. These activities produce harmful effects on targeted subjects that manifest in two forms. The primary form of damage consists of immediate losses resulting from physical damage and dysfunction of affected networks, creating the need for urgent repairs [44]. The secondary form of damage, which represents a significantly more serious security risk, includes indirect losses resulting from the slowdown or complete suspension of institutional or organizational operations, increased financial expenditures required to improve protection mechanisms, erosion of user trust, and compromise of confidential information [44].

[2] Ransomware-as-A-Service (RaaS) is a specific business model in which highly sophisticated groups develop ransomware and lease it to other cyber criminals, charging them a fixed price or a percentage of the successfully executed extortion.

Building resilience of critical infrastructure to cyber threats requires a combination of proactive, preventive, and strategically coordinated measures. The experience of Ukraine provides compelling evidence that cyber defence is not lost in advance, and that systematic preparation can significantly increase overall resilience. After a series of destructive cyber attacks during the conflict in Donbas in 2014, Ukrainian institutions invested substantial effort in strengthening their defensive capacities through targeted personnel training, development of technical skills, engagement of private companies specialized in cyber security, and intensified cooperation with partner states [45]. These measures contributed to the development of a level of resilience that enabled the protection of key infrastructure and the maintenance of the functionality of state systems even during the first years of extensive warfare in cyberspace.

Around two thirds of local self-government units in Serbia do not have a single employee who has undergone any form of cyber security training in the past three years [46]. Considering that the presence of illegal software on computers used by employees in the state administration of the Republic of Serbia is extremely frequent [46], and given the continuous neglect of proper cyber hygiene, including visits to risky internet domains and interaction with phishing messages via official computers, it is necessary to conclude that the overall level of information security is critically low. The most direct way to address these problems is through continuous training and awareness-raising campaigns among employees in state administration and in companies whose work is critical for the stability of the state, since individuals are the weakest link in cyber defence [47].

The protection of critical infrastructure requires an integrated, collaborative approach that includes both public–private partnerships and regional and international cooperation. Cooperation with the global private sector, led by specialized leaders in network security, has proven highly effective in combating sophisticated cyber threats, as their expertise and innovative capacities often exceed the available resources of state actors [48]. This form of engagement is crucial, since parts of critical infrastructure are privately owned or managed. However, establishing public–private partnerships with foreign entities in democratic societies must be subject to rigorous risk assessment [49]. Therefore, it is imperative to precisely define the circumstances and arrangements for the use of expert knowledge from international companies, ensuring the preservation of national sovereignty [50].

Cyber security in small states, perhaps more than in any other field of security, requires active cooperation with major powers that possess substantial capacities for defending their own digital domains. The example of Montenegro, which relied on NATO's capabilities—namely those of the United States and France, in its defence, demonstrates that an autonomous approach to critical infrastructure protection against cyber threats is not a sustainable solution [51]. The legal and financial frameworks provided by NATO and the EU, as well as the expertise of their specialized cyber security units, can serve as a strong incentive for deepening regional cooperation, both at the level of the Western Balkans and at the level of Europe, in this increasingly important domain [45]

4 Conclusions

The emergence of cyberspace has radically transformed the contemporary security environment, introducing it as the fifth dimension of operational activity and generating new, asymmetric forms of threats that often originate from non-state cyber actors. In the context of the increasing militarization of this domain and the growing digitalization of state apparatuses and critical infrastructure, strengthening the resilience of critical infrastructure has become the highest security priority. This results from the fact that the potential incapacitation of key systems—encompassing physical and cyber resources vital to state functioning—can cause significant and long-term socio-economic and security repercussions.

The analysis of case studies of cyber attacks on the critical infrastructure of Western Balkan states (Montenegro, Albania, and Serbia) confirms that small states operate within an exceptionally complex security environment. Although each of these cases contains certain specificities, they all point to a common structural deficit: insufficiently developed cyber resilience as the foundation of effective defence. The case of Montenegro most clearly demonstrates the limitations of national capacities and the indispensable role of international cooperation—in this instance, support from NATO allies—in mitigating the damage. At the same time, the ambivalence surrounding the attacker's status and the impossibility of reliable attribution illustrate a fundamental challenge that complicates the provision of an adequate response. Conversely, the case of Albania provides a clear demonstration of the use of cyber operations as a mechanism of direct geopolitical pressure, with consequences that involve not only the devastation of critical infrastructure but also the theft of personal data of a large number of citizens, which undermines public trust and compromises national security over the long term. The case of Serbia, a state that does not belong to any military-political alliance, further emphasizes the vulnerability that arises from domestic systemic shortcomings. The attacks on Elektroprivreda Srbije (EPS) and the compromise of other institutions show that insufficient institutional preparedness, weak cyber hygiene, and lack of transparency significantly increase systemic vulnerability, the effects of which are further exacerbated by the absence of external support. The exposure of Western Balkan states to cyber threats is additionally intensified by the coordinated activity of cyber actors linked to Russia and Iran, which increases the effectiveness of cyber attacks and amplifies the resulting damage.

In light of these findings, it can be concluded that building the resilience of critical infrastructure requires an integrated and multidimensional approach. This primarily entails substantial investments in the strengthening of human resources and continuous training of state administration employees, since they constitute the weakest link in cyber defence. The protection of critical infrastructure requires a collaborative model of public–private partnership, as well as rigorous risk assessment when cooperating with foreign entities in order to preserve state control over key security mechanisms. Finally, regional and international cooperation within the frameworks offered by the European Union and NATO represents essential support for small states, providing them with the necessary expertise and financial structures for enhancing collective defence. By comprehensively strengthening technical capacities and cultivating a culture of cyber security, the states of the region can secure vital functions and stability within an increasingly unstable digital environment.

References

1. Putnik, N.: Sajber rat i sajber mir. Univerzitet u Beogradu—Inovacioni centar Fakulteta bezbednosti; Akademska misao, Beograd (2022)
2. Keković, Z., Ninković, V.: Zaštita kritične infrastrukture: Sistemski pristup. Centar za analizu rizika i upravljanje krizama, Beograd (2020)
3. Aradau, C.: Security that matters: Critical infrastructure and objects of protection. Secur. Dialogue **41**(5), 491–514 (2010)
4. Lehto, M.: Cyber-attacks against critical infrastructure. In: Lehto, M., Neittaanmäki, P. (eds.) Cyber security: Critical infrastructure protection, pp. 3–42. Springer (2022)
5. Alcaraz, C., Zeadally, S.: Critical infrastructure protection: Requirements and challenges for the 21st century. Int. J. Crit. Infrastruct. Prot. **8**, 53–66 (2015)
6. Rinaldi, S.M., Peerenboom, J.P., Kelly, T.K.: Identifying, understanding, and analyzing critical infrastructure interdependencies. IEEE Control. Syst. Mag. **21**(6), 11–25 (2001)
7. United Nations Office for Disaster Risk Reduction (2017) The Sendai framework terminology on disaster risk reduction: Resilience. https://www.undrr.org/terminology/resilience
8. Zubieta, L.L.: *Resilience framework for critical infrastructures* (Doctoral dissertation). Tecnun Universidad de Navarra, Donostia-San Sebastián (2013)
9. Vielhaber, D., Bleek, P.C.: Shadow wars. Nonproliferation Rev. **19**(3), 481–491 (2012). https://doi.org/10.1080/10736700.2012.734194
10. Rid, T.: Cyber war will not take place. J. Strateg. Stud. **35**(1), 5–32 (2012). https://doi.org/10.1080/01402390.2011.608939
11. Veljković, S.: Possibility of applying the rules of international humanitarian law to cyber warfare. Law—Theory Pract., **41**(3), 17–28 (2024)
12. National Security Agency (2014) *Hearing of the House (Select) Intelligence Committee: "Cybersecurity threats: The way forward"* [Transcript]. NSA. Available at: https://www.nsa.gov/Press-Room/Speeches-Testimony/Article-View/Article/1620360/hearing-of-the-house-select-intelligence-committee-subject-cybersecurity-threat/
13. Office of the Director of National Intelligence (2019) *Worldwide threat assessment of the U.S. Intelligence Community.* Available at: https://www.dni.gov/files/ODNI/documents/2019-ATA-SFR—SSCI.pdf
14. Van Puyvelde., D.: Hybrid war – does it even exist? NATO Review. (2015). http://www.nato.int/docu/review/2015/Also-in-2015/hybrid-modern-future-warfare-russia-ukraine/EN/
15. Balkan Insight (2025) Balkan countries must respond proactively to evolving cyber threats: Report. Available at: https://balkaninsight.com/wp-content/uploads/2025/02/RDRAF-JAN25.pdf
16. Organized Crime and Corruption Reporting Project (2022) Cyberattacks hit Montenegro's state infrastructure. Available at: https://www.occrp.org/en/news/cyberattacks-hit-montenegros-state-infrastructure
17. Center for Strategic and International Studies (2024) Russia aims at Montenegro. Available at: https://www.csis.org/analysis/russia-aims-montenegro
18. Der Spiegel.: Montenegro sieht ausreichende Hinweise für russische Cyberattacke. (2022). Available at: https://www.spiegel.de/netzwelt/montenegro-sieht-ausreichende-hinweise-fuer-russische-cyberattacke-a-2b48d900-9389-4fb7-9c1c-d3e99ba72486
19. Reuters.: Montenegro Says FBI Will Help Investigate Cyber Attacks. Reuters. (2022). Available at: https://www.reuters.com/world/montenegro-says-fbi-will-help-investigate-cyber-attacks-2022-08-31/

20. Cybersecurity and Infrastructure Security Agency.: *#StopRansomware: Cuba ransomware.* (2023). Available at: https://www.cisa.gov/news-events/cybersecurity-advisories/aa22-335a
21. Radio Slobodna Evropa.: FBI predao Crnoj Gori nalaze o sajber napadima u toj zemlji. (2023). Available at: https://www.slobodnaevropa.org/a/istraga-napadi-vlada-sad/32220699.html
22. Tech Crunch.: Cybercriminals who targeted Ukraine are actually Russian government hackers, researchers say. (2023). Available at: https://techcrunch.com/2023/05/15/cybercriminals-who-targeted-ukraine-are-actually-russian-government-hackers-researchers-say/
23. Radio Slobodna Evropa.: Ruske službe izvele sajber napad na Vladu Crne Gore, tvrdi Agencija za bezbjednost. (2022). Available at: https://www.slobodnaevropa.org/a/ruski-hakeri-napad-vlada-crne-gore/32005931.html
24. Zaimi, G.: Iran's Balkan front: The roots and consequences of Iranian cyberattacks against Albania. Middle East Inst. (2022). Available at: https://mei.edu/publications/irans-balkan-front-roots-and-consequences-iranian-cyberattacks-against-albania
25. Oghanna, A.: How Albania became a target for cyberattacks. Foreign Policy. (2023). Available at: https://foreignpolicy.com/2023/03/25/albania-target-cyberattacks-russia-iran/
26. Cybersecurity and Infrastructure Security Agency.: Iranian state actors conduct cyber operations against the Government of Albania. (2022). Available at: https://www.cisa.gov/news-events/cybersecurity-advisories/aa22-264a
27. Kalinowski, M.: Homeland Justice operations against Albania (2022) Cyber Law Toolkit. (2024). Available at: https://cyberlaw.ccdcoe.org/wiki/Homeland_Justice_operations_against_Albania_(2022)
28. Internet Crime Complaint Center.: 2022 internet crime report. (2022). Available at: https://www.ic3.gov/Media/PDF/AnnualReport/2022_IC3Report.pdf
29. Reuters.: Albania cuts Iran ties over cyberattack, U.S. vows further action. Reuters. (2022). Available at: https://www.reuters.com/world/albania-cuts-iran-ties-orders-diplomats-go-after-cyber-attack-pm-says-2022-09-07/
30. Ozturk, T.: Albania blames Iranian-backed group for cyberattack on its statistical institute. Anadolu Agency. (2024). Available at: https://www.aa.com.tr/en/europe/albania-blames-iranian-backed-group-for-cyberattack-on-its-statistical-institute/3137301
31. Balkan Insight.: Battle for Balkan cybersecurity: Threats and implications of biometrics and digital identity. (2023). Available at: https://balkaninsight.com/2023/06/30/battle-for-balkan-cybersecurity-threats-and-implications-of-biometrics-and-digital-identity/
32. Higgins, A.: A NATO minnow reels from cyberattacks linked to Iran. The New York Times. (2023). Available at: https://archive.is/20230225162210/www.nytimes.com/2023/02/25/world/europe/albania-iran-nato-cyberattacks.html
33. Vicens, A.J.: Albania says Iranian hackers hit the country with another cyberattack. *CyberScoop.* (2022). Available at: https://cyberscoop.com/iranian-cyberattack-albania-homeland-justice/
34. Mapping Media Freedom.: Alert 25187. (2022). Available at: https://www.mapmf.org/alert/25187
35. Balkan Insight.: Iranian hackers leak database of Albanian criminal suspects. (2022). Available at: https://balkaninsight.com/2022/10/03/iranian-hackers-leak-database-of-albanian-criminal-suspects/
36. Balkan Investigative Reporting Network.: Battle for Balkan cybersecurity: Threats and implications of biometrics and digital identity. (2023). Available at: https://balkaninsight.com/2023/06/30/battle-for-balkan-cybersecurity-threats-and-implications-of-biometrics-and-digital-identity/
37. Group-IB.: The Qilin ransomware: Analysis and protection strategies. (2023). Available at: https://www.group-ib.com/blog/qilin-ransomware/
38. Cyberint.: Qilin ransomware: Get the 2025 lowdown. (2025). Available at: https://cyberint.com/blog/research/qilin-ransomware/
39. Balkan Insight.: Latest cyber-attacks in Serbia raise fresh questions about defence. (2024). Available at: https://balkaninsight.com/2024/01/16/latest-cyber-attacks-in-serbia-raise-fresh-questions-about-defence/

40. Radio Slobodna Evropa.: Hakersko „ ogoljavanje" poslovanja Hidroelektrane Bajina Bašta u Srbiji. (2024). Available at: https://www.slobodnaevropa.org/a/hidroelektrana-bajina-basta-hakerski-napad/33026495.html
41. Vreme.: Poverljivi podaci hidroelektrane "Bajina Bašta" mesecima dostupni na internetu. (2024). Available at: https://vreme.com/en/razno/poverljivi-podaci-hidroelektranebajina-basta-mesecima-dostupni-na-internetu/
42. РТС (2023) ЕПС био мета хакерског напада, систем и подаци безбедни. Available at: https://www.rts.rs/vesti/drustvo/5330910/eps-bio-meta-hakerskog-napada-sistem-i-podaci-bezbedni.html
43. Bogićević, A.: The influence of non-state cyber actors in conventional armed conflicts: A case study of the War in Ukraine. In: Blagojević, S., Trifković, D. (eds.) VojNa 2025: International scientific conference on military sciences, pp. 304–310. Military Academy (2025)
44. Салій, АЯ: Особливості відповідальності за кібератаки в країнах ЄС. Науковий вісник Ужгородського національного університету, Серія: Право **81**(2), 330–334 (2024)
45. Krasznay, C.: The role of civilian cybersecurity companies in military cyber operations. Land Forces Acad. Rev. **29**(113), 1–10 (2024). https://doi.org/10.2478/raft-2024-0001
46. NALED: (2023) Interni podaci (Excel format).
47. Gorgulenko, V.: Cyber warfare in modern military conflicts: Best practices, trends and patterns of development. Mod. Inf. Technol. Sphere Secur. Def., **50**(2), 11–28. (2024). https://doi.org/10.33099/2311-7249/2024-50-2-11-28
48. Kim, R.: Public–private partnerships in national cybersecurity. Mich. Technol. Law Rev.. (2022). https://mttlr.org/2022/01/public-private-partnerships-in-national-cybersecurity/
49. Christensen, K.K., Petersen, K.L.: Public–private partnerships on cybersecurity: A practice of loyalty. Int. Aff. **93**(6), 1435–1452 (2017). https://doi.org/10.1093/ia/iix189
50. Bateman, J., Beecroft, N., Wilde, G.: What the Russian invasion reveals about the future of cyber warfare. Carnegie Endow. Int. Peace. (2022). https://carnegieendowment.org/2022/12/19/what-russian-invasion-reveals-about-future-of-cyber-warfare-pub-88667
51. Haner, J.K., Knake, R.K.: Breaking botnets: A quantitative analysis of individual, technical, isolationist, and multilateral approaches to cybersecurity. J. Cybersecur. **15**(1), 1–15 (2021). https://doi.org/10.1093/cybsec/tyab003

Cybersecurity Incidents in the Water Sector

Nikola Milivojević and Vladimir Milivojević

Abstract Cybersecurity threats to the water sector have intensified as utilities adopt digital technologies for operational efficiency. This paper analyzes five major incidents between 2021 and 2025, highlighting vulnerabilities in both operational technology (OT) and enterprise IT systems. Findings reveal that most breaches exploited basic weaknesses such as insecure remote access, default credentials, and poor network segmentation rather than advanced exploits. Attack motives ranged from financial gain and political signaling to psychological impact, illustrating a diverse threat ecosystem involving criminal groups, hacktivists, and state-aligned actors. Consequences included service disruptions, reputational damage, and heightened regulatory scrutiny, even when physical harm was avoided. Comparative analysis underscores that smaller utilities face acute resource and expertise gaps, while larger entities remain exposed through IT systems critical to business continuity. Key lessons emphasize multi-factor authentication, IT–OT segmentation, secure remote access, continuous monitoring, and workforce training. Policy recommendations call for enforceable cybersecurity baselines, funding for under-resourced utilities, and international threat intelligence sharing. Research priorities include lightweight anomaly detection for legacy OT, cyber–physical modeling, and socio-technical response studies. Strengthening resilience requires coordinated action across policy, practice, and research.

Keywords Cybersecurity · Water sector · Comparative analysis

N. Milivojević (✉) · V. Milivojević
Jaroslav Černi Water Institute, Belgrade, Serbia
e-mail: nikola.milivojevic@jcerni.rs

V. Milivojević
e-mail: vladimir.milivojevic@jcerni.rs

Z. Keković et al. (eds.), *Critical Infrastructure: Risk and Crisis Management in the New Reality*, Advanced Sciences and Technologies for Security Applications,
https://doi.org/10.1007/978-3-032-20948-1_11

1 Introduction

Over the past decade, the cybersecurity threat landscape has expanded rapidly, affecting critical infrastructure systems across the globe. While sectors such as finance, energy, and transportation have received substantial attention and investment in cyber resilience, the water sector remains comparatively under-protected, despite its fundamental role in public health and environmental sustainability [1, 2]. Recent cyber incidents—including targeted attacks on SCADA systems, ransomware campaigns, and ideologically motivated intrusions—have exposed fragilities in the digital backbone of water utilities [3, 4], elevating cybersecurity to a central concern for operators, regulators, and policymakers.

Digitalization has brought significant benefits to water utilities—remote monitoring, predictive maintenance, optimized energy use, and real-time quality control—but has also expanded the attack surface, particularly through the integration of legacy operational technology (OT) with modern information technology (IT) systems [5, 6]. Many utilities, especially smaller or decentralized ones, struggle to maintain adequate cybersecurity controls, lacking dedicated personnel, modern equipment, and formal incident response procedures [7]. As adversaries grow more sophisticated and diverse, utilities must modernize while ensuring that security, privacy, and resilience are not compromised.

Case studies of real-world incidents offer insights beyond abstract threat modeling. By examining actual breaches and campaigns, researchers and practitioners can identify systemic weaknesses, operational blind spots, and recurring vulnerabilities [8, 9]. These cases also show how technical failures interact with organizational decisions and human behavior, revealing lessons for detection, containment, recovery, and long-term resilience.

This paper adopts a comparative case study–driven methodology to analyze evolving cybersecurity risks in the water sector. It focuses on five major incidents between 2021 and 2025: the attempted chemical manipulation at the Oldsmar water treatment plant (USA), the multi-national CyberAv3ngers campaign targeting SCADA components, the disruption of telemetry systems in County Mayo (Ireland), and the ransomware attack on the American Water utility, supplemented by the 2025 Norwegian dam incident. Each case is examined in terms of attack vector, operational impact, response capability, and broader implications. A comparative synthesis and discussion on policy, practice, and research aim to support more evidence-based and context-aware cyber risk management in the water domain.

2 Cybersecurity Landscape in the Water Sector

Water utilities worldwide are undergoing digital transformation to improve operational efficiency, service reliability, and regulatory compliance. Core technologies include SCADA systems, programmable logic controllers (PLCs), remote telemetry units (RTUs), and Industrial Internet of Things (IIoT) devices, supporting real-time control of abstraction, chemical dosing, pumping, reservoir management, leakage detection, and pressure optimization [10, 11].

Modern SCADA architectures typically feature a central control room connected to distributed assets via wired and wireless links. Human–machine interfaces (HMIs) support visualization, command issuance, and alarm handling. These OT systems are increasingly integrated with enterprise IT environments for analytics, asset management, remote access, and reporting [12]. While this convergence brings efficiency gains, it also enlarges the attack surface, especially when legacy components are brought online without adequate hardening [13].

Many small and medium utilities operate systems never designed with cybersecurity in mind. Legacy PLCs and RTUs may lack authentication, encryption, or secure boot; devices often retain default credentials, expose unnecessary ports, and sit in flat networks. Remote access—introduced for vendor support or off-hours operation—is frequently implemented with weak protection, such as poorly secured VPNs or exposed remote desktop interfaces [14].

The resulting cybersecurity posture is heterogeneous and often reactive. Larger, investor-owned utilities may have dedicated cybersecurity staff, IT–OT network segmentation, and incident response plans, while smaller public or municipal utilities operate with limited budgets, minimal cyber expertise, and aging infrastructure. Responsibilities are frequently fragmented among IT generalists, operations staff, and vendors, which can delay detection and response [15].

Threats are evolving in parallel. Water utilities face financially motivated ransomware groups, hacktivists, insiders, and state-sponsored actors seeking geopolitical leverage. Adversaries routinely scan for internet-exposed SCADA assets, exploit known OT vulnerabilities, and use phishing to pivot from IT networks into OT environments [16].

Regulatory frameworks are advancing—for example, the EU's NIS2 Directive and U.S. guidance from EPA and CISA—but enforcement is uneven and local implementation often under-resourced [17, 18]. Many utilities still lack real-time monitoring, intrusion detection, and consistent cybersecurity training.

In sum, digital systems bring substantial value but also introduce complex cyber risks that must be actively managed. Persistent gaps in visibility, governance, technical controls, and workforce capacity—especially in smaller organizations—underscore the need for coordinated policy, operational, and research efforts to ensure digitalization strengthens rather than undermines water service reliability and safety.

The selection of the case studies was guided by several considerations that ensured the cases would provide a broad and meaningful picture of cybersecurity risks in the water sector. First, events that occurred between 2021 and 2025 are chosen, since that period reflects the most recent and relevant developments in both operational technology and enterprise IT environments. This timeframe provides relevant information for highlighting the rapid evolution of threats and responses in the sector.

Geographic diversity was another important factor in the selection of case studies. Cases from the United States, Ireland, Norway, and a globally distributed campaign are included, which highlights that cyber threats to water utilities are not confined to one region but are truly international in scope. Also, organizational scale was considered, contrasting small municipal or rural utilities with large national providers. This way it is possible to assess the impact of resource limitations in smaller entities, while comparing it to IT-OT practices in larger organizations which face risks through complex IT systems that are critical to business continuity.

The cases were also chosen to represent a variety of attack types and motives. Some incidents involved direct manipulation of operational technology, while others targeted enterprise IT systems through ransomware or disrupted telemetry data. The motives ranged from financially driven extortion to politically motivated signaling and psychological pressure. By including this spectrum, it is demonstrated that the threat ecosystem is diverse, involving criminal groups, hacktivists, and state-aligned actors.

Finally, the impact of the incidents is considered. Some incidents posed immediate risks to public health, others disrupted customer services, and still others were designed to create fear or political leverage without causing physical damage. This range of consequences ensured that the analysis may extract lessons not only about technical vulnerabilities but also about organizational readiness, public communication, and regulatory response.

The following case studies were selected and will be presented in the following text:

1. Oldsmar Water Treatment Plant, Florida (2021)
2. CyberAv3ngers Campaign (2023–2025)
3. County Mayo, Ireland (2023)
4. American Water Utility (2024)
5. Norwegian Dam Incident (2025)

3 Case Studies

3.1 Oldsmar Water Treatment Plant, Florida (2021)

The February 5, 2021 cyber incident at the Oldsmar Water Treatment Plant in Florida remains a defining event for water-sector cybersecurity. Serving roughly 15,000 residents, the small facility used a SCADA system to manage chemical dosing and operations, with remote access provided by TeamViewer [19]. On the day of the attack, an operator observed an unexpected remote session. At about 1:30 PM, the intruder increased the sodium hydroxide (NaOH) dosage from 100 ppm to 11,100 ppm. The operator immediately reversed the change, preventing contaminated water from entering the distribution system.

Investigations indicated insecure remote access via TeamViewer without multi-factor authentication, shared credentials, and a workstation directly connected to SCADA, running outdated software and lacking proper segmentation [3]. There was no intrusion detection or systematic logging; discovery relied entirely on operator vigilance. The incident revealed systemic weaknesses common to small utilities: weak authentication, flat IT–OT networks, unsupported software, poor credential management, and low cybersecurity awareness. Although no physical harm occurred, the ease with which a critical dosing parameter was altered triggered concern at state and federal levels. The utility disabled remote SCADA access [20] and engaged the FBI and DHS. CISA and other agencies issued alerts to utilities nationwide [21], and public briefings reassured residents that water quality had not been compromised. Oldsmar became a catalyst for sector-wide discussions on remote access hardening and cyber investment, contributing to policy debates and federal funding decisions, including measures under the IIJA [22].

Key lessons include eliminating shared accounts, enforcing multi-factor authentication, segmenting IT and OT networks, introducing continuous monitoring, and embedding cybersecurity into daily operations. Oldsmar demonstrated how modest technical skill coupled with lax practices can pose direct public health risks.

3.2 CyberAv3ngers Campaign (2023–2025)

The CyberAv3ngers campaign, active from 2023 through 2025, represents a coordinated wave of politically motivated cyberattacks against water and energy infrastructure. Attributed to Iran-linked actors, the campaign signaled an escalation in using OT-targeted cyber operations to support geopolitical narratives and exert psychological pressure [23]. The group focused on Israeli-manufactured Unitronics PLCs, widely deployed in water treatment, irrigation, and municipal utilities. Many of these devices were accessible on the public internet with default credentials, making them easy targets. CyberAv3ngers reportedly used scanning tools, credential stuffing, and remote logins to compromise PLCs, then disrupted operations or defaced HMIs with slogans referencing the Gaza conflict and pro-Iran messages.

A widely reported incident occurred in November 2023 at the Municipal Water Authority of Aliquippa, Pennsylvania, where a Unitronics PLC at a remote booster station displayed "YOU HAVE BEEN HACKED. DOWN WITH ISRAEL." Similar events followed in other U.S. states and in countries such as Ireland and Albania. In many cases, utilities shut down affected sites as a precaution, even when no damage occurred. The campaign's distinguishing features were its scale, consistency, and overt ideological framing. Rather than a single-utility compromise, it blended psychological warfare, propaganda, and operational disruption, leading analysts to describe it as state-aligned hacktivism. Technically, it exposed a systemic lack of vendor-specific hardening. Many devices ran unpatched firmware, retained default administrative settings, and were placed in poorly segmented networks. Resource-constrained utilities often lacked monitoring or intrusion detection, allowing attackers to automate scanning and defacement at scale.

In 2024, researchers observed more sophisticated tooling, including malware samples labelled IOControl, designed to interact with Unitronics PLCs, modify logic, trigger unauthorized commands, and collect process data [24]. Although not destructive, IOControl demonstrated deeper understanding of industrial protocols and OT workflows. CISA, Israel's National Cyber Directorate, and private companies issued joint advisories urging utilities to disconnect exposed PLCs, change default passwords, and implement strict access controls. Many small operators, however, struggled to act quickly due to limited staffing and situational awareness. The campaign's physical impact was limited, but its psychological and political effects were significant. It showed how modest, widely deployed OT components can be weaponized in grey-zone conflicts and underscored the need for secure configuration management, supply chain accountability, OT-specific cyber awareness, and international cooperation.

3.3 County Mayo, Ireland (2023)

In late 2023, a cyberattack on a small water utility in County Mayo, Ireland, illustrated how even rural systems with limited scale can suffer significant operational disruption [25]. The target was a remote water supply and treatment operation serving about 160 residential properties. The utility employed a SCADA-based architecture with RTUs managing pumps, reservoir levels, and alarms. Infrastructure had partial modernization—cloud-based alerts and web dashboards—but lacked intrusion detection, endpoint protection on field devices, and centralized logging [26].

Attackers reportedly gained unauthorized access to the telemetry control system via a compromised password and weak remote access configuration. They disrupted communications between the central interface and reservoir-level sensors, causing false "full" readings and halting pumping operations [25]. Around 160 customers lost service for more than 24 h while teams manually inspected and reconfigured equipment. Public statements stressed that water quality remained safe, but concerns grew about detection and reporting capability [27].

The incident occurred amid broader OT-focused activity in Europe and against the backdrop of NCSC warnings about exposed industrial systems. Media speculated links to CyberAv3ngers, though attribution remained unconfirmed. The Mayo case exposed recurring weaknesses of small utilities: limited cybersecurity staffing, low awareness of internet exposure, minimal segmentation, and reliance on vendor defaults. It also showed that data integrity attacks—without malware—can cause effective denial-of-service in digital control environments. Response involved collaboration between local authorities, national cybersecurity teams, SCADA vendors, and Ireland's EPA. This cross-agency effort was key to restoring service, securing telemetry, and initiating longer-term risk assessments. Mayo County Council subsequently committed to improving SCADA monitoring, access controls, and staff cyber awareness.

Overall, the case underlines that critical infrastructure fragility can stem from configuration errors and weak controls, not only from advanced actors or complex malware. For small utilities, basic cyber hygiene, continuous remote access auditing, and external support structures are essential to resilience.

3.4 American Water Utility (2024)

In late 2024, a cyberattack on a major U.S. water utility—one of the country's largest water and wastewater service providers—demonstrated how digital threats increasingly affect enterprise IT environments with indirect but serious implications for service continuity and public trust [28].

In October 2024, customers experienced outages in the online billing portal, delayed invoices, and overloaded call centers. The company confirmed a significant intrusion affecting billing systems, customer databases, and internal communications [29]. Water treatment and distribution systems remained operational, but back-office services were taken offline, and some administrative processes shifted to manual modes.

Forensic analysis suggested a ransomware attack initiated via phishing against a privileged user. The attackers moved laterally and encrypted critical data. External cybersecurity firms and federal agencies assisted with containment and recovery; ransom payment, if any, was not publicly disclosed. A key feature of this case was that OT systems, including SCADA and PLCs, were not directly compromised. Existing IT–OT segmentation and read-only firewalls helped prevent the attack from propagating into control networks [30].

Business impacts, however, were considerable. Customer service disruptions lasted over two weeks; billing cycles were delayed or canceled; and the company faced reputational damage, regulatory questioning, and potential legal liabilities related to data exposure and service delays. The incident highlighted tight coupling between revenue collection, customer trust, and digital service availability. Despite having cybersecurity policies, a security team, and continuity plans, the utility remained vulnerable to social engineering and common IT weaknesses. In response, it accelerated deployment of endpoint detection and response (EDR), expanded phishing training, reviewed privilege models, and updated incident response plans for IT-focused scenarios. The event also prompted closer coordination between IT and OT security teams and more rigorous exercises for cyber-physical continuity.

Nationally, the case reinforced that IT systems are integral to critical service delivery and must be protected alongside OT. It underscored the value of OT segmentation while showing that business systems can themselves be critical points of failure.

3.5 Norwegian Dam Incident (2025)

In April 2025, pro-Russian hackers briefly seized control of a hydropower dam's digital control system in Bremanger, western Norway, remotely opening a gate and altering water discharge for roughly four hours before detection [31, 32]. Authorities stated the operation aimed to demonstrate reach and create psychological pressure rather than cause catastrophic flooding. Reports indicated the gate discharged around 500 L per second and that attackers later posted a video of the compromised interface on Telegram [33].

The incident highlights persistent IT–OT interface risks in dam operations, particularly remote access pathways and monitoring gaps that allow malicious commands to persist. Norwegian officials framed it as part of broader hybrid activities by Russian state-aligned actors against European energy, water, and transport assets. Analysts also noted companion activity in 2025 at a small hydropower facility in Poland, where pro-Russian hacktivists allegedly disrupted turbine operations [33, 34].

Operationally, the case shows how limited, reversible manipulations of water-control assets can achieve outsized psychological and political effects while avoiding escalation. Key lessons include tightening remote access control and session monitoring, deploying process-aware anomaly detection (e.g., rate-of-change checks on gates/valves), and rehearsing joint response between operators, regulators, and national cyber agencies.

4 Comparative Analysis of Case Studies

The case studies illustrate the evolving cyber threat landscape in the water sector across different geographies and organizational sizes. Despite differences in scale, sophistication, and impact, they reveal recurring weaknesses in attack surfaces, response capabilities, and preparedness.

Across all cases, attackers exploited basic cybersecurity failures: weak or unmanaged remote access, default or shared credentials, and poor network segmentation. In Oldsmar, TeamViewer without multi-factor authentication exposed chemical dosing controls; in the CyberAv3ngers campaign, publicly exposed Unitronics PLCs often retained default logins; County Mayo suffered from insecure telemetry and weak authentication; American Water's intrusion began with phishing that compromised a privileged IT account. Norwegian dam also suffered from insecure remote access. These examples show that many successful intrusions rely on preventable lapses rather than advanced zero-day exploits.

Attack types and motives nevertheless differ. Oldsmar and County Mayo targeted operations by manipulating SCADA or telemetry data. CyberAv3ngers pursued politically motivated, cross-border OT defacements and signaling. American Water was a financially driven ransomware attack on enterprise IT. This spectrum reflects a threat ecosystem that spans nation-state-aligned actors, hacktivists, and profit-seeking criminal groups.

Technical sophistication also varies: Oldsmar involved simple remote desktop abuse; CyberAv3ngers evolved from defacement to purpose-built malware (IOControl); County Mayo relied on data falsification without malware; American Water followed a typical ransomware playbook with phishing, lateral movement, and encryption. In every case, the absence of layered defenses and rapid detection amplified the impact of comparatively simple techniques.

Institutional readiness largely determined outcomes. Oldsmar avoided physical harm because an operator noticed and reversed the changes in real time. American Water contained the intrusion to IT because OT networks were segmented and incident response playbooks existed. County Mayo, lacking in-house expertise, depended on external support for recovery and long-term mitigation. During the CyberAv3ngers campaign, many utilities only learned they were compromised when external agencies or researchers notified them, underscoring uneven cybersecurity capacity between large and small operators.

Public and regulatory reactions ranged from national concern and new CISA guidance (Oldsmar) to international advisories and vendor mitigations (CyberAv3ngers), regional risk assessments (Mayo), and regulatory scrutiny over transparency and continuity planning (American Water). Public communication was consistently important in shaping perceptions and reinforcing the need for coordination and resilience.

Comparing organizational scale makes clear that smaller utilities like Oldsmar and County Mayo often operate with minimal cybersecurity staffing, limited budgets, and aging infrastructure, while larger entities such as American Water possess stronger governance but remain exposed through complex enterprise IT environments. The CyberAv3ngers campaign further showed that even small, remote devices can become targets when geopolitics enters the threat landscape. No utility is too small to be attacked, and no organization is too large to be immune.

In summary, the comparative analysis yields five key insights: (1) attackers predominantly exploit basic misconfigurations and weak credentials; (2) motivations range from political signaling and disruption to financial extortion; (3) detection and containment are most effective where automated monitoring and trained staff coexist; (4) resilience depends as much on planning, training, and collaboration as on technology; and (5) cybersecurity must be treated as a central pillar of operational continuity and public safety. Together, these cases underscore the urgent need for sector-wide investment, standardization, and targeted support to close persistent security gaps and prepare for future threats (Table 1).

Table 1 Comparison of case studies

Aspect	Oldsmar (USA, 2021)	CyberAv3ngers (Global, 2023–2025)	County Mayo (Ireland, 2023)	American Water (USA, 2024)	Norwegian Dam (Norway, Apr 2025)
Attack type	Attempted SCADA manipulation	Coordinated ideological OT attacks	Telemetry data disruption	Enterprise ransomware attack	OT manipulation of dam gate/discharge
Primary target	Chemical dosing system (NaOH)	Unitronics PLCs (SCADA/RTUs)	Remote reservoir telemetry system	Customer billing & internal IT systems	Hydropower gate/valve control (SCADA/PLC)
Entry vector	Unsecured remote desktop (TeamViewer)	Publicly exposed devices with default credentials	Weak remote access/ password management	Phishing email; privileged account compromise	Remote access compromise
Technical complexity	Low—no malware used	Moderate to High—custom malware (IOControl)	Low—no malware or automation	Moderate—typical ransomware toolkit	Low–Moderate—direct actuator commands; psyop signaling
Impact on operations	High potential; neutralized manually	Minimal physical; moderate psychological/ political	Temporary service outage for ~ 160 homes	IT disruption; OT unaffected due to segmentation	Unauthorized discharge; no downstream damage; high psychological impact
Detection method	Manual operator observation	External researcher/agency discovery	Local operator & vendor troubleshoot	SOC/IT system monitoring	Opertor/SCADA anomaly noticed; forensic review
Response capability	Immediate manual intervention	Mixed; many unaware until notified	Reactive; relied on external support	Structured incident response & containment	Operator intervention + national coordination

(continued)

Table 1 (continued)

Aspect	Oldsmar (USA, 2021)	CyberAv3ngers (Global, 2023–2025)	County Mayo (Ireland, 2023)	American Water (USA, 2024)	Norwegian Dam (Norway, Apr 2025)
Public/regulatory response	High; prompted federal alerts	International coordination and advisories	Moderate; national attention	High; regulatory and public scrutiny	High; national statements and NATO-area scrutiny
Organizational scale	Small municipal utility	Small–medium utilities worldwide	Small rural utility	Large national utility	Single hydropower facility within national operator ecosystem
Key lessons	MFA, segmentation, and real-time human oversight are critical	Secure configurations, vendor hardening, and awareness	Basic cyber hygiene and external support are vital	IT–OT segmentation and business continuity planning are essential	Process-aware alarms; secure remote access; multi-agency response

5 Implications for Policy, Practice, and Research

The case studies presented, highlight the multifaceted and evolving nature of cybersecurity risk in the water sector. These events offer important implications across three domains: policy formulation, operational practice, and academic and applied research. Together, they underscore the need for a coordinated, evidence-based, and forward-looking approach to strengthening cyber resilience in critical water infrastructure.

5.1 *Policy Implications*

From a policy perspective, the case studies reinforce the urgent need to formalize cybersecurity expectations for water utilities at both national and international levels. Regulatory gaps are particularly visible among small and medium-sized utilities, where cyber maturity is often low, and compliance with voluntary frameworks is inconsistent or absent. The Oldsmar and County Mayo cases illustrate that utilities of any size can be targeted, while the CyberAv3ngers campaign shows that geopolitically motivated threat actors are willing to exploit even low-tier devices in globally distributed attacks.

As a result, governments must adopt a minimum baseline of enforceable cybersecurity standards for all water utilities—regardless of scale. This includes mandatory implementation of basic controls such as network segmentation, multi-factor authentication, secure remote access protocols, and regular vulnerability assessments. Funding mechanisms and technical assistance programs must also be established to support under-resourced utilities in achieving compliance. At the international level, cross-border cooperation and threat intelligence sharing should be strengthened through platforms such as the EU Agency for Cybersecurity (ENISA), the U.S. CISA, and sector-specific Information Sharing and Analysis Centers (ISACs).

Furthermore, policy frameworks should incentivize proactive incident disclosure, without fear of reputational damage or punitive measures, to support a more transparent and learning-oriented security culture across the sector. The American Water case showed that public-facing disruptions can affect trust and regulatory scrutiny, highlighting the importance of pre-established communication protocols aligned with national incident reporting legislation.

5.2 *Practice Implications*

At the operational level, the case studies highlight the importance of both technological defenses and organizational readiness. Despite differences in context, a recurring pattern is the exploitation of preventable misconfigurations, such as open remote desktop services (Oldsmar), default PLC credentials (CyberAv3ngers), and weak access controls (County Mayo). This indicates that many incidents could have been prevented or mitigated by implementing basic cyber hygiene practices.

Utilities must integrate cybersecurity into day-to-day operations, not as a parallel IT function, but as an intrinsic component of operational safety and reliability. This includes:

- Developing and exercising incident response plans that span both IT and OT domains.
- Conducting regular security audits of SCADA, telemetry, and remote-control interfaces.
- Establishing asset inventories to track exposed devices and manage patch cycles.
- Enabling centralized monitoring and alerting, even in resource-constrained environments.
- Implementing role-based access control and deprivileging high-risk accounts.

Moreover, workforce development is essential. Operators and technicians must receive training to recognize cybersecurity red flags, such as unauthorized system access or anomalous process behavior, as shown in Oldsmar. In small utilities, the ability to collaborate with third-party responders (e.g., vendors, national CERTs, or regional cybersecurity units) must be institutionalized and practiced regularly.

5.3 Research Implications

From a research standpoint, the case studies identify significant gaps in tools, methodologies, and data availability for advancing cyber-physical resilience in the water sector. Most notably, there is a need for integrated modeling environments that can simulate the interaction between cyberattacks and physical water systems, particularly for studying the consequences of attacks on pump control, chemical dosing, or telemetry-driven automation.

Future research should also be explored:

- Development of lightweight anomaly detection tools tailored for legacy OT environments.
- Interdisciplinary frameworks combining hydraulic engineering, systems control, and cybersecurity risk analysis
- Methods for prioritizing mitigation investment in small and medium utilities based on risk impact modeling.
- Enhanced techniques for automated device discovery and exposure assessment, using secure and non-disruptive scanning.
- Evaluation of socio-technical response dynamics, including human error, organizational culture, and communication effectiveness during cyber events.

Additionally, there is a clear need for shared datasets and incident repositories to support empirical research. As most utilities are hesitant to disclose cyber incidents publicly, anonymized case databases—potentially hosted by regulators or research consortia—could provide a critical foundation for learning and innovation.

6 Conclusions

The case studies examined—from Oldsmar and CyberAv3ngers to County Mayo, American Water, and the Norwegian dam—show that cyber threats to the water sector are concrete, diverse, and increasingly entangled with geopolitical dynamics. Attacks range from simple misconfigurations and default credentials to coordinated campaigns targeting OT devices at scale.

Many successful intrusions exploit well-known weaknesses: insecure remote access, default or shared credentials, flat networks, and insufficient monitoring. Even when physical damage is avoided, as in Mayo, American Water, or Norway, the resulting service disruptions, manual workarounds, reputational impacts, and regulatory responses confirm that consequences extend far beyond technical remediation.

The analysis highlights a structural imbalance in preparedness: large utilities benefit from more mature governance and architectures yet remain susceptible to IT-focused attacks and social engineering, while smaller utilities often lack the resources and expertise to implement even basic protections. This uneven capacity creates systemic fragility, enabling both opportunistic and state-linked actors to exploit "too small to matter" assets and major providers alike.

Three overarching imperatives emerge. First, cybersecurity must be treated as core operational risk management, on par with water quality and asset integrity. This requires embedding controls such as multi-factor authentication, secure remote access, segmentation, and logging into everyday practice. Second, regulatory frameworks, funding, and technical assistance must be aligned with the realities of small and medium utilities through enforceable baselines, shared services, and structured support. Third, sustained research and innovation are needed in cyber–physical modeling, lightweight monitoring for legacy OT, and the socio-technical aspects of incident detection and response.

As digitalization deepens and geopolitical tensions persist, water utilities can no longer assume they are too mundane or remote to be targeted. Lessons from Oldsmar, CyberAv3ngers, County Mayo, American Water, and the Norwegian dam must translate into concrete action: stronger baselines, better-prepared organizations, coherent policy, and a culture that learns from near-misses as well as crises. Only through such coordinated, multi-level efforts can the sector ensure that digital technologies enhance security, resilience, and public trust rather than undermine them.

References

1. Laubshtein, Y.: Protecting water and sanitation infrastructure from cyberthreats: a cybersecurity study for latin America and the Caribbean. Nowersztern, A., Basani, M., Melean, F., Cohen Mizrav, H. (Eds.). (2023). https://doi.org/10.18235/0004876
2. European Union Agency for Cybersecurity.: 2023 consolidated annual activity report. Publ. Off. Eur. Union. (2024). https://data.europa.eu/doi/https://doi.org/10.2824/44117.
3. Hassanzadeh, A., Rasekh, A., Galelli, S., Aghashahi, M., Taormina, R., Ostfeld, A., Banks, M.K.: A review of cybersecurity incidents in the water sector. J. Environ. Eng. **146**(5), 03120003 (2020)
4. U.S. GAO.: Critical infrastructure protection: EPA urgently needs a strategy to address cybersecurity risks to water and wastewater systems. (2024). https://www.gao.gov/products/gao-24-106744
5. Stouffer, K., et al.: NIST SP 800–82r3 guide to operational technology (OT) Security. NIST Spec. Publ. (2023). https://doi.org/10.6028/NIST.SP.800-82r3
6. McCarthy et al.: Cybersecurity for the water and wastewater sector—a practical reference design for mitigating cyber risk in water and wastewater systems, NIST project description. (2023). https://www.nccoe.nist.gov/sites/default/files/2023-06/securing-water-and-wastewater-utilities-project-description-final.pdf.
7. Dragos Inc.: OT Cybersecurity Year in Review. (2024). https://hub.dragos.com/ot-cybersecurity-yir-2023
8. Boyer, S.: SCADA: Supervisory Control and Data Acquisition, 4th ed., ISA (2016)
9. Barrett, B.: The threat to the water supply is real—and only getting worse, wired. (2021). https://perma.cc/XAY2-LEAW.
10. U.S. Environmental Protection Agency. (2023). Water Sector Cybersecurity Program Case Study: Medium Drinking Water System, EPA 817-F-23–004
11. European Union Agency for Cybersecurity. (2023). ENISA threat landscape report (January 2021 to March 2023). European Union Agency for Cybersecurity
12. Knapp, E.D., Langill, J.T.: Industrial Network Security: Securing Critical Infrastructure Networks for Smart Grid, SCADA, and Other Industrial Control Systems, 2nd edn. Syngress, Waltham, MA, USA (2015)
13. Tuptuk, N., Hazell, P., Watson, J., Hailes, S.: A systematic review of the state of cyber-security in water systems. Water **13**, 81 (2021)
14. CISA (2018). Cybersecurity Best Practices for Industrial Control Systems, Cybersecurity and Infrastructure Security Agency, U.S. Department of Homeland Security, Washington, DC, USA. https://www.cisa.gov/sites/default/files/publications/Cybersecurity_Best_Practices_for_Industrial_Control_Systems.pdf
15. Cybersecurity of the municipal water sector: background and issues for congress. (2025). https://www.congress.gov/crs-product/R48556
16. CISA, EPA, and FBI (2024). Top cyber actions for securing water systems, fact sheet. https://www.cisa.gov/sites/default/files/2024-03/fact-sheet-top-cyber-actions-for-securing-water-systems.pdf.
17. European Commission, Directive (EU) 2022/2555 of the European Parliament and of the Council, Official Journal of the European Union, L 333/80, (2022)
18. Greenberg, A.: A hacker tried to poison a Florida city's water supply, officials say. (2021). Wired.com. https://www.wired.com/story/oldsmar-florida-water-utility-hack/
19. U.S. EPA (2023). Evaluating cybersecurity during public water system sanitary surveys
20. Joint Cybersecurity Advisory (2021). Ongoing Cyber Threats to U.S. Water and Wastewater Systems.
21. U.S. Congress (2021). Infrastructure Investment and Jobs Act (IIJA), H.R.3684
22. Greenberg, A.: CyberAv3ngers: The Iranian Saboteurs Hacking Water and Gas Systems Worldwide, Wired Magazine. (2024). https://www.wired.com/story/cyberav3ngers-iran-hacking-water-and-gas-industrial-systems/

23. Israel National Cyber Directorate (2025). 2024 Cybercrime Activity Report
24. Martin, A.: Two-day water outage in remote irish region caused by pro-iran hackers, the record. (2023). https://therecord.media/water-outage-in-ireland-county-mayo
25. Christian, B.: Securing remote monitoring systems in water distribution networks: key insights and strategies, WaterOnline. (2024). https://www.wateronline.com/doc/securing-remote-monitoring-systems-in-water-distribution-networks-key-insights-and-strategies-0001
26. Trevor, Q.: Anti-Israeli hackers leave 180 Mayo homes without water in cyber-attack, BreakingNews.ie. (2023). https://www.breakingnews.ie/ireland/anti-israeli-hackers-leave-180-mayo-homes-without-water-in-cyberattack-1561775.html
27. Shipkowski, B.: American water, the largest water utility in US, is targeted by a cyberattack, US News. (2024). https://apnews.com/article/american-water-cyberattack-36423062dbce05c9aa70ef8aa07810cb
28. Reuters.: Water utility American Water disconnects computers following 'cybersecurity incident'. (2024). https://www.reuters.com/technology/cybersecurity/water-utility-american-water-disconnects-computers-following-cybersecurity-2024-10-08/
29. American Water Works Association (2017). Process Control System Security Guidance for the Water Sector, Version 2.0, Denver.
30. McLaughlin, M.G., Sanger, K., Adigwe, A.R., Yeung, T., Eapen, D.K.: Navigating troubled waters: cybersecurity threats to the water industry, buchanan ingersoll & rooney PC (2024)
31. Nerijus, A.: Norway spy chief blames Russian hackers for dam cyber sabotage in April, Reuters. (2025). https://www.reuters.com/technology/norway-spy-chief-blames-russian-hackers-dam-sabotage-april-2025-08-13/
32. Burrows, E.: Norwegian police say pro-Russian hackers were likely behind suspected sabotage at a dam, AP News. (2025). https://apnews.com/article/russia-norway-dam-sabotage-cyberattack-16673f35c17aacf5ed871918136bdf6f
33. Lemos, R.: Water systems under attack: Norway, Poland Blame Russia Actors, Dark Reading. (2025). https://www.darkreading.com/ics-ot-security/water-systems-attack-norway-poland-russia-actors
34. Riotta, C.: Russian hackers accused in wave of water sector cyberattacks, bankinfosecurity/information security media group. (2025). https://www.govinfosecurity.com/russian-hackers-accused-in-wave-water-sector-cyberattacks-a-29264

AI-Generated “Fake News” as an Instrument of Crisis Induction: A New Typology

Marko Selaković, Marijana Krkić, and Nenad Filipović

Abstract The development of ICT has created an information environment in which vast amounts of data flow through digital space, algorithms shape exposure to content, and an increasing share of material is generated by artificial intelligence, raising new risks for critical infrastructure, public safety, and democratic processes. Riding this wave, the gatekeeping practice of mainstream media imperceptibly shifts toward gatewatching, at times contributing to the spread of crises. Technically, the recognition and detection of AI-generated “fake news” has become a hot topic, with startups and major technology companies investing heavily in solutions. This paper examines AI-generated “fake news”, used here as a shorthand for synthetic disinformation and related deceptive content, that function as inductors of crises. Drawing on the BANI (Brittle, Anxious, Nonlinear, Incomprehensible) perspective, the study analyzes a corpus of 217 research articles on AI-generated disinformation and semi-structured interviews with state officials and decision-makers to (i) identify crisis-inducing types of AI-generated “fake news”, (ii) compare them with an established fake-news typology, and (iii) propose an updated framework tailored to AI-generated content and crisis contexts. Findings indicate six crisis-inducing types: (1) deepfakes; (2) false statements or notorious misinformation; (3) content with a fictitious source; (4) favorable reports or statements; (5) conspiracy theories; and (6) satire without clear labeling. The paper outlines implications for detection pipelines, critical-infrastructure protection, and journalistic practice, and suggests directions for future research on linking AI-generated “fake news” to specific crisis categories, escalation dynamics, and early-warning and response playbooks.

M. Selaković (✉)
S P Jain School of Global Management, Dubai International Academic City, Dubai, UAE
e-mail: marko.selakovic@spjain.org

M. Krkić
Faculty of Political Sciences, University of Belgrade, Belgrade, Serbia

N. Filipović
Faculty of Engineering, University of Kragujevac, Kragujevac, Serbia
e-mail: fica@kg.ac.rs

Z. Keković et al. (eds.), *Critical Infrastructure: Risk and Crisis Management in the New Reality*, Advanced Sciences and Technologies for Security Applications,
https://doi.org/10.1007/978-3-032-20948-1_12

Keywords "Fake News" · Disinformation · Artificial intelligence · Critical infrastructure

1 Introduction

The digital information ecosystem of the last decade is marked by brittleness, anxiety, nonlinearity, and incomprehensibility, elements consolidated in the BANI framework (Brittle, Anxious, Nonlinear, Incomprehensible), proposed as a complement and correction to VUCA to capture the dynamics of contemporary chaos and unreliable feedback loops [1]. Concurrently, the role of traditional editorial "gatekeepers" in mass media has evolved toward gatewatching, collective monitoring and curation of news flows from networked sources, often with insufficient verification, which under certain conditions may escalate crisis narratives [2, 3].

Within this context, AI-generated "fake news", from textual narratives to audiovisual deepfakes, introduces novel risk vectors for critical infrastructure, public safety, electoral processes, and corporate reputation. Reviews and surveys from recent years suggest that leaders and the public increasingly perceive AI-enabled disinformation as a salient short-term global risk and a significant driver of trust erosion. Building on these concerns and the accompanying presentation that motivated this study, the paper (i) operationalizes a new typology of AI-generated "fake news" that function as inductors of crises; (ii) compares it to the Selaković et al. [4] typology; and (iii) discusses implications for risk management, infrastructure security, and media practice.

2 Historical and Conceptual Foundations

2.1 "Fake News" is not New

"Fake news" is not a recent phenomenon [5]. In theory, its emergence traces back to the earliest writing systems [6]. A classic modern illustration is the 1938 radio adaptation of H. G. Wells's The War of the Worlds, whose dramatic format reportedly frightened around a million listeners in the United States [7, 8]. Although the magnitude of the panic caused has been debated, the episode remains a symbol of how media form and context shape public perception. As emphasized by Goldman and Lopez [8], "fake news" has real-world consequences, which is precisely why it warrants rigorous study; theorizing should therefore begin by clarifying the concept.

2.2 The Definitional Debate

Despite the term's longevity, scholars have not yet agreed on a single, broadly accepted definition. Numerous literature sources describe "fake news" as news-like information in newspapers, broadcast media, or social platforms that is intentionally based on falsehoods or uses deceptive framing.

Institutional guidelines follow this narrative. For example, University of Michigan describes "fake news" as material that is not accurate, stories that are invented, not grounded in facts and sources, often deployed as propaganda to steer audiences toward incorrect conclusions or produced to chase clicks on web portals [9]. University of Oregon offers a deliberately narrow definition: information that is clearly and demonstrably fabricated, packaged and distributed to appear as legitimate news, and intended to deceive consumers into believing it is real [10]. This narrower view helps distinguish "fake news" from other misleading forms (satire, poor journalism, or opinion).

Certain authors frame "fake news" as false but verifiable items, fabrications built from real-world scaffolding. They are created intentionally [11]. The aim of "fake news" is to evoke emotion, with the intent to deceive readers and persuade rather than inform [5, 8, 12]. In the media-studies canon, a widely cited effort by Allcott and Gentzkow [13] defines "fake news" as media texts that are intentionally or verifiably false and likely to mislead readers; in discussing the term's usage, they note it has also been applied, sometimes over-broadly, to disinformation, manipulation, journalistic errors, and even satire or parody [14].

2.3 Canonical Taxonomies: Six and Seven-Part Models

Allcott and Gentzkow's [13] classification includes six practical forms frequently discussed in teaching and policy contexts: (1) unintentional errors; (2) rumors; (3) conspiracy theories; (4) satire; (5) false statements; (6) reports favoring a particular option. In parallel, Wardle's [15] influential First Draft scheme distinguishes seven categories: (1) satire/parody; (2) misleading content; (3) fabricated content; (4) content with an invented source; (5) false content; (6) false context/miscontextualized content; (7) manipulated content [15, 16]. These models, though framed for the broader mis/disinformation landscape, are highly compatible with the AI-specific extensions developed in this paper.

2.4 Truth Decay and Platform Dynamics

Dynamic changes in the information environment, 24-h news cycles and scale-optimized social platforms, amplify opinion, accelerate mass dissemination, and blur boundaries between evidence and assertion [17]. These dynamics magnify the reach and speed of deceptive content, including AI-generated variants.

2.5 Digitization, Clickbait, and Incentive Structures

The digitization of media has substantially changed newsroom routines and incentives. With digitization, the world entered the platform era that "thousands of sites serve merely as bulletin boards for someone's opinion, without genuine interest in accuracy or verification" [18]. This phenomenon has been magnified with the rise of clickbait: web content engineered to maximize clicks via sensational or affective cues, frequently yielding misbeliefs or misinterpretations [19]. This pattern is also observed by Beisecker and Schlereth [20] in their consumer-perception work. Those authors further argue that portals often resort to sensationalism and emotionally charged headlines to attract attention, distinguishing between wholly fabricated statements and rhetorical questions that function as suggestive claims, which may or may not be false. Complementing this, large-scale diffusion research shows that false content can travel further and faster than true content [21], while simple accuracy prompts can improve trustwothiness and reduce sharing of false headlines [22].

2.6 Politicization and Variation in Usage

Usage of the term "fake news" varies with the intent and context. Surčulija Milojević and Kulić [14] document how political actors and institutions deploy the label strategically. For instance, during the Trump era, critical outlets (notably CNN) were sometimes branded "fake news" or even "enemies of the people" [23]. In contrast, ahead of the 2018 European Parliament elections, the European Commission signaled Russia as a principal disseminator of "fake news" in Europe and pledged to "wage war" on such content [24]. These cases illustrate how normative judgments about misinformation are embedded in political struggle, which is another reason to anchor analysis in clear operational definitions.

To minimize definitional drift and politicization, this paper adopts consistent usage of misinformation—false content without intent, disinformation—false content with deceptive intent, and malinformation—true content used out of context to cause harm, and treats "fake news" as a non-technical umbrella label [25]. Thus, nowadays the term "fake news" has dual meaning: it can be a genre of disinformation, or a label for political content [5]. Given its prominence in public and policy discourse, the term "fake news" is retained in the article's title, abstract, and in references to existing typologies as a descriptive shorthand that reflects common usage and facilitates discoverability. In our own conceptual and empirical analysis, however, we rely primarily on the more precise notions of mis-, dis-, and malinformation, and use "fake news" explicitly in quotation marks to signal its status as a quoted, non-technical label.

3 Theoretical Background: BANI, Gatewatching, and Prior Typologies

A useful lens for understanding contemporary informational disruptions is the BANI perspective, which characterizes the environment as brittle, anxious, non-linear, and incomprehensible. In such conditions, marked by information overload and algorithmic intermediation, systems become prone to amplified cascades and unpredictable outcomes, making the rapid, nonlinear diffusion of disinformation especially consequential for infrastructures and markets [1]. This lens aligns with the observed erosion of traditional gatekeeping and the rise of "gatewatching," where hierarchical editorial control gives way to collaborative curation. Social platforms intensify this shift by enabling low-friction circulation and optimizing for engagement, thereby increasing exposure to AI-synthetic content that is fast to produce, highly scalable, and often plausibly realistic [2, 3].

Within this BANI-informed, gatewatched ecosystem, existing typologies provide a baseline for comparison and extension. A prominent line of work proposes matrices organized around degree of factuality, deceptive intention, and financial motivation, offering clear axes along which to position evolving subtypes of synthetic content [4, 11]. These baselines help isolate the crisis-inducing potential of AI-specific variants, such as high-plausibility, low-cost manipulations that synchronize public behavior or undermine institutional trust, while keeping analytical continuity with earlier frameworks. Together, the BANI lens, the gatewatching shift, and baseline typologies form an integrated scaffold for diagnosing where and how synthetic disinformation is most likely to propagate and destabilize critical social and technical systems.

4 Methodology

4.1 Research Design

This study follows a two-stage qualitative design: (1) a systematic review of 217 scholarly works relevant to AI-generated "fake news" with extraction of types and crisis effects; and (2) semi-structured interviews with state officials and decision-makers regarding geopolitical uses of such content and infrastructure vulnerabilities. The resulting list of types is benchmarked against the Selaković et al. [4] typology.

Procedure and analysis:

- Corpus: 217 academic articles and technical reports (2017–2025).
- Coding: open and axial coding along criteria of modality (text/audio/video), source/attribution, intention, financial motivation, target (public/corporations/infrastructure), and escalation potential.

- Validation: triangulation with interview insights; member checking on a subsample (energy-sector official, public-administration representative, corporate risk manager).
- Comparison: mapping to the three axes (factuality, intention, financial motivation) per Selaković et al. [4] classification.

4.2 Results: A New Typology of AI-Generated "Fake News" that Induce Crises

Based on the corpus review and interviews, six AI-relevant types are confirmed, with operational definitions and typical crisis effects:

1. Deepfakes

Definition: realistically generated audio/video that attributes statements or actions to individuals who did not make or perform them.

Crisis effect: destabilization of political processes and erosion of trust in institutions and crisis messaging; exposure to infrastructure-related deepfake narratives is linked with increased distrust toward government.

Early-warning indicators: lip-sync/blink anomalies, phonetic inconsistencies, metadata gaps; need for verification via independent sources.

2. False statements / notorious misinformation

Definition: content containing factually incorrect claims, generated with the intention of presenting falsehoods as facts.

Crisis effect: panic and confusion, especially in public-health, financial, and security contexts.

3. Content with a fictitious source

Definition: attribution of a news-like item to non-existent persons, institutions, or documents, or the misuse of an AI persona.

Crisis effect: undermining legitimacy of institutions and media; disrupting the crisis-communication source-verification chain.

4. Favorable reports or statements

Definition: AI-synthetic content that privileges one option/actor (e.g., in campaigns) while suppressing risks and counter-evidence.

Crisis effect: asymmetric informing of the public; potential for reputational crises and compromised decision integrity.

5. Conspiracy theories

Definition: narratives offering untrue or incomplete explanations of events or persons within a frame of a hidden agenda.

Crisis effect: polarization and delegitimization of institutions and media; a spiral of cynical skepticism.

6. Satire without clear labeling

Definition: news-like humorous content without unambiguous signals that it is satirical, leading to misinterpretation.

Crisis effect: public confusion, misguided responses in emergencies; possible legal ramifications.

Alignment with Selaković et al. [4] concept: The six types align with the three axes, factuality–intention–financial motivation, but explicitly introduce AI-specific modalities such as deepfake audio/video; synthetic persona/source) and foreground crisis-induction as a selection criterion. The categories are also interoperable with the six- and seven-part models previously discussed [3].

5 Discussion

5.1 *Diffusion Dynamics and the Speed–accuracy Dilemma*

AI tools can reduce the time from creation to distribution of persuasive content. Gatewatching environments without strong editorial intermediation heighten the risk of false-positive amplification. This shifts the burden from reactive corrections to proactive verification and pre-arranged protocols. In practice, the dilemma should be translated into decision variables with thresholds for detection and verification times, acceptable false-positive rates, and escalation triggers, as detailed before.

5.2 *Trust and Institutional Communication*

Emerging experimental and literature evidence indicates that exposure to realistic manipulations, including but not limited to deepfakes, can reduce trust in institutions, although effects vary across contexts. Cognitive reflection and related deliberative tendencies are consistently associated with more accurate veracity judgments and lower susceptibility to false claims [22]. Complementing this, syntheses and trials on media literacy and prebunking report small-to-moderate, reliable gains in resilience to misinformation without indiscriminately eroding trust in true content [26, 27]. For crisis-message design, the translation is straightforward: it is recommended to pair verifiable evidence, such as logs, third-party attestations, or a live dashboard, with reasoning aids that make the logic testable and include prebunk cues about manipulation strategies relevant to the incident, such as false context or invented sources.

5.3 Public Risk and Perceptions

At the system level, AI-enabled mis/disinformation is now ranked among the top near-term global risks. The World Economic Forum's Global Risks Report 2024 lists "misinformation and disinformation" as the top short-term risk (two-year horizon); the 2025 edition reiterates the threat to societal cohesion and governance in an election-dense period [28, 29]. Public-opinion work echoes this salience: polling summarized by the Alan Turing Institute in 2024 reports broad concern about deepfakes affecting election results, highlighting widespread perceived vulnerability [30].

These risk perceptions have accelerated efforts toward standardized verification. The C2PA standard, also known as Content Credentials, specifies interoperable provenance manifests that bind assets to signed claims and attestations, enabling tamper-evident source and edit history across images, audio, and video; adoption allows public institutions to publish evidence-rich rebuttals with machine-readable provenance, as documented by the Coalition for Content Provenance and Authenticity, 2023. In parallel, NIST's AI Risk Management Framework catalogs governance and technical controls relevant to synthetic-content risks [31]. For crisis communication, combining provenance signaling through C2PA manifests with transparent artifacts such as time-stamped logs and third-party attestations, alongside media-literacy programming, operationalizes a defense-in-depth approach aligned with both standards and behavioral evidence.

6 Practical Framework for Detection, Response, and Prevention

6.1 Detection

Effective detection starts from the premise that misinformation spreads across multiple channels and jumps platforms quickly; monitoring should therefore run as a continuous data stream across news APIs, social platforms, and closed or messaging channels, with semantic claim-tracking and multimodal analytics covering text, image, audio, and video [32–36].

Recommendations include the following: set alert thresholds empirically, not by feelings: define critical keywords associated with crisis triggers, estimate a baseline for each, and trigger an alert on a deviation. Weight the signal by source credibility, such as a verified outlet versus an anonymous account, and by modality, with video above image above text.

Once triggered, run a fixed verification pipeline:

1. Source triangulation: locate the first sighting, cross-check with two independent confirmations, record a chain of custody (who/when/what).

2. Metadata forensics: hash checks; EXIF/content credentials; timestamps; audio spectra; for video, keyframe analysis and manipulation signals (cuts, lip-sync, frame anomalies).
3. Technical checks: phonetic inconsistencies, reverse image search, archive matching (recycled footage).
4. Expert confirmation: internal subject-matter experts (grid/water/security) or partner labs; decision as likely true / likely false / indeterminate.

Staff this through a Trust & Safety war-room:

- Incident Lead/Strategist (prioritization, decisions)
- Media Desk (press outreach, statements, Q&A)
- Social Desk (platform operations, moderation, community)
- Tech/Forensics (metadata + multimodal analysis)
- Legal Liaison (publication/privacy risk)

Track metrics: Mean Time to Detect (MTTD), Mean Time to Verify (MTTV), false-positive rate, source coverage (% of priority channels monitored), and recall@K for critical phrases. Review thresholds and keyword sets weekly and after every incident.

6.2 Response

Response depends heavily on speed with offered evidence. The golden-hour rule means issuing an initial statement within 60 min that includes: (a) a clear status line, (b) a proof bundle, (c) a link to a live evidence source (if available), and (d) the timeline for the next update. If full verification isn't complete, publish a prebunk: flag the manipulation techniques in circulation and show how audiences can verify it.

Further, it is recommendable to use audience segmentation to avoid cross-talk. Partnerships can be helpful in borrowing credibility and velocity: pre-arranged fact-checking networks, university audio/video forensics labs, and platform contacts for labeling/rate-limiting have the potential to be helpful in setting timely, factually true, and accurate crisis response. After publication of the response, it is also important to measure impact: decay rate of the false narrative, share of voice, correction reach, sentiment shift, and unresolved inbound questions.

6.3 Prevention

Prevention includes technical, procedural, and educational measures. In critical sectors (energy, water, health), establishing fallback public-alert channels (web, SMS, radio) with two-way authentication is useful. Stakeholders can use such systems to confirm authenticity (signature/provenance), while authorities can securely receive submissions from the field.

With platforms, it is helpful to formalize rapid-response protocols for interim labeling during verification, distribution throttling, and set archival flags when old "fake news" resurface in a new context or timeline.

Finally, it is critically important to build media literacy and prebunking into routine practice: short, frequent modules for communities and schools focused on common AI patterns (deepfakes, invented sources, fake "official notices"). Keep the modules active (reverse image search, basic video keyframe checks, domain/author verification, reading content credentials), and equip professionals with a prompt card: Who asserts this? Where is the evidence? What would falsify it? Run trainings and workshops on regular basis to refine thresholds, scripts, and publication protocols.

It is extremely helpful to institutionalize a learning culture: every incident ends with an after-action review (what was detected, how fast, what worked/failed) and a process patch (updated thresholds, new FAQ items, additions to the proof pack). Prevention thus becomes continuous improvement, not a one-off initiative.

6.4 Composite Classification for Case Analysis

To operationalize a case-based content analysis, this paper proposes a composite scheme that synthesizes elements from Allcott & Gentzkow [13] and Wardle [37] and aligns them with the AI-focused crisis typology above. The categories are defined as follows:

1. Unintentional errors, mistakes (e.g., typos, misreadings) that alter context or meaning without intent to mislead.
2. False statements or information, factually incorrect claims created with intent to present a falsehood as fact.
3. Content with an invented source, reliance on claims attributed to fictitious persons or nonexistent documents.
4. Reports or statements favoring a particular option, selective framing that privileges one side while omitting salient counterevidence.
5. Manipulative content, uses mostly accurate data or claims but frames and interprets them in deceptive ways to yield false conclusions.
6. Rumors, (partial) information that circulates especially in crises, with uncertain provenance and low verifiability.
7. Conspiracy theories, untrue or incomplete accounts presented as part of a hidden plan.
8. Satire, fictional "news" meant as satire without clear labeling, making misinterpretation likely.

This grouping supports transparent classification and traceability from raw claims to categories, while remaining compatible with the AI-specific lens (identifying when any category is amplified, authored, or augmented by AI).

7 Limitations and Future Research

This study's insights are bounded by its qualitative design and specific institutional settings, which limits generalizability. Cross-linguistic and cultural variation further complicates detection and comparison, and current pipelines underperform in smaller languages and locally specific contexts, increasing false negatives and misclassification. Future work should therefore combine quasi-experimental and longitudinal designs to test causal mechanisms and track dynamics over time, and prioritize domain-specific, multilingual datasets with evaluation benchmarks tailored to local culture, context, discourse, and slang. A structured bridge to practice is also needed: map AI-generated "fake-news" subtypes onto concrete crisis categories (energy, financial, reputational, public-health), estimate escalation thresholds for each, and validate them in realistic simulations and field settings. Building on this baseline, a broader program can strengthen external validity, bolster performance where detection is weakest, and translate conceptual advances into operational early-warning and response playbooks.

8 Conclusion

Within a BANI environment, AI-generated content constitutes a systemic threat: it can trigger crisis events, compromise infrastructure governance, and undermine democratic processes. AI-generated "fake news", used here as shorthand for synthetic disinformation and related deceptive content, amplifies existing vulnerabilities in information systems by lowering production costs, increasing plausibility, and accelerating diffusion across gatewatched, platform-driven ecosystems. In such conditions, treating synthetic content merely as a communication nuisance underestimates its potential to destabilize critical social and technical systems.

The typology proposed in this paper, deepfakes, notorious falsehoods, fictitious sources, favorable statements, conspiracy theories, and unlabeled satire, builds on existing work but explicitly foregrounds AI modalities and crisis-induction capability as selection criteria. By aligning these six types with the factuality-intention-financial motivation axes of Selaković et al. [4] and integrating them with canonical six- and seven-part taxonomies, the paper offers an interoperable bridge between the broader mis/disinformation literature and AI-specific risk analysis. The composite eight-category scheme for case analysis further supports transparent classification, traceability from raw claims to categories, and practical deployment in real-world monitoring and investigation pipelines.

Beyond classification, the study advances a pragmatic framework for detection, response, and prevention. Continuous multimodal monitoring, empirically calibrated alert thresholds, and structured verification pipelines shift organizations from ad hoc reactions to repeatable, auditable crisis routines. When combined with provenance signaling (content credentials), evidence-rich rebuttals, and targeted prebunking and media-literacy interventions, these measures form a defense-in-depth approach that is compatible with emerging technical standards and behavioral evidence.

Ultimately, the findings underscore that AI-generated "fake news" is not only a technical challenge but a governance problem. Effective mitigation requires coordinated action across infrastructure operators, public institutions, media organizations, platforms, and civil society. By linking AI-enabled disinformation subtypes to concrete crisis scenarios and offering operational guidance for early-warning and response playbooks, this paper aims to support that coordination. If taken seriously and expanded through sector-specific simulations and longitudinal evaluation, the proposed framework can help reduce escalation risks, strengthen institutional resilience, and protect both critical infrastructure and the democratic public sphere.

References

1. Cascio, J.: Facing the age of chaos (BANI framework). Open the Future. (2020)
2. Bruns, A.: Gatewatching: Collaborative online news production. Peter Lang. (ISBN 9780820474328). (2005)
3. Bruns, A.: Gatewatching and news curation: Journalism, social media, and the public sphere. Peter Lang (2018). https://doi.org/10.3726/B13293
4. Selaković, M.: Typology of business-related fake news online: A literature review. GATR J. Manag. Mark. Rev., **5**(4), 234–243. (2020). https://doi.org/10.35609/jmmr.2020.5.4(5)
5. Selaković, M., Tarabasz, A., Ljepava, N.: Evolucija dezinformacija: Analiza digitalne transformacije lažnih vijesti. Medijske Stud., *15*(30), 114–131 (2024). https://doi.org/10.20901/ms.15.30.6
6. Marcus, J.: Mesoamerican writing systems: Propaganda, myth, and history in four ancient civilizations. Princeton University Press (1992)
7. Cantril, H.: The invasion from Mars: A study in the psychology of panic. Trans. Publ. (Original work published 1940) (2005). https://doi.org/10.4324/9781315132730
8. Tandoc, E.C., Jr., Lim, Z.W., Ling, R.: Defining "fake news": A typology of scholarly definitions. Digit. J. **6**(2), 137–153 (2018). https://doi.org/10.1080/21670811.2017.1360143
9. University of Michigan Library. (n.d.). Fake news, fact checking, and misinformation. https://guides.lib.umich.edu/fakenews
10. University of Oregon Libraries. (n.d.). Evaluating news sources: What is fake news? https://researchguides.uoregon.edu/fake-news
11. Selaković, M.: Countering business-related fake news online: An examination of the application of situational crisis communication theory (Doctoral dissertation). ProQuest Dissertations & Theses Global. (2021). https://www.researchgate.net/publication/361743020_Countering_Business_Relaed_Fake_News_Online_An_Examination_of_the_Application_of_Situational_Crisis_Communication_Theory
12. Belloir, N., Ouerdane, W., Pastor, O.: Characterizing fake news: A conceptual modeling-based approach. In Ralyté, J., Chakravarthy, S., Mohania, M., Jeusfeld, M.A., Karlapalem, K. (Eds.), Concept. Model. (ER 2022), Lect. Notes Comput. Sci. **13607**, pp. 115–129 (2022). Springer. https://doi.org/10.1007/978-3-031-17995-2_9
13. Allcott, H., Gentzkow, M.: Social media and fake news in the 2016 election. J. Econ. Perspect. **31**(2), 211–236 (2017). https://doi.org/10.1257/jep.31.2.211

14. Surčulija Milojević, J., Kulić, M.: Lažne vesti" ili dezinformacije: (Samo)regulatorni okvir i izazovi u praksi. Fakultet političkih nauka, Univerzitet u Beogradu. (2020). https://rfpn.fpn.bg.ac.rs/bitstream/handle/123456789/1007/fulltext.pdf?isAllowed=y&sequence=1
15. Wardle, C.: The seven types of mis- and disinformation. First Draft (2017)
16. Kulić, M.: Savremeno novinarstvo: Od ideologije profesije do novinarskih žanrova. Univerzitet u Beogradu, Fakultet političkih nauka (2022)
17. Kavanagh, J., Rich, M.D.: Truth decay: An initial exploration of the diminishing role of facts and analysis in American public life. RAND Corporation (2018). https://doi.org/10.7249/RR2314
18. Craig, R.: Online journalism: Reporting, writing, and editing for new media (2nd ed.). Cengage Learning (2010)
19. Rubin, V.L.: Disinformation and misinformation triangle: A conceptual model for "fake news" epidemic, causal factors and interventions. J. Doc. **75**(5), 1013–1034 (2019). https://doi.org/10.1108/JD-12-2018-0209
20. Beisecker, S., Schlereth, C., Hein, S.: Shades of fake news: How fallacies influence consumers' perception. Eur. J. Inf. Syst. **33**(1), 41–60 (2024). https://doi.org/10.1080/0960085X.2022.2110000
21. Vosoughi, S., Roy, D., Aral, S.: The spread of true and false news online. Sci. **359**(6380), 1146–1151 (2018). https://doi.org/10.1126/science.aap9559
22. Pennycook, G., Rand, D.G.: Lazy, not biased: Susceptibility to partisan fake news is better explained by lack of reasoning than by motivated reasoning. Cognition **188**, 39–50 (2019). https://doi.org/10.1016/j.cognition.2018.06.011
23. CNN. (2018). Coverage of "enemy of the people" rhetoric and "fake news" labeling. https://transcripts.cnn.com/show/rs/date/2018-08-19/segment/01
24. The Guardian. https://www.theguardian.com/technology/2018/oct/23/risk-of-interference-in-mep-elections-growing-eu-commission-facebook-scandal-monitoring
25. Lewandowsky, S., Ecker, U.K.H., Cook, J.: Beyond misinformation: Understanding and coping with the "post-truth" era. J. Appl. Res. Mem. Cogn. **6**(4), 353–369 (2017). https://doi.org/10.1016/j.jarmac.2017.07.008
26. Roozenbeek, J., van der Linden, S.: Fake news game confers psychological resistance against online misinformation. Palgrave Commun. **5**, 65 (2019). https://doi.org/10.1057/s41599-019-0279-9
27. Van der Linden, S., Roozenbeek, J., Compton, J.: Inoculating against fake news about COVID-19. Harv. Kennedy Sch. Misinformation Rev., **1**(2) (2020). https://doi.org/10.37016/mr-2020-015
28. World Economic Forum. (2024). Global risks report 2024. https://www.weforum.org/reports/global-risks-report-2024/
29. World Economic Forum. (2025). Global risks report 2025. https://www.weforum.org/reports/global-risks-report-2025/
30. The Alan Turing Institute. (2024). 9 in 10 concerned about deepfakes affecting election results. https://www.turing.ac.uk/news/9-10-concerned-about-deepfakes-affecting-election-results
31. National Institute of Standards and Technology. (2023). AI risk management framework (AI RMF 1.0). https://doi.org/10.6028/NIST.AI.100-1
32. Starbird, K., Wilson, T.: Cross-platform disinformation campaigns: Lessons learned and next steps. HKS Misinformation Rev., **1**(1). (2020). https://doi.org/10.37016/mr-2020-002
33. Resende, G., Messias, J., Melo, P., Vasconcelos, M., Benevenuto, F., Sousa, H., Almeida, J.M.: (Mis)information dissemination in WhatsApp: Gathering, analyzing, and countermeasures. Companion Proceedings of the 2019 World Wide Web Conference (WWW '19). (2019). https://doi.org/10.1145/3308558.3313688
34. Alam, F., Cresci, S., Chakraborty, T., Silvestri, F., Dimitrov, D., Da San Martino, G., Shaar, S., Firooz, H., Nakov, P.: A survey on multimodal disinformation detection. Proceedings of the 29th International Conference on Computational Linguistics (pp. 6625–6643). International Committee on Computational Linguistics. (2022). https://aclanthology.org/2022.coling-1.576/ ACL Anthology

35. Google Developers. (n.d.). Fact Check (ClaimReview) markup for Search. https://developers.google.com/search/docs/appearance/structured-data/factcheck Google for Developers
36. Larraz, I.: Semantic similarity models for automated fact-checking: ClaimCheck as a claim matching tool. Profesional de la Información, **32**(3), e320321. (2023). https://revista.profesionaldelainformacion.com/index.php/EPI/article/download/87284/63433.revista.profesionaldelainformacion.com
37. Wardle, C., Derakhshan, H.: Information disorder: Toward an interdisciplinary framework for research and policymaking. Counc. Eur. (2017). https://rm.coe.int/information-disorder-toward-an-interdisciplinary-framework-for-researc/168076277c

Using AI in the Evaluation and Monitoring of the Scientific Proposals

Nenad Filipovic

Abstract This comprehensive initiative of Science Fund of the Republic of Serbia (SFRS) aims to leverage artificial intelligence to fundamentally transform Serbia's research and innovation infrastructure by establishing an advanced AI-assisted evaluation and monitoring system designed to enhance the efficiency, accuracy, and transparency of scientific funding processes. Implementing AI-driven automation can streamline proposal reviews by rapidly analyzing submissions for completeness, relevance, and quality, thus allowing funding agencies to quickly identify and prioritize qualified candidates for grants and support. Predictive analytics, utilizing extensive historical data on funded projects and research outcomes, can be employed to identify emerging trends, forecast the potential societal impact or return on investment of proposed projects, and optimize resource allocation by focusing on areas with the highest potential benefits. The system can also assist in strategically prioritizing research domains aligned with societal needs and national development goals, such as health, climate change mitigation, technological advancement, and sustainable development, ensuring that funding efforts address pressing issues effectively. In line with Horizon Europe's emphasis on excellence in research and societal impact, Serbia's Science Fund of the Republic of Serbia (SFRS) will support this transition by fostering a culture of responsible AI adoption, capacity building, and multi-stakeholder collaboration involving academia, industry, government, and civil society. It is important to highlight that SFRS support critical infrastructure, which includes key large economic sectors vital for society and the economy. These sectors encompass energy, transportation, healthcare, digital infrastructure, water supply, and other essential services. Ensuring their security and resilience is crucial for supporting societal needs and national development goals. Focusing on these areas in research and projects helps strengthen societal resilience and contributes to long-term priorities such as stability, economic growth, healthcare, and digital transformation.

N. Filipovic (✉)
Science Fund of the Republic of Serbia, Belgrade, Serbia
e-mail: fica@kg.ac.rs

Faculty of Engineering, University of Kragujevac, Kragujevac, Serbia

Z. Keković et al. (eds.), *Critical Infrastructure: Risk and Crisis Management in the New Reality*, Advanced Sciences and Technologies for Security Applications,
https://doi.org/10.1007/978-3-032-20948-1_13

Keywords Science Fund of the Republic of Serbia · Artificial intelligence · Proposal evaluation and monitoring · Horizon Europe · Critical infrastructure

1 Introduction

The advent of artificial intelligence (AI) is reshaping various sectors, including research and innovation. In Serbia, the Science Fund of the Republic of Serbia (SFRS) is implementing a comprehensive initiative aimed at integrating AI into its scientific funding processes. This paper explores the potential of AI to enhance the efficiency, accuracy, and transparency of grant evaluations, ultimately transforming the research landscape in Serbia. As nations strive to address pressing societal needs, integrating AI in funding decisions is critical [1, 2].

Global policies on AI in scientific funding evaluations vary significantly, with a strong emphasis on maintaining human oversight. In the European Union, the Horizon Europe framework urges researchers to avoid using generative AI in the peer review process, ensuring evaluations remain human-led. Similarly, the NIH in the US prohibits using AI in peer reviews, emphasizing confidentiality and integrity. While the UKRI permits limited AI use by applicants with transparency, it restricts assessors from using such tools [3]. These guidelines reflect a cautious approach, prioritizing ethical considerations and data integrity. As regions like Canada, Australia, and others formulate their AI policies, a global consensus on best practices remains critical to enhance trust and transparency in scientific funding processes [4, 5].

2 Methods

The proposed AI-assisted evaluation and monitoring system will employ a variety of methodologies designed to enhance the efficacy of grant proposal assessments and foster innovation.

AI-Driven Automation will be at the core of this system, enabling a seamless and efficient analysis of grant proposals. By developing sophisticated algorithms, the system will assess submission completeness, relevance, and overall quality. Employing machine learning techniques, the system will rank proposals according to pre-established criteria, ensuring that the evaluation process is systematic and objective. This automation not only streamlines the review process but also significantly reduces the administrative burden on evaluators, allowing them to focus on more complex aspects of proposal analysis [6].

Predictive Analytics will further enrich the evaluation framework by leveraging historical data from previously funded projects. By identifying trends and forecasting potential outcomes, the system can provide insights into the anticipated societal impacts and returns on investment for various projects. Utilizing advanced statistical models, the system will assess not only the merit of individual proposals but also their alignment with strategic goals, helping funding bodies make informed decisions based on data-driven projections.

In addition to technical methodologies, **Stakeholder Engagement** is critical for the system's success. To ensure that the AI-assisted evaluation aligns with actual societal needs, the involvement of academia, industry leaders, government representatives, and civil society is essential. Regular workshops and consultations will be conducted to gather comprehensive input from these diverse stakeholders, fostering a collaborative approach to system design and implementation.

Infrastructure Support is another critical consideration, targeting the evaluation of vital sectors such as energy, transportation, healthcare, and digital infrastructure. The system will assess both risks and opportunities in these areas, providing valuable insights to inform funding decisions. This strategic focus will ensure that resources are directed to areas with the potential for maximum societal benefits.

Moreover, the system will underscore the importance of **Transparency** by asking applicants to disclose their use of AI in their research and applications. Creating an environment conducive to transparency will empower researchers, encouraging them to share information without fear of repercussions. Alongside transparency, agencies must conduct **Due Diligence** to evaluate AI tools based on standards of quality, integrity, data protection, confidentiality, and intellectual property rights. This careful scrutiny ensures that the selected tools meet high ethical standards [7].

Monitoring the evolving AI landscape is also crucial. Agencies must stay proactive in their involvement with developing AI technologies, adapting their practices to align with advancements in the field. This vigilance will help maintain the relevance and effectiveness of the evaluation process [8].

Finally, ethical considerations are paramount. Agencies should adhere to the guidelines put forth by the European Research Area, ensuring that their operations are grounded in principles of research integrity and trustworthy AI. To support this ethical framework, **Training** programs should be promoted and funded to educate researchers on the responsible use of AI. These initiatives will equip them with the knowledge and skills necessary to navigate the complexities of AI in their work.

By integrating these methodologies, the proposed AI-assisted evaluation and monitoring system will not only improve the efficiency and effectiveness of grant assessments but also align with ethical standards and societal priorities, driving innovation and positive outcomes in funded projects [9, 10].

2.1 IDEA Program

The IDEA Program is, to date, the largest program of the Science Fund that supports 105 projects across all scientific fields. The supported projects are based on excellent ideas that may significantly impact the future development of science and research, as well as other areas of society. As of the third quarter of 2025, out of the 105 supported projects, 101 have completed implementation, while 4 projects are in the final stages of project activities. A total of 1,042 researchers from 98 research organizations in the

Republic of Serbia, along with 18 researchers from the diaspora, are participating in the program. The total budget for the program amounts to 30 million euros, financed from combined sources: the SAIGE project and the budget of the Republic of Serbia. In the upcoming period, the completion of all supported projects in the final phase of implementation is expected, along with the preparation of a Final Report on the program and the promotion of project results.

2.2 IDENTITIES Program

The IDENTITIES Program supports projects in the field of social and humanities sciences that explore various types of identities and how they shape our society. Within this program, 16 projects have been supported, involving 115 researchers from 29 research organizations. During the reporting period, the implementation of the last 4 projects was completed, concluding the realization of all projects within the program. The total budget of the program is 2 million euros, financed by the budget of the Republic of Serbia. In the coming period, the completion of all supported projects in the final phase of implementation, the preparation of a Final Report on the program, and the promotion of project results are expected.

2.3 Green Program for Science and Industry Cooperation

The Green Program for Science and Industry Cooperation is a program of the Science Fund aimed at developing scientific solutions that will find practical application in reducing environmental pollution through collaboration with industry. In the third quarter of 2025, implementation of all 20 projects financed under this program was completed. The total budget for the program is 4 million euros, funded by the budget of the Republic of Serbia. This program has supported 16 projects involving 231 researchers from 38 research organizations. Documentation for the second phase of program implementation was prepared by May 30, 2025, after which the program proposal was sent to the Government of the Republic of Serbia for approval. The second phase of the program will focus on commercializing products resulting from scientific research and developing cooperation with partners from the industry.

2.4 PRISMA Program

Within the PRISMA Program, the Science Fund of the Republic of Serbia finances 97 projects involving 948 researchers from 264 research organizations. The PRISMA Program supports basic and applied research projects across all scientific fields. Projects within this program do not have predetermined themes and are realized under six subprograms. The projects are in their second year of implementation. The total budget for the program is 25 million euros, financed through the budget of the Republic of Serbia and funds from the SAIGE project (World Bank and IPA funds).

2.5 PROMIS 2023 Program

Under the PROMIS 2023 Program for outstanding projects by young researchers and early-career scientists, the Science Fund of the Republic of Serbia finances 30 projects involving 157 researchers from 42 research organizations. PROMIS 2023, like the previous phase of the PROMIS program, is intended for outstanding projects of young researchers and scientists in early career stages across all scientific areas. The total budget of the approved projects amounts to 4 million euros, financed through the budget of the Republic of Serbia and funds from the SAIGE project. The projects are in their second year of implementation.

2.6 Proof of Concept Program

The Proof of Concept Program is intended for researchers whose scientific solutions can serve as a basis for commercial products and services. Within this program, the Science Fund finances 52 projects, involving 232 researchers from 27 research organizations. The total budget for the program amounts to 1 million euros, funded through the SAIGE project. In addition to financial support, the Proof of Concept Program also provides business support, requiring all research teams to select a business advisor from an official list of approved business advisors by the Science Fund, who will assist them in the technology transfer process. Of the 52 projects financed under this program, the implementation of 50 projects has been completed, while the final reporting process is ongoing.

3 Results

Artificial Intelligence (AI) is rapidly reshaping various industries globally, including the field of Monitoring and Evaluation (M&E) for scientific projects. As organizations and development programs increasingly aim to boost their effectiveness, AI emerges as a crucial tool that optimizes M&E processes. This technology significantly enhances the speed and quality of data analysis, reporting, and decision-making, ultimately resulting in improved outcomes for both development and humanitarian initiatives.

To fully understand the impact of AI on M&E in scientific projects, it's important to grasp the fundamental principles of this field. Monitoring and Evaluation refers to systematic processes designed to assess the performance of projects, programs, or policies, fostering accountability and effectiveness. Monitoring involves the ongoing collection and analysis of data to track progress toward defined objectives, while evaluation focuses on assessing outcomes and impacts to gauge the overall success of interventions. Traditionally, M&E relied heavily on manual data collection methods—including surveys and field visits. However, with the increasing volume and complexity of data, organizations are turning to AI to streamline and enhance various aspects of the M&E process.

AI revolutionizes M&E in several significant ways. One of the key advantages is the enhancement of data collection and analysis. Traditional methods can be both time-consuming and prone to error, but AI technologies automate many of these processes, relieving the workload on field staff and enabling real-time data collection. For example, Natural Language Processing (NLP) enables organizations to analyze unstructured data from platforms like social media and various reports, providing a more comprehensive understanding of their situational context. Additionally, machine learning algorithms can identify trends and patterns in data that may go overlooked by human analysts, offering predictive insights that inform strategic adjustments and improve program effectiveness.

Another transformative aspect of AI is its ability to facilitate real-time data monitoring. Unlike traditional M&E methods that depend on periodic assessments, AI allows for continuous monitoring, enabling organizations to swiftly respond to emerging challenges. For example, AI-powered sensors and Internet of Things (IoT) devices can track environmental conditions or health indicators in real time, giving M&E teams the immediate feedback necessary for making data-driven decisions. This timely intervention is especially crucial in public health, where AI can alert authorities to potential disease outbreaks.

In addition to enhancing data collection and monitoring, AI simplifies the reporting and visualization processes, which can be labor-intensive in traditional systems. Automated reporting tools can generate reports, charts, and dashboards based on real-time data, substantially reducing the time analysts typically spend preparing information for stakeholders. These AI-generated reports can be tailored to emphasize key performance indicators, making it easier for stakeholders to grasp project progress and outcomes. Furthermore, AI can detect anomalies or discrepancies in the data, prompting detailed investigations into areas needing attention.

Moreover, AI-driven predictive analytics enable organizations to foresee future program outcomes. By analyzing historical data, predictive models can identify potential challenges and opportunities, allowing for proactive strategy adjustments. For instance, AI may assess the likelihood of success for a new intervention by evaluating data from similar past initiatives, informing better resource allocation and targeting for beneficiaries. This forward-thinking approach helps design programs that are more effective and maximize their impact.

Finally, AI enhances stakeholder engagement and transparency, essential components of successful M&E. By making data more accessible and comprehensible, AI ensures that all parties involved in a program have access to relevant information and can provide feedback. AI-powered platforms can create personalized dashboards for stakeholders, enabling them to monitor progress and identify issues pertinent to their specific interests and roles. This level of transparency fosters greater trust and collaboration among stakeholders, empowering them with the information needed to make informed decisions.

In summary, AI is dramatically transforming the landscape of Monitoring and Evaluation by modernizing traditional practices and providing innovative solutions that improve data collection, analysis, and reporting processes. As organizations continue to integrate AI technologies into their M&E frameworks, the potential for enhanced effectiveness and accountability increases, leading to more impactful development and humanitarian efforts.

3.1 Data Privacy and Security

In the age of artificial intelligence, the reliance on large volumes of data raises significant concerns regarding data privacy and security. As organizations implement AI systems, it is crucial that they prioritize compliance with data protection regulations, such as the General Data Protection Regulation (GDPR) and others relevant to their regions or sectors. Such compliance helps to protect the rights of individuals whose data is being collected and processed. Organizations must therefore ensure that robust security measures are in place to safeguard sensitive information from potential breaches or unauthorized access. This includes employing encryption protocols to protect data both in transit and at rest, implementing stringent access controls to limit who can view or manipulate data, and conducting regular security audits. By fostering a culture of data privacy and security, organizations can not only bolster their defenses against cyber threats but also build trust with stakeholders, ensuring that individuals feel safe sharing their data.

Furthermore, organizations must keep in mind that data privacy and security are not one-time considerations, but ongoing responsibilities. As technology evolves and new threats emerge, organizations should stay informed about the latest best practices and emerging regulations related to data security. Implementing regular training for staff about data handling practices is equally important. This training can empower employees to recognize potential threats, such as phishing attempts or inadequate data handling procedures. By instilling a strong awareness of data privacy and security throughout the organization, entities can create a fortified framework that supports their AI initiatives while respecting the privacy rights of all individuals involved.

3.2 Bias and Fairness

The integration of AI into decision-making processes introduces essential considerations around bias and fairness. AI algorithms are only as reliable as the data upon which they are based; if this data contains biases or reflects historical inequities, the outcomes produced by AI can perpetuate these disparities. For instance, biased training data can lead to discriminatory practices in funding allocations or program assessments, thus undermining the very goals of fairness and equity that organizations strive to achieve. To counteract this issue, it is imperative that organizations actively seek to ensure that the datasets used to train AI models are diverse and representative of the populations they serve.

Moreover, organizations ought to continually assess the algorithms employed and refine them as necessary to promote fairness. This might involve implementing machine learning techniques that specifically seek to identify and mitigate biases within the data. Additionally, incorporating feedback from diverse stakeholder groups, including underrepresented communities, can provide invaluable insights into potential biases and help organizations adjust their models accordingly. By committing to fairness in AI, organizations can enhance the credibility of their evaluations and reinforce their commitment to equitable outcomes in their programs.

3.3 Capacity and Skills

Successfully implementing AI within Monitoring and Evaluation (M&E) frameworks demands a workforce equipped with specialized skills and knowledge. Organizations must recognize that investing in human capital is just as crucial as investing in technological tools. This may entail providing training programs for current staff focused on AI principles, data science, and analytical methodologies. Upskilling team members not only enhances their capabilities but also fosters an organizational culture that values continuous learning and adaptation to new technologies.

In addition to training existing staff, organizations may benefit from hiring specialists with expertise in AI and data analytics. These professionals can bring fresh perspectives and innovative approaches to M&E practices, thus enabling organizations to leverage AI technologies more effectively. Furthermore, enhancing the organization's technology infrastructure is essential to support the integration of advanced AI tools and systems. This may involve investing in cloud-based solutions, data warehouses, or analytical platforms that can handle the computational demands of AI algorithms. By establishing a strong foundation of skilled personnel and adequate technological resources, organizations are better positioned to maximize the impact of AI in their M&E processes.

3.4 *Cost and Resource Allocation*

While the potential advantages of AI in M&E are promising, the associated costs can present challenges, particularly for organizations with limited budgets. The initial investment required for implementing AI solutions, coupled with ongoing maintenance, can strain financial resources. Therefore, organizations must engage in careful planning and resource allocation to ensure that they can capitalize on the benefits of AI without jeopardizing their financial stability. Conducting a thorough cost–benefit analysis allows organizations to determine which AI applications are likely to yield the highest return on investment.

Additionally, organizations may want to prioritize AI projects that align closely with their strategic objectives and target areas where AI can provide the most meaningful impact. Exploring funding opportunities, partnerships, and grants can also ease the financial burden associated with adopting AI technologies. Collaborations with academic institutions, non-profits, or government agencies may provide access to resources and expertise that bolster AI initiatives while sharing the costs involved. By thoughtfully navigating the financial landscape of AI implementation, organizations can successfully integrate these technologies into their M&E practices to enhance effectiveness and achieve better outcomes for the populations they serve.

Addressing challenges related to data privacy and security, bias and fairness, capacity building, and cost management is essential for the successful integration of AI in Monitoring and Evaluation. By proactively tackling these considerations, organizations can harness the power of AI to transform their M&E processes, ultimately leading to improved effectiveness and accountability in their programs.

4 Discussion

The integration of artificial intelligence (AI) into the funding process brings several notable advantages. One prominent benefit is the automation of proposal evaluations, which significantly reduces the time needed for reviews. This efficiency enables funding agencies to prioritize qualified candidates more effectively. Additionally, AI-driven processes offer clear metrics for decision-making, fostering greater trust among stakeholders. By targeting sectors that are essential for societal welfare, the Scientific Funding and Research System (SFRS) can ensure that research funding aligns with national development strategies. Promoting responsible AI practices will also equip stakeholders with the skills and knowledge necessary to adapt to changing technologies.

However, alongside these benefits, several challenges must be confronted. Key issues such as data privacy, algorithmic bias, and the need for continuous evaluation of AI systems require attention. Establishing a robust framework for ethical AI use is vital to build confidence among both researchers and the public.

The inclusion of AI in the funding landscape significantly enhances the efficiency and effectiveness of grant evaluations. The ability to automate proposal evaluations drastically reduces review times, allowing funding agencies to allocate resources more effectively and prioritize candidates. This streamlining not only speeds up the evaluation timeline but also permits funders to consider a larger number of proposals, ultimately promoting a more equitable selection process.

Furthermore, AI-driven processes provide clear metrics that reinforce decision-making. By relying on data-driven insights, stakeholders can enhance their trust in the funding process. Increased transparency regarding how proposals are scored and assessed bolsters the accountability of funding agencies, assuring both researchers and the public that decisions are based on objective criteria rather than subjective biases. This clarity can further motivate researchers to engage with funding opportunities, as they are assured that their proposals will be evaluated fairly.

Another significant advantage of integrating AI, particularly for the SFRS, is the potential to align research funding with national development strategies. By concentrating on sectors critical for societal welfare—such as health, education, and environmental sustainability—the SFRS can ensure that funded research makes meaningful contributions to broader societal goals. This strategic alignment maximizes the impact of research investments and highlights the crucial role of funded projects in addressing urgent national challenges.

Moreover, promoting responsible AI practices within the funding ecosystem will provide stakeholders with the necessary competencies to keep pace with technological advancements. Training programs focused on the ethical use of AI can empower both researchers and funding agencies to effectively navigate the complexities of AI integration. By imparting knowledge about best practices, organizations can cultivate a culture of responsible innovation that takes ethical considerations into account.

Nonetheless, it is essential to recognize the challenges presented by AI integration. Issues such as data privacy and algorithmic bias can undermine trust in AI systems if not adequately addressed. Organizations must prioritize implementing stringent data protection measures to safeguard sensitive information and comply with regulatory requirements. In addition, tackling algorithmic bias is critical to ensure fairness in funding decisions, which may involve conducting regular audits and updates to AI systems to guarantee they are trained on diverse and representative datasets.

Finally, establishing a continuous evaluation framework for AI systems is crucial for maintaining the integrity of the funding process. By consistently assessing the performance and impact of AI tools, organizations can identify potential issues early and make necessary corrections. Developing a strong framework for ethical AI usage will be essential for instilling confidence among researchers and the general public. By ensuring that AI technologies are employed responsibly and transparently, funding agencies can enhance their credibility, leading to more effective funding outcomes and societal advancements.

5 Conclusion

The initiative by the Science Fund of the Republic of Serbia not only signifies a transformative step in the modernization of research funding through the integration of artificial intelligence (AI) but also greatly enhances transparency within the funding process. This increased transparency has profound implications for various stakeholders, including researchers, policymakers, and the general public, fundamentally altering their engagement and trust in the funding system.

By utilizing AI to streamline proposal evaluations, the Science Fund ensures that funding decisions are based on objective criteria. This objectivity is crucial for fostering a fair and equitable environment where researchers can feel assured that their work is evaluated without bias. The clear articulation of how decisions are made allows stakeholders to better understand the rationale behind funding allocations, promoting a sense of fairness and accountability. When researchers see that their proposals are evaluated transparently, they are more likely to engage actively in the funding process, knowing their ideas are given due consideration.

For policymakers, increased transparency aids in aligning research funding with national priorities and societal needs. When the decision-making processes are open and accessible, it encourages a collaborative atmosphere, enabling better communication between researchers and policymakers. This collaboration can lead to more relevant research outcomes that are attuned to pressing national challenges.

The public also stands to benefit from this transparency. By having insights into how funding decisions are made, citizens can hold the Science Fund accountable, ensuring that taxpayer money is used efficiently and effectively. This level of oversight promotes public trust in both the funding body and the research community, as people can see the tangible impacts of funded projects on their society.

Moreover, enhanced transparency allows for the sharing of data regarding funded projects, enabling a broader discourse about research processes and findings. Stakeholders can track the progress and outcomes of funded initiatives, creating opportunities for feedback and discussion that can lead to further innovation and improvement in research practices.

In conclusion, the emphasis on transparency in the Science Fund's initiative significantly impacts stakeholders by fostering trust, encouraging engagement, and promoting accountability. This commitment to openness not only bolsters the integrity of the funding process but also cultivates a more dynamic and responsive research environment. As the initiative unfolds, continuous evaluation of its transparency mechanisms will be vital to maximizing its effectiveness and ensuring that it meets the needs of all involved parties.

References

1. European Commission.: Horizon Europe: the EU Research and Innovation Programme (2021–2027) (2020)
2. Lacity, M.C., Willcocks, L.P.: Robotic process automation at Telstra: scoping and making the business case. J. Inf. Technol. Teach. Cases **8**(2), 71–82 (2018)
3. Kitchin, R.: Thinking critically about and researching algorithms. Inf. Commun. Soc. **20**(1), 14–29 (2017)
4. Russell, S., Norvig, P.: Artificial Intelligence: a Modern Approach. Pearson (2016)
5. Moor, J.H.: The ethics of digital theft: a conceptual framework. Comput. Soc. (2006)
6. AlSabah, K.F., Alazzaz, A., Ali, A.: The role of artificial intelligence in public administration: a catalyst for transformation. J. Public Adm. Res. Theory **32**(1), 45–62 (2022)
7. Liu, B.C., Martin, G.R., Thorne, J.A.: The impact of automation on funding decision-making processes. Res. Policy **48**(2), 123–134 (2019)
8. McCarthy, P.J., Keller, F., Ross, S.: Artificial intelligence for innovation in funding processes. Innov.: Manag. Policy Pract. **22**(3), 289–302
9. Thompson, K.S.: Transparency in research funding: the importance of unbiased proposal assessments. Sci. Public Policy **48**(5), 637–648 (2021)
10. Weitzner, E.G., Van Doren, S., Ly, M.: Aligning research funding with societal needs for impactful outcomes. J. Appl. Res. Technol. Innov. **4**(1), 15–29 (2018)

CBRN Terrorism in the Age of AI: Prevention Strategies, Response Capabilities and Potential Risks

Marina Dabetić, Ivana Perović, and Zoran Keković

Abstract With the rapid advancement of deep learning (DL), machine learning (ML), and artificial intelligence (AI) more broadly, asymmetric threats are evolving at an unprecedented pace and becoming increasingly unpredictable. Emerging technologies and expanding knowledge bases may be exploited for the preparation of chemical and biological agents, even by individuals without specialized training. At the same time, AI offers substantial advantages by strengthening prevention systems, enhancing safety, and providing cognitive and operational support to frontline responders. The paper uses a literature review to gain insight into relevant research in this area in order to examine the advantages and challenges that the future holds in the field of CBRN terrorism, with a focus on potential advantages of emerging technology in prevention, detection, and response.

Keywords Artificial intelligence (AI) · CBRN terrorism · Dual-use goods · Prevention · Response · Resilience

1 Introduction

Although chemical, biological, radiological, and nuclear (CBRN) attacks have constituted a relatively small proportion of terrorist activities and military operations, their potential use continues to pose a significant concern for policymakers and the public [1]. Advancements in artificial intelligence (AI), including machine learning

M. Dabetić (✉) · I. Perović
Department of Physical Chemistry, Vinča Institute of Nuclear Sciences - National Institute of the Republic of Serbia, University of Belgrade, Belgrade, Serbia
e-mail: fmarina@vin.bg.ac.rs

I. Perović
e-mail: ivanaperovic@vin.bg.ac.rs

Z. Keković
Faculty of Security Studies, University of Belgrade, Belgrade, Serbia
e-mail: zkekovic@fb.bg.ac.rs

Z. Keković et al. (eds.), *Critical Infrastructure: Risk and Crisis Management in the New Reality*, Advanced Sciences and Technologies for Security Applications,
https://doi.org/10.1007/978-3-032-20948-1_14

(ML) [2, 3] and deep learning (DL) [3, 4], enhance CBRN detection and crime scene analysis through automated data processing and advanced forensic support; however, these same technologies exhibit pronounced dual-use potential and may be exploited to modify, accelerate, or diversify threat modalities, thereby contributing to the transformation of the contemporary threat landscape. In the domain of chemical, biological, radiological, and nuclear (CBRN) terrorism, AI acts as a powerful dual-use technology: while it can significantly support prevention, detection, and response mechanisms [5], it presents a dual challenge: its implementation can open new vulnerabilities while also helping malicious actors build new capabilities [6, 7]. Such free AI-powered sources, open-source datasets, and automated knowledge generation tools increase the risk that non-state actors or individuals without requisite domain knowledge may misuse the technology to design, model, or optimize hazardous agents and delivery pathways [8]. Meanwhile, AI creates new opportunities to build on situational awareness, assist with decision making, and improve the frontline operation of responders. A primary concern regarding emerging technologies is the increasing accessibility of information and technical expertise that could facilitate CBRN attacks. Recent progress in large language models (LLMs) has further complicated the CBRN terrorism landscape by enabling rapid and simplified access to scientific and technical content that was previously difficult to obtain [8]. Although much of this information is publicly available, AI-driven tools substantially lower the barriers to its retrieval, interpretation, and practical application. This shift reduces dependence on traditional intermediaries or specialized networks, enabling malicious actors to independently acquire guidance related to the development or use of CBRN materials [6].

At the same time, Intelligent surveillance systems, predictive analytics, autonomous detection systems, and optimized response protocols can play an important role in enhancing national resilience against CBRN terrorism [2, 4, 9]. While scholarly interest in the consequences of AI on CBRN attacks is increasing, the impact of AI for CBRN terrorism is under-researched, particularly with respect to misuse, developing threat vectors, and the balance between innovation and safety [10]. This paper is meant to explore prevailing trends and provide an overview of AI for both enhancing and mitigating CBRN terrorism risks to examine present research practices and their impacts. Using a systematic review of existing literature, this study explores prevention strategies, technological benefits, response capacity, and growing points of vulnerability, providing the background of the AI-based future security environment.

The authors conducted a structured and transparent literature review in Scopus, following the basic principles of the PRISMA-P approach to ensure clarity in the selection and analysis of relevant studies.

2 Methodology

A structured literature review of peer-reviewed publications was conducted using the Scopus database as the primary source. The review followed the core principles of the PRISMA-P approach to ensure transparency and clarity in the identification, selection, and evaluation of relevant studies. The search strategy incorporated multiple keywords, including "artificial intelligence," "machine learning," "deep learning," "CBRN terrorism," "dual-use technologies," "prevention," and "emergency response."

Articles published in English and indexed as peer-reviewed journal papers or conference proceedings were considered. Other papers that were not directly associated with AI in the context of CBRN terrorism, as well as duplicate records and non-scientific sources, were excluded. The final collection of studies was qualitatively analyzed to identify key trends, technological capabilities, and potential risks associated with AI-enabled CBRN terrorism. This search strategy was designed to align with the aims of the study by capturing literature related to both the beneficial applications of AI in CBRN prevention, detection, and response, as well as research addressing emerging risks, dual-use concerns, and evolving threat vectors.

3 Findings: Benefits of Artificial Intelligence in the CBRN Context

AI improves CBRN preparedness and response by enabling faster and more accurate situational assessment, enhanced decision support, and safer operational planning [1, 2]. In combination with AI-based consensus analysis and predictive analytics, these capabilities enhance operational effectiveness and strengthen the overall resilience of CBRN emergency management systems [1, 4].

3.1 AI for Prevention and Detection of CBRN Threats

In the prevention and detection phase, AI enables the early identification of potential CBRN threats through advanced data processing, pattern recognition, and anomaly detection [2, 10]. Machine learning algorithms enhance the analysis of sensor data, environmental monitoring systems, and forensic indicators, allowing for faster and more reliable detection of hazardous substances and abnormal events [1, 4]. Predictive analytics further support preventive decision-making by modeling potential exposure pathways and identifying high-risk scenarios, thereby strengthening situational awareness and reducing the likelihood of escalation [9].

Beyond real-time detection, AI contributes to the integration of heterogeneous data sources, including satellite imagery, open-source intelligence, and historical incident databases, enabling a more comprehensive understanding of evolving threat patterns [2, 10]. Such capabilities improve early warning systems and support risk prioritization by distinguishing between routine anomalies and indicators of intentional or accidental CBRN release [1, 4]. In addition, AI-assisted monitoring can enhance border security and critical infrastructure protection by supporting the identification of illicit trafficking routes, unauthorized movements of hazardous materials, and deviations from established safety thresholds [9].

By augmenting human expertise, AI-driven tools reduce cognitive burden on analysts and first responders, enabling more timely and informed preventive actions while minimizing false positives that could otherwise lead to unnecessary disruption or resource allocation [2, 10]. However, the effectiveness of AI in prevention and detection remains contingent upon the quality of training data, interoperability of monitoring systems, and the integration of these technologies within existing legal, organizational, and operational frameworks [1, 4, 9].

3.2 AI-Supported Response and Operational Capabilities

AI plays a critical role in strengthening response capabilities during CBRN incidents by enhancing situational awareness, operational coordination, and responder safety. AI-enabled reconnaissance drones and autonomous unmanned aerial vehicles (UAVs) provide real-time environmental data, rapid incident assessments, and hazard identification without exposing personnel to hazardous conditions, thereby significantly reducing operational risk [11, 12] These platforms support dynamic scene assessment and facilitate informed decision-making during the early and most uncertain phases of CBRN incidents.

AI-driven decision-support systems further assist responders by integrating real-time sensor inputs with historical incident data, predictive models, and operational protocols, enabling more effective coordination, prioritization, and allocation of resources under time-critical conditions [5]. Such systems support command-and-control functions by improving information flow across response units and enhancing interoperability among agencies involved in CBRN emergency management.

Empirical studies report a high level of acceptance of these technologies among CBRN-specialized personnel, primarily due to improvements in analytical accuracy, speed of information delivery, and clarity of data visualization, while also emphasizing the importance of maintaining appropriate levels of human oversight, trust, and autonomy in operational decision-making [11]. These findings highlight that AI-supported response systems are most effective when deployed as decision aids that complement, rather than replace, human judgment in complex and high-risk operational environments [13].

While these AI-supported response capabilities significantly enhance operational effectiveness and responder safety, their increasing integration into CBRN emergency management systems simultaneously introduces new technical, organizational, and governance-related challenges, which are examined in the following section.

4 Emerging Challenges in Countering CBRN Threats

Despite significant advances in prevention, preparedness, and response capabilities, countering CBRN threats remains a complex and evolving challenge [1, 6]. A persistent concern highlighted by security experts is the assessment that the use of CBRN materials by non-state actors represents not a question of possibility, but rather a question of timing, reflecting enduring vulnerabilities in the control and protection of hazardous materials [14]. These challenges are further intensified by rapid technological developments that have altered the traditional risk profile of CBRN threats by lowering barriers to access, development, and deployment. Emerging technologies affect two critical dimensions of CBRN risk: the knowledge and development domain—through tools such as LLMs, additive manufacturing, and advanced biotechnologies including CRISPR—and delivery mechanisms, particularly unmanned aerial systems and drones, which expand the range and accessibility of potential attack vectors [3, 8, 9]. Together, these dynamics complicate early detection and risk assessment, underscoring the need for adaptive CBRN risk management strategies that integrate technological innovation with robust governance frameworks, international cooperation, and sustained capacity-building efforts [5, 10].

Table 1 synthesizes the analyzed literature by categorizing key studies according to their primary focus, core findings, and thematic relevance, illustrating the distribution of AI-enabled benefits, risks, and emerging threat vectors in the CBRN terrorism context.

From the literature mentioned above, AI has been framed more and more not just as a technology enabling the prepared-and-responded-to operations of the CBRN, but also as a transformational phenomenon altering operational landscape of CBRN terrorism. In contrast to conventional CBRN risk discussions centered on accidents or state actors, the reviewed literature addresses both the advantages and risks of AI in the terrorism context, emphasizing its dual role in enhancing CBRN prevention and response while also enabling terrorist and non-state actors to adapt, innovate, and circumvent traditional security barriers.

According to some authors, AI tools may change fundamental features of CBRN terrorism through lowering technical thresholds, expediting the phases of planning and executing, and increasing access to advanced knowledge, which has once been a domain restricted to experts [3, 8]. From this perspective, AI not only amplifies forces to create and deploy new threat vectors within terrorist networks, but also drives them to build new environments with different approaches for agents to be designed, deliver and operate in, to conceal themselves operationally [15, 16]. On the other hand, AI also provides in the literature evidence that it significantly strengthens counterterrorism capabilities in CBRN. Machine learning based detection systems, forensic attribution tools and decision-support platforms enhance a security agencies' ability to identify, attribute and prevent intentional events of CBRN more rapidly and accurately [2, 4, 10]. These capabilities are directly applicable to terrorism settings where detecting, attribution, and interrupting early are essential for averting mass

Table 1 Overview of selected studies on AI and CBRN terrorism

Author(s), year	Type of study	Primary focus	Key contribution/findings	Thematic category
Wilson et al. 2024 [1]	Journal article	CBRN forensics & counterterrorism	Integrates military forensics with CBRN counterterrorism operations	Detection & response
Kantarcıoğlu 2025 [4]	Journal article	CBRN forensic sciences	AI applications in forensic identification and attribution	Detection & response
Sandbrink 2023 [3]	Journal article	Biological AI misuse	Differentiates risks of LLMs vs biological design tools	Dual-use & risks
Barrett et al. 2024 [7]	Policy report	AI governance & dual-use	Proposes a combined benchmark and red-team approach for assessing dual-use CBRN and cyber risks	Governance & risk management
Miotto 2024 [8]	Policy analysis	LLM misuse	Explores terrorist use of LLMs for CB terrorism	Dual-use & risks
Seyedin et al. 2025 [5]	Journal article	Frontline protection	AI tools enhancing safety and decision-making of responders	Response capabilities
Kegyes et al. 2024 [2]	Journal article	ML-based decision support	ML framework supporting decision-making in CBRN protection	Detection & response
Koblentz 2020 [20]	Journal article	Emerging technologies & CBRN terrorism	Analyses how emerging tech (incl. AI) reshapes CBRN threat vectors	Dual-use & risks/emerging Threats
Houser and Dong 2025 [11]	Systematic review	AI & terrorism	Comprehensive review of AI convergence with terrorism	Dual-use & risks
Popa 2025 [16]	Literature review	CBRN detection	Overview of AI-based detection methodologies	Detection & response
Biron et al. 2024 [12]	Conference paper	Autonomous CBRN drones	Examines trust–control balance in military drone autonomy	Emerging threats
Tripathi and Ajaykumar 2025 [19]	Book chapter	AI risk assessment	Argues for the development of AI-specific risk assessment models to address emerging CBRN disasters and incidents	Prevention

(continued)

Table 1 (continued)

Author(s), year	Type of study	Primary focus	Key contribution/findings	Thematic category
Pannu et al. 2025 [17]	Journal article	Biological AI dual-use	Identifies high-consequence dual-use capabilities of biological AI models and argues for pre-deployment evaluation to mitigate biosecurity and biosafety risks	Dual-use & risks
Ajaykumar 2024 [10]	Conceptual analysis	Emerging CBRN technologies	Examines emerging tech in CBRN threats	Emerging threats
Farhat et al. 2024 [6]	Journal article	Regional preparedness	AI-enhanced Delphi model for CBRN preparedness in MENA	Prevention & preparedness
Galatas 2020 [15]	Monograph	Prevention of CBRN terrorism	Prevention of access to CBRN materials by terrorists	Prevention & preparedness
Bruzzone et al. 2022 [19]	Conference paper	Crisis decision-making	Simulation-based decision support	Detection & response
Kumar et al. 2025 [18]	Book chapter	AI in biotechnology	Discusses AI-driven bioengineering processes	Dual-use & risks
Szklarski 2020 [22]	Final project report	CBRN terrorism overview	Overview of improvised CBRN use	Threat landscape

death and secondary attacks. Of particular significance, the studies examined indicate that the marriage of AI and CBRN terrorism is a dynamic one that is situated within the tension between creativity and exploitation. If anything, the growing integration of AI in combat operations provides greater preparedness, resilience and response efficacy, but can also create new vulnerabilities that terrorists can exploit where governance, oversight and risk mitigation tools are not robustly developed [7, 16, 18]. This dual-use dilemma locates AI at the core of the future progress of CBRN terrorism rather than a peripheral technological cudgel.

5 Discussion

Our review shows that this evolution in this context highlights that employing AI in the CBRN domain has been a qualitative change. AI will reshape the defensive and offensive aspects of CBRN security, specifically when seen in relation to terrorism and non-state actors [15, 16, 18]. Instead of more classical CBRN risk frameworks, which are limited to accidental releases, industrial disasters or state programs, the literature reviewed above shows the impact that AI capabilities have on asymmetric threat behavior. A key insight gained from the reviewed articles is that AI changes the operational logic of CBRN terrorism. AI helps shrink decision cycles, facilitates rapid experimentation, and provides access to more sophisticated analytic tools, thus removing traditional obstacles that had limited terrorist actors. This is not an automatic creation of a future-looking piece of AI-designed CBRN weapons but rather a broadening of operational alternatives involving enhanced concealment, planning proficiency, and flexibility. And consequently, the threat environment is becoming more fluid, less predictable and difficult to attribute, adding complexity to early warning and interdiction. Meanwhile, the existing literature shows that AI greatly augments counterterrorism capabilities in a CBRN setting. Advanced detection systems, forensic attribution tools, and decision-support platforms allow security agencies to identify anomalous behaviors, place attribution on incidents more precisely, and take preemptive action at earlier stages in the attack. These capabilities are especially important in the case of terrorism scenarios wherein the ability to identify, disrupt, and mediate quickly is vital to halting large scale casualties and other secondary effects. This is why AI serves as a force multiplier for preparedness, response coordination, and resilience-building for our security institutions. However, the discussion in the sources reviewed across the literature shows consistently that a widening gap between technological capability and governance preparedness is evident. With rapid developments in AI-enabled detection and response tools in the industry, oversight mechanisms, ethics frameworks, and risk mitigation frameworks are lagging. This gap is even more acute in the world of terrorism, as the threat landscape is rife with opportunities for cyber attackers taking advantage of regulatory asymmetries, jurisdictional fragmentation, and worldwide availability of digital tools. The lack of standardization in the management of risks associated with AI-based CBRN also increases the potential for the systems used primarily to secure organizations to inadvertently increase the threat surface. An obvious second theme to be stressed in the discussion about the fragmentation of research and policy approaches. And technical studies often draw attention to operational optimizations and performance gains, whilst policy-oriented analyses generally concern governance and ethical challenges, which often don't integrate well. The compartmentalization of approaches prevents the formulation of holistic approaches to mitigating the risks and strengths associated with AI-enabled CBRN security. Combating this gap necessitates interdisciplinary partnership among AI developers, security professionals, policy makers, and lawyers.

6 Conclusion

The review shows that employing AI in the CBRN domain has been a qualitative change. AI will transform both defensive and offensive features of CBRN security, specifically when seen in relation to terrorism and non-state actors. AI can shrink decision-making loops, enable quick testing, make analytical tools more sophisticated and hence, make it easier to overcome the traditional problems which had prevented terrorist actors. Sophisticated detection systems, forensic attribution tools, and decision-support platforms enable security organizations to detect anomalous actions, assign more accurate attribution to incidents and take preemptive actions earlier in the attack. Such capabilities are particularly relevant for terrorism, and the capability to recognize, disrupt and mediate quickly is necessary to stop mass casualties and other collateral damage at the earliest. That is why AI works as a force multiplier for preparedness, response coordination, and resilience for our security institutions. Nevertheless, the conversations from sources cited throughout the paper found in the literature constantly demonstrate that the distance between technical capacity and governance readiness is growing. Rapid development of AI-enabled detection and response tools in the industry, however, is prompting a shortage of oversight mechanisms to monitor them, ethics frameworks for using such tools as well as frameworks of risk mitigation. This divide is particularly pronounced in the context of terrorism, where the increasing accessibility of advanced knowledge, biotechnological tools, and AI-enabled analytical capabilities lowers traditional technical barriers and enables non-state actors to exploit emerging technologies without the need for extensive resources or institutional support.

References

1. Wilson, L., Roffey, P., Lothridge, K., Rowan, S.S.A., Crispino, F.: Intersection of CBRN, military forensic science, and counterterrorism. Forensic Sci. Int. **364**, 112218 (2024)
2. Kegyes, T., Süle, Z., Abonyi, J.: Machine learning-based decision support framework for CBRN protection. Heliyon **10**(4) (2024)
3. Sandbrink, J.B.: (2023). Artificial intelligence and biological misuse: differentiating risks of language models and biological design tools. https://doi.org/10.48550/arXiv.2306.13952
4. Kantarcıoğlu, V.D.: Use of Artificial Intelligence Technologies in CBRN Forensic Sciences. Adli Bilimler ve Suç Araştırmaları **7**(2), 161–178 (2025)
5. Seyedin, H., Moslehi, S., Tavan, A., Narimani, S.: The role of artificial intelligence in protecting frontline forces in CBRNE incidents. Progr. Disaster Sci. 100443
6. Farhat, H., Alinier, G., Bajow, N., et al.: Preparedness and response strategies for chemical, biological, radiological, and nuclear incidents in the Middle East and North Africa: an artificial intelligence-enhanced Delphi approach. Disaster Med. Public Health Prep. **18**, 244 (2024). https://doi.org/10.1017/dmp.2024.160
7. Barrett, J., Mukherjee, S., Arnold, J.:A framework for assessing and managing dual-use hazards of AI foundation models. Berkeley Center for Responsible AI Report (2024). https://vcresearch.berkeley.edu/news/framework-assessing-and-managing-dual-use-hazards-ai-foundation-models
8. Miotto, N.: The potential terrorist use of large language models for chemical and biological terrorism. European Leadership Network (2024). https://europeanleadershipnetwork.org/commentary/the-potential-terrorist-use-of-large-language-models-for-chemical-and-biological-terrorism/
9. Egan, J., Rosenbach, E.: Biosecurity in the age of AI: what's the risk? Belfer Center for Science and International Affairs Report (2023). https://www.belfercenter.org/publication/biosecurity-age-ai-whats-risk
10. Ajaykumar, S.: Emerging technologies in the development and delivery of CBRN threats (2024). https://www.orfonline.org/research/emerging-technologies-in-the-development-and-delivery-of-cbrn-threats

11. Houser, T., Dong, B.: The convergence of artificial intelligence and terrorism: a systematic review of the literature. Stud. Conflict Terror. 1–24 (2025)
12. Biron, B., Schraml, S., Bauer, G., Seböck, W.: Military perspectives on CBRN drone autonomy: balancing trust and control. In: International Conference on Information Technology in Disaster Risk Reduction, pp. 195–209. Springer Nature Switzerland, Cham (2024)
13. Dabetić, M., Keković, Z., Ćurčić, M., Obrovski, B.: CBRN Incidents and critical infrastructure: the role of first responders in an integrated emergency management and resilience building in Serbia. In: Building Resilience in a Dynamic Critical Infrastructure Landscape, pp. 145–154. Publisher, Springer (2025)
14. Dabetić, M., Ćurčić, M., Veljković, S.: Phases of integrated risk management in CBRN risk mitigation and recovery. In: 12th International Conference on Electrical, Electronics and Computer Engineering (IcETRAN), pp. 185–188, Čačak (2025)
15. Galatas, I.: Prevention of CBRN materials and substances getting into the hands of terrorists. International Centre for Counter-Terrorism (ICCT), The Hague, Netherlands (2020)
16. Popa, M.: Literature review on recent technological methodologies in detection and identifying CBRN agents. Land Forces Acad. Rev. **30**(2), 202–210 (2025)
17. Pannu, J., Bloomfield, D., MacKnight, R., Hanke, M.S., Zhu, A., Gomes, G., et al.: Dual-use capabilities of concern of biological AI models. PLoS Comput. Biol. **21**(5), e1012975 (2025)
18. Kumar, K., Sharma, R., Ranjan, K., Pandey, D.K.: Artificial intelligence in biotechnology. In: Artificial Intelligence and Biological Sciences, pp. 192–218. CRC Press (2025)
19. Tripathi, P., Ajaykumar, S.: CBRN disasters and incidents: the need for AI risk assessment models. India's AI Imperative **123** (2025)
20. Koblentz, G.D.: Emerging technologies and the future of CBRN terrorism. Wash. Q. **43**(2), 177–196 (2020)
21. Bruzzone, A.G., Massei, M., Reverberi, A., Giovannetti, A., De Paoli, A., Pedemonte, M., Rovelli, R.: Models & solutions for strategic decision makers during CBRN crises in industrial & urban environments. In: 12th International Defence and Homeland Security Simulation Workshop, DHSS 2022, vol. 5 (2022). https://doi.org/10.46354/i3m.2022.dhss
22. Szklarski, Ł., Maik, P., Wieczorek, A.: European system for improved radiological hazard detection and identification (SU-DRS04-2019-2020): final project report. Research and Innovation Action, Horizon, European Commission, Brussels (2020)

Effects of AI-Driven Financial Sector Transformation

Nikola Vidović

Abstract The accelerating integration of artificial intelligence (AI)—particularly generative AI—is reshaping the financial services ecosystem through automation, data-driven intelligence, and new business models. This study explores how AI investments influence structural transformation in finance, focusing on cost efficiency, improved decision-making, regulatory alignment, and long-term strategic outcomes. A mixed-method approach, combining quantitative analysis of financial performance with qualitative case studies across banking, insurance, auditing, and fintech, provides a comprehensive perspective on AI's economic and operational impact. Findings highlight that the expansion of digital transactions has created opportunities for advanced analytics and predictive modeling. Financial institutions increasingly deploy AI to detect fraud, assess credit risk, and enhance customer experience through personalized "next best action" systems. The study emphasizes that AI's return on investment extends beyond short-term financial gains, encompassing sustainable innovation, talent development, and responsible governance. Evidence suggests that organizations systematically investing in AI achieve superior adaptability, resilience, and innovation capacity. If effectively governed, generative AI could add between 0.5 and 3.4% points annually to productivity growth while contributing to inclusive and human-centered financial systems. The paper contributes to both scholarship and practice by providing an evidence-based framework for aligning AI investments with long-term institutional and societal value.

Keywords Financial sector · Artificial intelligence · Return on investment

N. Vidović (✉)
Faculty of Security Studies, University of Belgrade, Belgrade, Serbia
e-mail: vidovicnikola.finance@gmail.com

Z. Keković et al. (eds.), *Critical Infrastructure: Risk and Crisis Management in the New Reality*, Advanced Sciences and Technologies for Security Applications,
https://doi.org/10.1007/978-3-032-20948-1_15

1 Introduction

Inspired by statistics and probability theory, artificial intelligence (AI) researchers have developed numerous algorithms that enable computers to learn from data, and the field of machine learning has become its own independent subdiscipline of AI, intentionally separate from symbolic AI [1]. As Vučenović et al. state [2], the scope of application of AI technologies and models is unlimited, given that in all areas of business there is the possibility of improving operational processes and activities through the prism of this technology, and it is certain that it is increasingly present in the financial industry. Financial firms, which have always relied heavily on statistics and quantitative analysis in their services, were naturally suitable for the development of strategic AI technologies [3].

Slavković et al. [4] point out that digital transformation has become an inseparable part of all business activities, as it represents a process in which digital technologies create disruptions that trigger strategic responses from organizations that want to change their paths to creating new value, while managing structural changes and organizational barriers. AI is revolutionizing the financial services landscape, offering transformative opportunities for efficiency, innovation, and improved user experience [5], and especially for increasing economic competitiveness in the financial industry, including the banking sector [6], the insurance sector [4, 7], accounting and auditing [5, 8, 9], and investment business entities and start-up firms. AI has emerged as a transformative force in improving the efficiency of financial markets around the world. By leveraging advanced computing techniques, including machine learning, natural language processing, and predictive analytics, AI enables financial institutions to automate processes, analyze vast amounts of data, and make data-driven decisions with unprecedented speed and accuracy. This efficiency revolution is reshaping the way financial institutions operate, driving increased productivity, reduced costs, and improved overall performance [10].

The application of AI technologies in central banks and financial institutions improves economic forecasting [11], financial stability and risk analysis, as well as the creation of macroeconomic and monetary models [12], thereby increasing the resilience of the financial system in the face of global complexity [13]. The introduction of digital services in central banking, together with process automation, leads to cost reduction, more accurate fraud detection and more efficient decision-making, while in the broader financial sector, AI improves communication with clients, automates loan approval and enables the development of more advanced analytical and risk management procedures. Banks and fintech companies, as carriers of innovation, transfer their business to the digital sphere by relying on AI, which increases performance, scalability and competitive advantages, while algorithmic techniques in insurance improve premium calculation and risk management [4]. The fintech ecosystem, based on AI, big data and cognitive computing [14, 15], contributes to the development of more diverse and accessible services, greater user trust and stability of the sector, while automation through chatbots reduces the burden on employees [16]. In the field of accounting and controlling, AI takes over routine tasks in data processing and report preparation, enabling the transformation of the profession and the development of new technical competencies with increased cooperation between people and algorithms [8, 17–19].

2 Transforming the Financial Sector by Applying AI Technologies

The underpinning influence and implementation of AI technologies in financial services dates back to the 1950s when early attempts were made to apply computer techniques to financial modelling and analysis [21]. Over the next decades, AI technologies, such as expert systems, neural networks and genetic algorithms, gradually gained strength in finance, enabling tasks such as credit scoring, risk assessment and portfolio optimisation. The growth of big data and advances in machine learning algorithms have further triggered the adoption of AI in finance, introducing an era of automated market, algorithmic trading strategies and predictive analysis. Looking from a technological perspective, AI encompasses a wide range of concepts and technologies that allow machines to perform tasks that traditionally require human intelligence [5, 22]. The digitisation of financial services—from the emergence of ATMs and online banking [23] to the widespread use of mobile applications—has transformed industry from all perspectives: Providers, customers, employees, regulators and other stakeholders. The application of AI to automate and expand tasks through the operational models of most organisations quickly established it as one of the most transformative technologies the world has ever seen. Banking, insurance, capital markets and wage transport are among the areas where their influence is expected to be most widespread [24] (Table 1).

According to the Zoting survey [25], the global value market size of generative AI in banking and financial services was $1,260.16 million in 2024, with a growth of $1,676.02 million in 2025, and is projected to reach $21.82 billion by 2034, at a compound annual growth rate of 33% (Fig. 1).

Regional adoption of AI technologies shows the dominance of North America, where financial centers such as New York and San Francisco lead the development of innovations, fintech systems and startups that are constantly improving generative AI models for the financial industry. European financial institutions are also actively applying AI to improve competitiveness, personalization of services, process automation and operational efficiency, supported by startup ecosystems and venture capital investments. At the same time, in the Asia-Pacific region, countries such as China, Japan and Singapore are investing significant resources in the development and research of AI, with a focus on employee support, improving risk assessment, fraud detection, user experience and financial forecasting. Overall, the greatest profit from AI technologies is recorded in North America, where leading financial companies generate more annual revenue than the combined financial sectors of France, Germany and Italy [11].

Table 1 Use of the AI model in the financial sector with the highest level of generated value

Industry	Function	Description	Assigned value
Banking	Sales and customer service	Customer support agents receive quick and comprehensive information on all aspects of products, policies, and procedures from various sources	Greater agent efficiency, increased accuracy of responses and aster response times
Capital markets	Client servicing/ investment management	Companies use AI models to create investment portfolios, provide financial guidance, and deliver real-time insights and trading recommendations to clients	Increased customer satisfaction and retention, competitive advantage
Payments	Fraud management and detection	Preventive fraud detection utilizes technologies capable of proactively identifying suspicious behavior or anomalies before fraudulent transactions occur	Enhanced customer protection from fraud, improved user experience by minimizing false positives
Insurance	Claims processing	Automation of claims and customer document handling streamlines data collection and validation	Improved workflow efficiency, simplified document collection and validation, greater agent productivity
Financial services	Risk management and underwriting	Prediction of fraudulent transactions, more efficient insurance processing, and advanced risk scoring models	Reduced internal and external risks, greater accessibility of established credit scoring and evaluation, faster underwriting processing time, improved data protection
	Technology development	Streamlining the software development lifecycle from code writing to automated testing as well as understanding and decommissioning outdated code environments	Improved workflow and accuracy, reduction of technical debt, shorter development cycles, increased efficiency

The impact of the integration of AI technologies with its dependent subsets of machine and deep learning, launched in the era of digital transformation of markets and global economic flows, is visible in financial markets and stock market operations, trading models, volatility forecasting, portfolio management [26], performance assessments and risk analyses [27], as well as default assessments, cryptocurrencies, financial derivatives, investment activity analysis and foreign exchange management [28]. AI technology generates profits for business entities in the financial industry

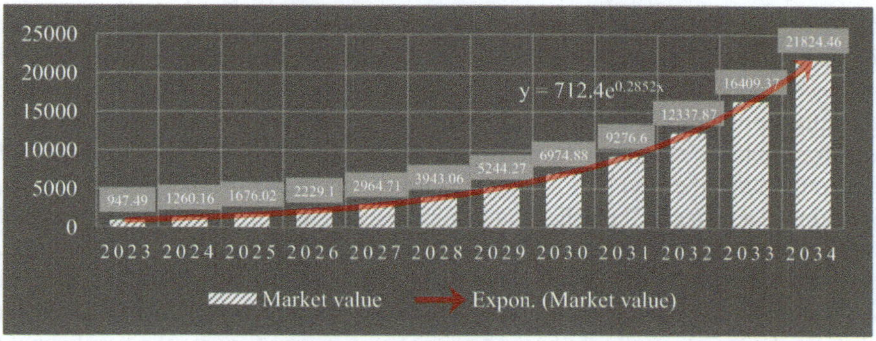

Fig. 1 Demonstration of the global value market size of generative AI in the banking and financial services sector from 2023 to 2034 (mill. $)

through predictive functions and contributes to market stability by minimizing information asymmetry and volatility, resulting in profitable investment systems and accurate performance assessments [28]. With the support of AI, financial institutions can now assess the possible outcomes and risks associated with alternative techniques by creating synthetic market situations and modeling different investment strategies. By evaluating past data and creating hypothetical future financial scenarios, generative AI can also help with financial forecasting. This allows businesses to more accurately predict stock prices, market trends, and consumer behavior, resulting in better decision-making. The market is driven by various factors, including increasing investment in AI, increasing product launches, increasing collaboration among market players, technological advancements, and the benefits offered by these techniques [26].

The integration of AI into the banking and insurance sectors enables rapid and extensive transformations of business processes by building a digital core that encompasses infrastructure and security [29], data and AI layers, and application platforms, while continuously adopting new technologies and business opportunities. In banking, AI automates customer service, loan approval, receivables collection, risk management, and regulatory compliance, reducing costs and increasing efficiency; at the same time, generative and predictive models enable the analysis of large data sets, the prediction of market trends, and the more precise management of credit, market, and operational risks. In the insurance sector, AI and Internet of Things (IoT) contribute to better risk assessment, loss prevention, and service personalization. The expected growth of the IoT insurance market is 40% in the forecast period 2021–2026 according to Slavković et al. [4]. Benefits include a better user experience [12, 17], more accurate fraud detection [26, 30], expanded access to financial services, and significant savings in regulatory and audit processes, while audit automation enables resource optimization [31, 32] and shortening of the time required for control [33–36].

While AI applications contribute to improving efficiency, analytical capabilities, and decision-making processes in the financial sector, they also present risks that are fundamentally different from traditional financial risks. In particular, algorithmic bias can arise as a result of incomplete or unrepresentative data, which affects decisions in lending, insurance, and fraud detection. In addition, the limited transparency of advanced models makes audit, oversight, and regulatory compliance difficult, while the reliance of multiple institutions on the same algorithmic structures, data sources, or infrastructure creates the potential for model and systemic risk. The development of generative and other complex AI systems further increases exposure to cyberthreats, including data integrity breaches and manipulation of model outputs. As the scope of automated decision-making increases, challenges arise in the areas of governance, accountability, and internal controls. These risks indicate that the application of AI is not just a technological improvement of existing processes, but a factor that changes the structure of risks in the financial system and requires the development of appropriate management and supervision mechanisms [37, 38].

3 Analysis of Realized Benefits and Development of the Financial Industry

In 2023, financial services companies invested $35 billion in AI technologies, with projected investments in banking, insurance, capital markets, and payments businesses, where investment capitalization is expected to reach $97 billion by 2027 [24]. A comprehensive analysis of AI applications in the financial sector has shown its ability to profoundly transform the field [23], offering innovative solutions ranging from customer service automation to financial advisory and risk management optimization [38, 39]. For global banking, McKinsey estimates that AI technologies could potentially deliver up to $1 trillion in additional value each year. Traditional AI and analytics are indicative of traditional machine learning (e.g., clustering) and statistical techniques (e.g., basic regression), while advanced AI technology is indicative of deep learning neural networks (e.g., convolutional neural networks). The potential for value creation is one of the highest in any industry (Fig. 2).

Companies are already beginning organizational changes aimed at creating future value through generative AI (GenAI). Research shows that starting from a comprehensive, strategic vision that includes security integration, access control, code reuse, and built infrastructure is more effective than standalone solutions, which allows for faster and cheaper introduction of additional functionality and creates a competitive advantage. Successful companies focus simultaneously on the adoption and expansion of solutions, as well as on the development of advanced technologies [40] (Fig. 3).

Radak and Urošević [41] show that the insurance sector has undergone significant transformations under the influence of digital transformation and AI, where advanced algorithms are used to solve business problems, optimize investment portfolios according to the Basel II standard, and prevent fraud. Pavlović [7] points out that the key to the successful implementation of intelligent chatbots in insurance is training on a large volume of data about products and customer interactions, which

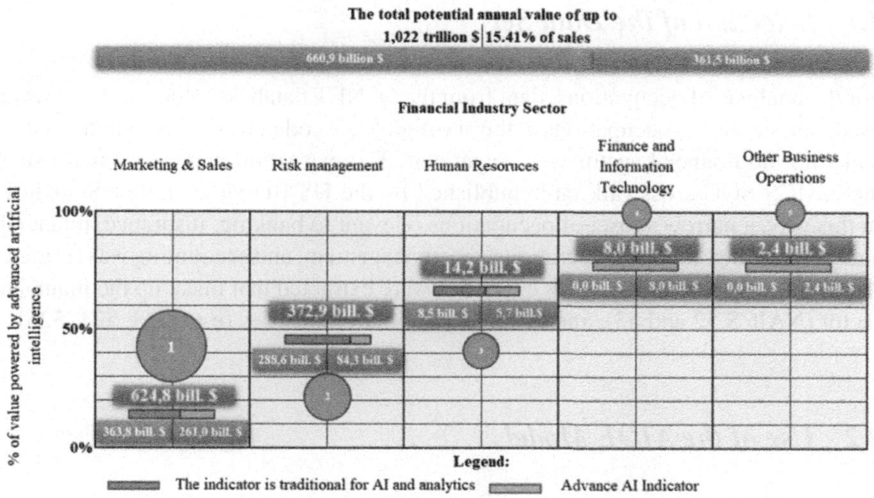

Fig. 2 Potential total annual value generated by AI indicators in the financial sector

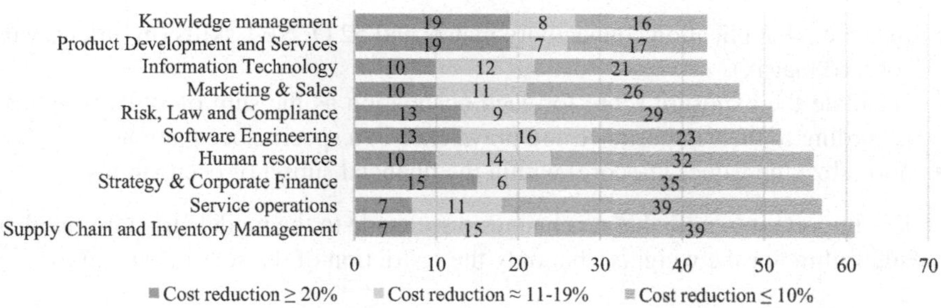

Fig. 3 Cost reduction within an entity by business sector using GenAI models in 2024 [40]

allows for more accurate content generation, encourages innovation, and increases the quality and efficiency of services.

4 Results

The methodology for measuring occupational and industry exposure in the financial sector in this paper is based on the AIOE/AIIE model developed by Felten et al. [42]. Unlike the original studies, in which exposure is calculated for the entire structure of financial occupations in the US, in this study the sector methodology is adapted by separating only the occupations and industry segments of which it is a component. classification (financial services, banking, insurance, accounting, financial consulting, etc.).

4.1 Selection of the Data Set

For the analysis of occupations, data from the O*NET database (Release 24.3) were used, but out of 774 occupations at the six-digit SOC code level, 97 occupations that make up the financial sector were separated. The extraction was performed using the NAICS-SOC crosswalk table published by the US Bureau of Labor Statistics. In this way, a narrow subset of occupations relevant to banking, insurance, financial analysis, trading, financial operations, risk management, and accounting was formed. At the industry level, 12 NAICS industries were extracted that make up the financial sector (NAICS 52 and 53), and in some cases, sub-industries (e.g., 523, 524, 525).

4.2 Use of the AIOE Model

The created AIOE model available on GitHub [42] was used in its original form, but the data was pre-filtered to include only relevant SOC codes. The model was run using Python scripts from the repository to:

- load the AI application connectivity matrix and 52 O*NET skills (mTurk crowd-sourced matrix),
- calculate the exposure score for each occupation as the sum of skills weighted according to their importance and prevalence in a specific occupation,
- normalize the values (z-score) within the financial subset of occupations.

It is important to note that no changes were made to the model structure (neither the algorithm nor the weights), but only the restriction of the set of occupations.

4.3 Adaptation of the Model to the Financial Sector

The adaptation consisted of:

- extraction of occupations that make up the financial sector;
- recalculation of the AIOE score only for this subset;
- scaling and normalization of the results of the internal financial recall, which allows for comparison of occupational exposures within the domestic sector (rather than in relation to the entire US workforce);
- interpretation of the results in the context of specific financial activities (analytics, risk modeling, insurance, credit processes, financial operations).

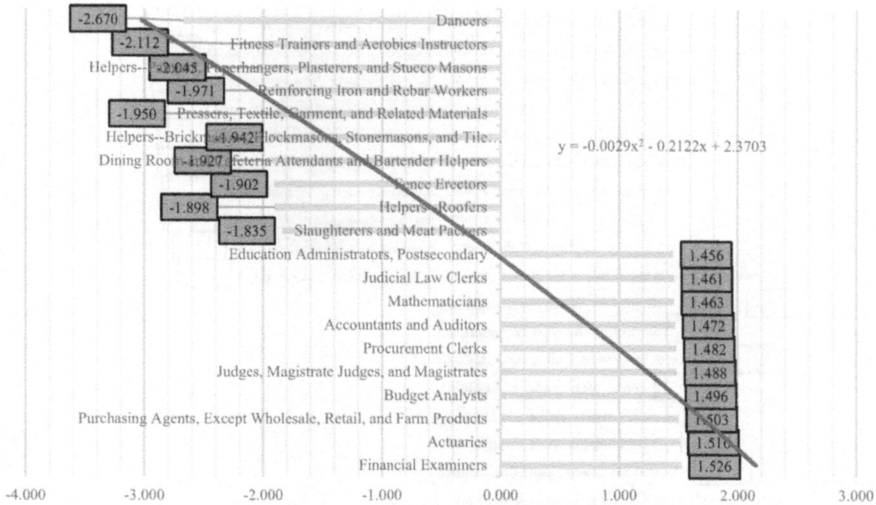

Fig. 4 Results of a comparative analysis of occupational exposure to the impacts of AI technology

4.4 *Construction of AIIE for the Financial Industry*

After calculating the AIOE for all occupations in the financial sector, an AIIE was also constructed based on employment by occupation in the financial industry (NAICS 52). Employment data from the BLS [42] were used. Industry exposure was calculated as a weighted average of AIOE values within each industry. In this way, Figs. 4 and 5 show the result of this adapted, restructured model.

4.5 *Advantage of This Approach*

This approach allowed [42] model to be applied to the banking and finance domain in a scientifically valid manner, while maintaining consistency with the original methodology, but also focusing on sector specificities that were not covered by the original research.

In the Figure above, we can see that the most exposed to advances in AI technologies are predominantly those professions and occupations in the financial sector, such as financial managers, appraisers, analysts, and actuaries, while the lowest values include non-office jobs that require a high degree of physical effort and are of a manual nature.

Just as with the ranked occupations, in the chart below we find that the most exposed industries tend to be those that require a high level of education, such as financial services, accounting, insurance, and legal services. On the other hand, the industries with the lowest scores tend to be industries that involve manual labor, such

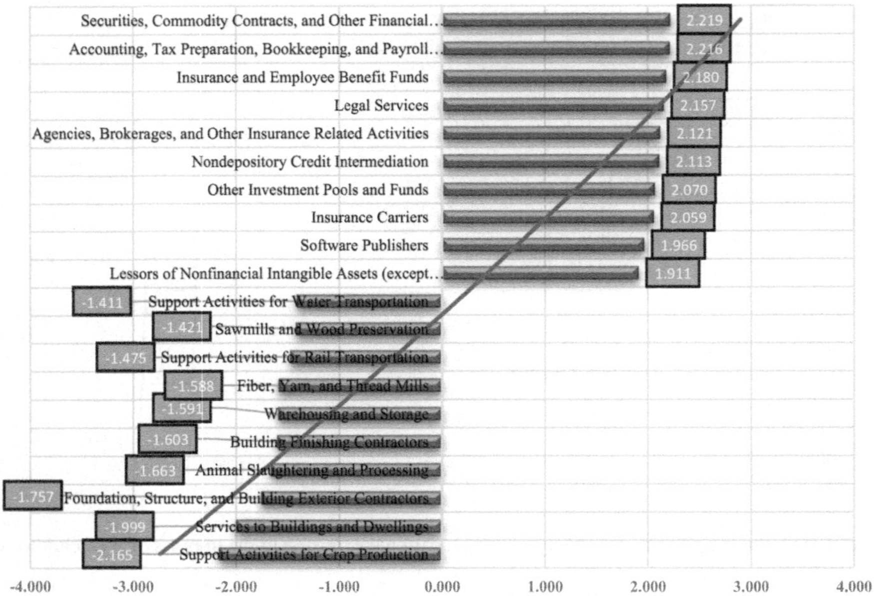

Fig. 5 Results of a comparative analysis of the exposure of industrial sectors to the effects of AI technologies

as crop support activities, construction and housing services, construction contracting services, and AI technologies are rapidly developing and advancing, and we see that they will have an even greater impact than what is known today on financial flows, the global economy, and macroeconomic movements in markets. From the perspective of economic feasibility, it is assumed that the activities of investing in AI technologies will begin to be automated when the price of the solution falls below the level of the salary paid to a human for performing that activity.

From the above, we conclude that the net effect of AI is that it replaces the workforce and increases productivity [43, 44], where we see both a reduced risk of labor shortages and pressure to reduce unit labor costs. This is particularly relevant in the eurozone, where the working-age population is projected to decline by 19% by the end of the century as a result of population aging [11].

5 Discussion

The return on investment that certain companies have had for years in new technologies, skills and ways of working has varied. The emergence of AI has generated the possibility of creating competitive advantages for companies that use it. Therefore, this research includes case studies of successful practices and positive solutions that have been achieved through the application of AI in the banking, insurance, accounting and auditing sectors, investment companies, fintech and start-up business entities, which have created new values for the industry, and have given rise to and developed models and applications that have internal and external application. The following tables show pragmatic examples, highlighting the functional descriptions of the models in question, as well as the benefits generated in business [9, 33].

The insurance, accounting and auditing sector is also recording significant advances in the approach to operationalizing business activities through AI models, which have generated both material gains and non-financial values for agencies and companies [5, 9, 12].

However, the highest rate of investment activity in innovations through AI is observed in the sector of leading global multinational companies such as "BlackRock", "JPMorgan Chase", "Vangard", "Goldman Sachs", but also pioneers in financial markets, which belong to the group of financial technology companies and start-up firms, such as "Monsoon Credit Technii", "Capital Floati", "Finomena" and the German "Kreditech" [3, 5, 8, 9, 12, 33].

The examples presented in Tables 2, 3 and 4 represent systematically selected cases of documented and publicly available successful implementations of AI in the financial sector, based on the analysis of official documents, technical publications and relevant academic sources [3, 5, 8, 9, 12, 33]. However, such analysis of secondary data implies inherent methodological limitations, primarily in terms of incomplete availability of information, variations in the quality of reporting and the selectivity of corporate publications, which most often present technological implementations in a positive context. Consequently, the examples presented do not include unsuccessful, insufficiently documented or internally archived projects, nor do they provide deeper insight into operational processes, problems during implementation or long-term sustainability of the model. Therefore, the results in the tables can be considered representative examples of well-documented successful practices, and not a comprehensive cross-section of all outcomes of AI implementation in the financial industry. Accordingly, the aim of the paper is not statistical generalization, but rather the analytical selection and critical consideration of examples that possess a sufficient degree of transparency, verifiability, and detail to enable qualitative analysis.

6 Development Prospects and Potential of Artificial Intelligence

The integration of AI into financial services has ushered in a new era of innovation, efficiency, and transformation [5]. Examples include AI's ability to support customer interactions, generate creative content for marketing and sales, and draft computer code based on natural language instructions, among many other tasks [45]. The advancement through generative models will have a significant impact across all industry sectors. Across the banking industry, for example, the technology could deliver an additional $200 billion to $340 billion per year in value if its use cases

Table 2 Case studies of successful applications and practices of AI in banking and monetary regulatory institutions

Sector	Bank/ institution	Developed AI model/application	Functional description	Business benefits
Banking	Banco Bilbao Vizcaya Argentaria (BBVA)	Generative AI, BBVA Valora, one-click loans, AI financial coach	Application of generative AI in operational transformation, digital marketing, personalized loan offers, data analytics, and customer-relationship enhancement. BBVA Valora provides recommendations for property purchase/rental, while the AI coach offers personalized financial guidance	Increase in customer base (150%), shortened onboarding process (from days to minutes), digitalized sales (70% digital), reduced costs (43% cost-to-income ratio), leading position in financial innovation
Monetary institution	European Central Bank (ECB)	PRISMA	A microdata analytics network enabling structured data processing, improving accuracy, and supporting inflation and price fluctuation forecasting across the Eurozone and individual countries	More accurate economic forecasts, more efficient monitoring of inflation drivers, improved monetary decision-making
		ATHENA	A supervisory and financial oversight platform. Uses NLP, optical character recognition, and translation tools to analyze suspicious financial transactions	Significant reduction in analysis time, public-sector savings, better insight into high-risk transactions, automated supervision
		HEIMDALL	A tool for assessing the suitability and fitness of banks. Uses machine reading, translation, process automation, and text analysis from questionnaires	Improved assessment quality, reduced manual workload, more time for complex cases

(continued)

Table 2 (continued)

Sector	Bank/ institution	Developed AI model/application	Functional description	Business benefits
		SupTech—GABI and NAVI	Supervisory tools for analyzing big data (GABI) and banking-system networks (NAVI)	Faster and more efficient supervision of the complex Eurozone banking industry, enhanced risk management, timely intervention
Banking	HSBC holdings plc	Cloud-based AI system (developed with Google Cloud) for anti-money laundering	AML system uses big-data analytics to detect anomalies in transactions and customer behavior. Analyzes networks and global transactions to identify potential financial crimes	Improved detection of financial misconduct, faster and more accurate transaction analysis, global oversight of customer activity, strengthened regulatory compliance
Monetary institution	Monetary authority of Singapore (MAS)	FEAT—Responsible AI framework	A framework that enables financial institutions to use AI ethically by assessing compliance with principles of fairness, ethics, accountability, and transparency in AI models, particularly in credit approvals and insurance	Reduced risk of discrimination, improved reliability of AI models, enhanced consumer protection, promotion of ethical AI adoption in finance

were fully implemented. In retail and consumer goods, the potential impact is also significant at $400 billion to $660 billion per year. Generative AI has the potential to change the anatomy of work, augmenting the capabilities of individual workers by automating some of their individual activities. Current generative AI and other technologies have the potential to automate work activities that currently absorb 60 to 70 percent of employees' time.

The use of AI techniques in the provision of financial services can increase efficiency, reduce costs, improve quality, raise customer satisfaction, and foster financial inclusion, largely due to the opportunities they offer to automate operational processes and increase analytical capacity. Techniques such as natural language processing or image recognition allow financial institutions to automate tasks that are repetitive or have little added value (e.g., answering frequently asked questions). This reduces the possibility of human error, increases productivity, and reduces the cost

Table 3 Case studies of successful application and practice of AI in the insurance, accounting and auditing sectors

Sector	Agency/ company	Developed AI model/ application	Functional description	Business benefits
Insurance	Generali	Mobile & web hub	A centralized omni-channel platform for mobile and web communication with clients. It unifies the customer experience across multiple markets and enables personalized interaction	Consistent and high-quality customer experience worldwide, improved collaboration between IT and design teams, support for long-term customer relationships
		Agent hub	An application designed to enhance interaction between insurance agents and clients, focused on human experience and personalized communication	Higher agent productivity, improved customer support, strengthened agent–client relationships, positive impact on sales and customer retention
Accounting & auditing	BiZDocs	AI document archiving system with RPA, ML and OCR	A digital archive solution for accounting and payroll documents, supported by robotic process automation, machine learning, and optical character recognition. Adapted to IoT environments and fully compliant with GDPR	Reduced paper documentation, faster document processing, regulatory compliance, better accessibility, and more efficient inter-company communication
Insurance	Ping an insurance	AI algorithms for claims processing and fraud detection	Integration of AI into claims handling, customer-relationship management, and fraud detection processes in insurance	Faster claims processing, improved fraud detection, higher customer satisfaction, more efficient resource management
		AI-based predictive analytics	AI models that predict insurance risks and support decision-making to optimize offerings and reduce losses	Enhanced risk management, increased profitability, stronger market competitiveness, improved accuracy in risk assessment

(continued)

Table 3 (continued)

Sector	Agency/ company	Developed AI model/ application	Functional description	Business benefits
Accounting & auditing	Viseeon Portugal	AI-based accounting technology model with network marketing	An integrated platform that uses AI for accounting advisory, training, productivity, and financial guidance, supported by network connectivity and shared IT infrastructure	Increased client growth, improved accounting services, continuous training, efficient resource management, and revenue generation through modern IT and AI infrastructure

of these tasks. As a result, customer satisfaction levels increase [45], as customers receive better service (shorter response times and greater service availability) and, potentially, all at a lower price [17].

Innovators have accumulated an average profit margin of 5.6% points higher than other companies in the same industry by investing in generative AI technology in the previous period, which leads to a 2.4-fold profit margin gap. The view [9] is that it will have a positive impact on other business entities, motivating them to proactively act and invest in relevant technologies. AI accelerates the need for change and enables a more extensive and bold form of re-innovation in companies. What makes AI different from other technological innovations is that it generates opportunities for conceptualizing every aspect of the organization and acts as a catalyst for business activities, contributing to growth, productivity, and agility.

7 Conclusion

The ROI indicator for generative AI applications by business entities in the financial services and products industry should not be limited to immediate cost and revenue gains, but should also consider the long-term strategic benefits for the financial sector, as well as society as a whole. A clear and empirical view of these benefits will help entities decide where to allocate resources, which projects to scale, and how to optimize their transformation path. It will also help to gain support from everyone in the bank for the holistic adoption of generative AI.

Not all implementations are economically viable at this time. However, the overall trend in generative AI implementation and consumption is towards lower costs and greater feasibility. The challenge for entities in the financial industry is to position themselves to take advantage of new usage and application modalities that are economically feasible and generate new value and benefits. Scaling AI is not just about immediate financial gains, but also about preparing for the future. Approaches to assessing ROI should extend beyond immediate economic returns and lay the

Table 4 Case studies of successful implementation and practice of AI in the investment company, fintech and start-up business entities sector

Sector	Bank/ institution	Developed AI model/application	Functional description	Business benefits
Investment financial institutions	JPMorgan Chase	AI fraud-detection algorithms, NLP, ML systems	Application of machine learning and natural language processing to improve fraud detection, personalized client support, and automated trading	Reduced false positives, improved customer support, more accurate risk management, enhanced client engagement
	Vanguard	AI-driven robo-advisors	Automated portfolio management, client-profile analysis and market-trend assessment via ML to create personalized investment strategies	Democratization of wealth-management services, increased accessibility, reduced advisory costs, expanded client base
	BlackRock	AI for market analysis and portfolio optimization	Use of ML algorithms to identify investment opportunities and optimize asset allocation based on client risk profiles	Improved returns, more efficient risk management, data-driven advanced investment strategies
	Goldman Sachs/Morgan Stanley	Algorithmic trading, predictive AI for markets and risk	Various AI platforms for automated trading, risk assessment, and market-movement prediction	Higher efficiency in trading and risk management, scalable solutions, improved market position through advanced algorithms
	Veritas Finance Ltd.	Responsible AI tool (FEAT framework, open-source)	Tool for ethical and responsible AI deployment in finance, aligned with FEAT principles, enabling fair and transparent evaluation and adoption of AI	Increased trust in AI, regulatory compliance, responsible innovation in finance

(continued)

Table 4 (continued)

Sector	Bank/ institution	Developed AI model/application	Functional description	Business benefits
Fintech/ Start-up	Capital float	AI for SME creditworthiness assessment	Similar to Monsoon Credit Tech—focus on risk analysis and improved credit decisioning for SME clients	Increased SME credit availability, improved portfolio risk management
	Monsoon Credit Tech	AI for SME creditworthiness assessment	Use of AI to analyze SME liquidity and solvency to reduce default risk	Reduced risk in SME financing, improved credit decision-making
	Kreditech (Germany)	Automated big-data and ML system	Real-time credit scoring and loan approval in domestic and international markets	Faster decision-making, global credit accessibility, higher efficiency in lending
	Finomena	AI/ML for client portfolio risk analysis	Decision-making system for loan approvals based on AI/ML-driven portfolio risk assessment	Improved risk assessment, more accurate lending decisions, reduction of non-performing loans
	Loan frame	AI/ML model for creditworthiness assessment	Integrated system for borrower-profile and credit-risk analysis	Faster and more accurate credit-potential assessment, improved lending decisions

foundation for continuous reinvention. It is therefore essential to define and track short-, medium-, and long-term goals, which include talent development, improved customer and employee experience, sustainability, and responsible use of AI [31]. The proliferation of digital payments and banking—which has helped generate financial transaction data that banks can use to track, predict, and better respond to consumer behavior—and the growing demand for online banking and financial information offerings [46] have opened up opportunities for AI implementation in retail banking and finance, where they identify the next best action for adopting AI for digital transformation [33]. One of the key milestones in the evolution of AI in finance is the development and implementation of neurogenetic tools, as they combine the principles of genetic algorithms with neural networks to optimize financial decision-making processes, enabling more accurate forecasting and efficient resource allocation [46].

The future of financial services is intrinsically linked to the continued evolution and adoption of AI technologies. Innovations driven by AI will shape industry dynamics, redefine business models, and empower financial institutions to deliver personalized, efficient, and secure services to customers in a rapidly changing environment. As AI becomes increasingly integrated into financial operations and decision-making processes, its impact on efficiency, risk management, and customer experience will be profound, driving industry-wide transformation, and reshaping the competitive landscape [5]. Accordingly, research has found that innovative generative AI could enable labor productivity growth of 0.1–0.6% per year by 2040, depending on the rate of technology adoption and the reallocation of work time to other activities. Combining generative AI with all other technologies, automation of work could add 0.5–3.4% points to annual productivity growth. However, employees in the financial industry will need support in learning new skills, adaptability and flexibility to change. If the transition of workers and other risks can be managed, generative AI could significantly contribute to economic growth and support a more sustainable, inclusive financial world [45]. The research has found that innovation is an imperative in today's business and at the same time a guarantor of the future. The market is extremely dynamic and turbulent, and accordingly, business entities must constantly look for ways to, on the one hand, meet the needs of increasingly demanding consumers, and on the other, to survive in the market and achieve better business performance, as they state [20], especially in the financial services sector.

References

1. Mitchell, M.: Artificial Intelligence: a Guide for Thinking Humans. Picador, New York (2020)
2. Vučenović, S., Vukmirović, G., Marić, R., Macura, N.: The potential application of artificial intelligence in retail business in the shadow of the controversy around technology. Econ. Ideas Pract. **55**, 1–12 (2024)
3. Mangunson, W.: Artificial Financial Intelligence, Harv. Bus. L. Rev. **337** (2020)
4. Slavković, A., Slavković, N., Zajić, G., Kostić, S.: Digital transformation, artificial intelligence and internet of things as a support for new insurance industry systems. Challenges and insurance market's responses to the economic crisis, Belgrade: University of Belgrade, Faculty of economics and business, Publishing centre, pp. 419–438 (2023)
5. Shubham, S., Dhamiwal, A.: Artificial intelligence in financial services. In: International Conference on AI in Cyber Security, Jaipur, India (2024)
6. Kovačević, A., Radenković, S., Nikolić, D.: Artificial intelligence and cybersecurity in banking sector: opportunities and risks. In: Contemporary Issues in Economics, Business and Management—EBM 2024, Kragujevac (2024)
7. Pavlović, B.: Application of ChatGPT in insurance. Challenges and insurance market's responses to the economic crisis, pp. 419–438. University of Belgrade, Faculty of economics and business, Publishing centre, Belgrade (2023)
8. Gonçalves, M.J.A., da Silva, A.C.F., Ferreira, C.G.: The future of accounting: how will digital transformation impact the sector? Informatics **9**(1), 19 (2022)
9. Azagury, J., Ashraf, M., Wright, O., Fang Grant, K., Moore, M.: Reinvetion in the age of generative AI, Accenture, www.accenture.com/research. Accessed 04 June 2025

10. Ochuba, N.A., Adewunmi, A., Olutimehin, D.O.: The role of AI in financial market development: enhancing efficiency and accessibility in emerging economies. Finance Account. Res. J. **6**(3), 421–436 (2024)
11. Cipollone, P.: Artificial Intelligence: a central bank's view (2024)
12. Martin, V.: Integrating artificial intelligence into central banking: opportunities, challenges and implications. J. Process Manag. New Technol. **12**(1–2), 49–60 (2024)
13. Rodrigues, A.R.D., Ferreira, F.A., Teixeira, F.J., Zopounidis, C.: Artificial intelligence, digital transformation and cybersecurity in the banking sector: a multi-stakeholder cognition-driven framework. Res. Int. Bus. Financ. **60**, 101616 (2022)
14. Patil, R., Bharathi, S.V.: A study on the business transformation, security issues and investors trust in fintech innovation. Cardiometry **24**, 918–932 (2022)
15. Kovačević, A.: Mašinsko Učenje i Sajber Bezbednost. Zbornik Radova Sa Konferencije Strateški i Normativni Okvir Republike Srbije Za Reagovanje Na Savremene Bezbednosne Rizike, pp. 158–171 (2023)
16. Tiron-Tudor, A., Donțu, A.N., Bresfelean, V.P.: Emerging technologies' contribution to the digital transformation in accountancy firms. Electronics **11**(22), 3818 (2022)
17. Fernández, A.: Artificial intelligence in financial services. Banco de Espana Article **3**, 19 (2019)
18. Rupić, I.B., Obradović, D.B., Rupić, B.: The importance of digitalization for the sustainability of accounting. Ecologica **29**(106), 201–208 (2022)
19. Bigham, T., Nair, S., Soral, S., Tua, A., Gallo, V., Lee, M., Mews, T., Fouche, M.: AI and risk management: innovating with confidence, Centre for regulatory strategy EMEA, Delloite LLP (2018)
20. Stanković, M., Anđelković, T., Mrdak, G., Stojković, S.: The trend of digitalization in the insurance market in the period after the pandemic caused by the Covid-19 virus. Knowl.-Int. J. **54**(1), 51–57 (2022)
21. Webb, A.: The Big Nine: how the Tech Titans and Their Thinking Machines Could Warp Humanity. PublicAffairs, New York (2019)
22. Lopez de Prado, M.: Machine Learning for Asset Managers. Cambridge University Press (2020)
23. Biswas, S., Carson, B., Chung, V., Singh, S., Thomas, R.: AI-bank of the future: can banks meet the AI challenge?. McKinsey & Company, Global Banking & Securities (2020)
24. World Economic Forum.: Artificial Intelligence in Financial Services. https://reports.weforum.org/docs/WEF_Artificial_Intelligence_in_Financial_Services_2025.pdf. Accessed 13 July 2025
25. Zoting, S.: Generative AI in Financial Services Market Size, Share and Trends 2025–2034. Precedence research (2025)
26. Hilpisch, Y.: Artificial intelligence in Finance: a Python-Based Guide. O'Reilly Media, Inc. (2021)
27. Svoboda, A.: The impact of artificial intelligence (AI) on portfolio optimization. Econ. Financ. **12**(3), 21–26 (2024)
28. Bahoo, S., Cucculelli, M., Goga, X., Mondolo, J.: Artificial intelligence in finance: a comprehensive review through bibliometric and content analysis. SN Bus. Econ. **4**(2), 23 (2024)
29. Cin., D., P., Fox, J., Nunn-Price, J., Sidhu, H.: State of Cybersecurity Resilience—How cybersecurity boosts enterprise reinvention to drive business resilience. Accenture. https://www.accenture.com/content/dam/accenture/final/accenture-com/document/Accenture-State-Cybersecurity.pdf#zoom=40. Accessed 11 June 2025
30. Maple, C., Szpruch, L., Epiphaniou, L., Staykova, K., Singh, S., Penwarden, W., Wen, Y., Wang, Z., Hariharan, J., Avramovic, P.: The AI revolution: opportunities and challenges for the finance sector. The Alan Turing Institute. arXiv:2308.16538 (2023)
31. Smith, K., Abbott, M.: Banking in the age of generative AI, Accenture Research analysis. https://www.accenture.com/content/dam/accenture/final/accenture-com/document-2/Accenture-Age-AI-Banking-New-Reality.pdf. Accessed 02 Aug. 2025
32. Svoboda, A.: The impact of artificial intelligence on the banking industry. J. Bank. Financ. Manag. **4**(1) (2023)

33. IDRBT—Institute for Development and Research in Banking Technology: AI in Banking. https://www.idrbt.ac.in/wp-content/uploads/2022/07/AI_2020.pdf. Accessed 06 July 2025
34. Kovačević, A.: The dark side of generative artificial intelligence: abuse and vulnerabilities. In: Artificial Intelligence Conference, Belgrade (2024)
35. Olumide, M.: AI and predicitive financial models: shaping the future of banking risk strategies (2025)
36. Jeremić, N., & Luka, S.: Artificial Intelligence—Challenges of use in auditing. Revizor J. Organ. Manag. Financ. Audit. **27**(107), 29–55 (2024)
37. Vidović, N., Beriša, H.: Economic aspects of cyber security: socio-financial consequences of cyber attacks. Int. J. Contemp. Secur. Stud. **1**(1), 149–162 (2025)
38. Vidović, N., Cvetković, V.M., Beriša, H., Milašinović, S.: Understanding ransomware through the lens of disaster risk: implications for cybersecurity and economic stability. Int. J. Disaster Risk Manag. **7**(1), 247–264 (2025)
39. Becerra-Vicario, R., Salas-Compás, B., Valcarce-Ruiz, L., Sánchez-Serrano, J.R.: The impact of artificial intelligence in the financial sector: opportunities and challenges. Int. J. Bus. Manag. Stud. **05**(10) (2024)
40. Singla, A., Sukharevsky, A., Yee, L., Chui, M., Hall, B.: The state of AI: how organisations are rewiring to capture value. https://www.mckinsey.com/capabilities/quantumblack/our-insights/the-state-of-ai. Accessed 10 Sep. 2025
41. Radak, V. & Urošević, B.: A review of machine learning algorithms for fraud detection. Challenges and insurance market's responses to the economic crisis, pp. 419–438. University of Belgrade, Faculty of economics and business, Publishing centre, Belgrade (2023)
42. Felten, E., Raj, M., Seamans, R.: Occupational, industry, and geographic exposure to artificial intelligence: a novel dataset and its potential uses. Strateg. Manag. J. **42**(12), 2195–2217 (2021)
43. Conic, N., & Demko-Richter, J.: The influence of the development of business intelligence tools on the function of controlling. In: Accounting Knowledge as a Factor of Economic and Social Progress, vol. 209 (2023)
44. Colić, S.: Industry 4.0 and its application in banking. In: Proceedings of the VII International Scientific Conference "Regional Development and Cross-Border Cooperation", Pirot, p. 45 (2024)
45. Chui, M., Hazan, E., Roberts, R., Singla, A., Smaje, K., Sukharevsky, A., Yee, L., Zemmel, R.: The economic potential of generative AI—The next productivity frontier. McKensey&Company. https://www.mckinsey.com/~/media/mckinsey/business%20functions/mckinsey%20digital/our%20insights/the%20economic%20potential%20of%20generative%20ai%20the%20next%20productivity%20frontier/the-economic-potential-of-generative-ai-the-next-productivity-frontier.pdf. Accessed 10 June 2025
46. Adeyelu, O.O., Ugochukwu, C.E., Shonibare, M.A.: Ethical implications of AI in financial decision-making: a review with real world application. Int. J. Appl. Res. Soc. Sci. **6**(4), 608–630 (2024)

Regulatory Challenges in the Field of Artificial Intelligence (Legal Regulation and Ethics)

Dražen Cerović and Aneta Spaić

Abstract Artificial intelligence (AI) has evolved over the past decade from a technological innovation into a key factor in social and legal transformations. The dynamic development of data processing algorithms, machine learning, and automated decision-making has necessitated new regulatory approaches to establish a balance between innovation and the protection of fundamental rights. The European Union recognized this need by adopting the AI Act, the first comprehensive legislative framework aimed at regulating the use of artificial intelligence based on the level of risk generated by such systems. This paper analyzes the main regulatory challenges in the field of AI, with a particular focus on the EU AI Act, its structure, implementation dynamics, as well as broader international implications and comparative regulatory models. The paper combines legal, technical, and ethical approaches to examine the complexity of balancing technological progress with social responsibility.

Keywords Artificial intelligence · EU AI act · Regulation · Technology ethics · AI system risks

1 Introduction

The development of artificial intelligence represents one of the most significant achievements of modern society. Although the idea of "intelligent machines" emerged as early as the mid-twentieth century, it is only the last decade that has brought practical and social implications requiring legal attention. From vehicles with autonomous control systems to facial recognition systems and automated decision-making in public policies, AI is increasingly assuming a role that encroaches on citizens' rights, privacy, and fundamental principles of democracy [1]. Confronted

D. Cerović (✉) · A. Spaić
Faculty of Law, University of Montenegro, Podgorica, Montenegro
e-mail: drazenc@ucg.ac.me

A. Spaić
e-mail: anetas@ucg.ac.me

Z. Keković et al. (eds.), *Critical Infrastructure: Risk and Crisis Management in the New Reality*, Advanced Sciences and Technologies for Security Applications,
https://doi.org/10.1007/978-3-032-20948-1_16

with this reality, states and international organizations are striving to establish a legally sustainable and technologically neutral regulatory framework. The European Union took a leading role in this process in 2021 by publishing the proposal for the Regulation on Artificial Intelligence (EU AI Act), thereby offering for the first time a comprehensive legal framework that classifies AI systems according to their level of risk and establishes mechanisms for controlling their use [2].

This paper analyzes the broader context of regulatory challenges, from insufficiently defined technical terms and boundaries of liability to issues of jurisdiction and international harmonization of rules. Particular attention is devoted to the dilemma between innovation freedoms and the protection of civil rights, as well as practical problems of implementing regulations in EU member states.

2 Legal Framework and Historical Genesis of AI Regulation

The idea of legally regulating technological progress is not new, but it emerges in a particularly complex context with artificial intelligence. While earlier forms of technical innovation, such as nuclear energy, biotechnology, or genetic engineering, were regulated based on clearly defined physical parameters, AI systems operate at the level of abstract algorithms and data, making their regulation significantly more demanding [3].

Historically, legal thought on artificial intelligence can be categorized into three phases, reflecting a progressive evolution from rudimentary technical simulations to sophisticated interdisciplinary frameworks addressing AI's profound societal implications [4]:

(a) Phase of technological neutrality (up to 2015)

During this period, the regulation of information technologies did not recognize AI as a distinct phenomenon. Legal frameworks relied on existing norms in data protection, liability, and digital services. It was generally considered that general legal principles could encompass new technologies without the need for specific regulations.

(b) Phase of normative fragmentation (2015–2020)

With the emergence of advanced machine learning systems and "deep learning" models, many states and organizations began developing partial regulations, most often in the context of privacy protection and algorithm transparency. Examples include the GDPR (EU General Data Protection Regulation), which introduces concepts such as "automated decision-making without human intervention" and "profiling," as well as OECD and UNESCO documents on AI ethics.

(c) Phase of systemic regulation (from 2021)

With the adoption of the EU AI Act proposal, the European Union inaugurates a new era, the first legally binding system that recognizes AI as a distinct category of technology with its own risks and responsibilities [5]. This shift from a reactive to a proactive risk management model, which anticipates mandatory oversight and certification measures for high-risk systems [6].

The fundamental characteristic of the contemporary approach is interdisciplinarity, the integration of law, informatics, ethics, and economics. Regulatory models no longer treat AI solely as a technical concept but as a social phenomenon that impacts labor rights, access to justice, education, and security. This transformation requires redefining a range of classical legal concepts, from subjectivity and fault to liability for machine decisions. For instance, if an algorithm makes a decision that causes harm, the question arises whether responsibility lies with the programmer, the user, or the organization that implemented the system [7]. Consequently, legal scholarship increasingly discusses "algorithmic accountability" and "legal transparency", concepts that seek to bridge principles of technical auditing and legal control.

3 EU AI Act—Structure, Objectives, and Key Provisions

The European Union became the first jurisdiction in the world to establish a comprehensive, horizontal legal framework for artificial intelligence with the adoption of the AI Act (Regulation (EU) 2024/1689) in May 2024. This regulation exhibits a dual character, manifesting in two key dimensions: *regulatory*, a regulatory instrument to protect fundamental rights, public safety, and market integration within the EU, and *strategic* -a strategic tool to promote innovation and the development of "safe and reliable" artificial intelligence [8]. Unlike national initiatives in the US, Japan, or China, which are predominantly focused on ethical guidelines or sector-specific regulations, the EU AI Act has binding legal force as a regulation directly applicable in all member states.

According to Article 1, the Act's objective is to "ensure that AI systems placed on the EU market are safe, ethically grounded, and in compliance with the Union's fundamental rights and values" [9]. The Act is based on a proportional risk approach, meaning different AI systems are subject to varying obligations depending on their potential harm to society.

Article 3(1) of the EU AI Act defines "artificial intelligence" as: "a system that uses machine learning, logic, statistical approaches, or other techniques to produce outputs such as predictions, classifications, recommendations, or decisions which influence physical or virtual environments." This definition is significantly broader than earlier technical interpretations (e.g., OECD or ISO standards), as it encompasses all forms of algorithmic decision-making with elements of autonomous reasoning [10]. One of the Act's key innovative elements is the four-tier risk classification, which forms the basis of the entire legislative model [11]:

(a) Prohibited systems (Unacceptable Risk)

This category covers AI practices deemed incompatible with fundamental rights. Prohibited practices include, among others: citizen social scoring systems by public authorities, manipulative techniques exploiting individual vulnerabilities, and real-time biometric identification in public spaces (except in specific security cases).

(b) High-risk systems (High Risk)

These systems are permitted but subject to strict regulation and oversight. They cover AI in areas such as employment, education, justice, critical infrastructure, medical devices, and public administration. Compliance assessment and entry into the EU high-risk systems database are mandatory.

(c) Limited risk (Limited Risk)

Systems impacting user experience (e.g., chatbots, recommendation systems). Transparency is key, users must be aware they are interacting with an AI system.

(d) Minimal risk (Minimal or No Risk)

This category includes most everyday applications, video games, image filters, or programs without real social consequences. They are not subject to any specific obligations.

Such classification enables a dynamic balance between protection and innovation: risky systems are strictly controlled, while innovative projects without serious consequences enjoy regulatory freedom. For high-risk systems, the Act prescribes detailed technical and organizational requirements: establishing data quality management systems, algorithm transparency and model training process documentation, human oversight of automated decision-making, cybersecurity, mandatory labeling and registration before market placement. AI system developers are required to conduct ex ante risk assessments, as well as post-market monitoring mechanisms. This introduces the concept of "lifecycle responsibility," encompassing all phases, from design to application and algorithm review [12].

The Act provides for the establishment of new regulatory bodies: European AI Board, National supervisory authorities in each member state, European Centre for Algorithmic Transparency (ECAT), operating under the European Commission.

These bodies are tasked with ensuring compliance, information exchange, and market oversight coordination. The system is similar to that of the GDPR but with a greater degree of technical control and testing. The Commission retains the right to update the list of prohibited and high-risk practices in line with technological developments [13].

3.1 Implementation Dynamics of the EU AI Act and Its Institutional Architecture

The EU AI Act entered into force on August 1, 2024, featuring a carefully phased implementation schedule to allow stakeholders time to adapt while prioritizing public safety. Prohibitions on unacceptable AI practices under Article 5 took effect just six months later, on February 1, 2025; rules governing general-purpose AI models, such as foundation models like GPT, followed 12 months after entry into force on August 1, 2025; obligations for high-risk systems apply after 24 months, from August 1, 2026; and additional specific requirements for foundation and generative AI systems activate 12 months post-entry into force, aligning with the general-purpose AI timeline on August 1, 2025 [14].

This timeline allows member states and industry to gradually establish necessary procedures and oversight capacities. This schedule enables businesses and public institutions to adapt internal processes, particularly regarding risk assessment and certification. Member states are required to form national AI bodies and establish market testing and monitoring mechanisms by mid-2025. Effective application of the EU AI Act requires a complex institutional structure and gradual implementation dynamics. The goal is to ensure market consistency, protection of fundamental rights, and stimulation of innovation.

Each EU member state must appoint a national AI supervisory authority to enforce the EU AI Act through an integrated framework of key tasks, ensuring seamless coordination with the European AI Board (EAIB): conducting rigorous market surveillance and compliance controls by overseeing AI producers and users, to verify adherence to technical and ethical standards; managing the approval and registration of AI systems through meticulous verification of technical documentation for enhanced transparency; handling incident and issue reporting by promptly coordinating with the EAIB on systemic failures, user rights violations, or emerging threats; and imposing stringent sanctions for non-compliance [9].

National bodies serve as the first line of regulation, while the European Commission and EAIB play roles in coordination and standards harmonization.

The European AI Board (EAIB) serves as a pivotal regulatory and coordination body at the European level, ensuring cohesive oversight and harmonized implementation of AI governance across member states. Its core competencies encompass issuing authoritative opinions on AI system classifications to clarify risk categories; coordinating national supervisory authorities for consistent enforcement and information sharing; providing comprehensive guidelines on the interpretation and practical application of AI Act provisions; and preparing delegated acts that define precise technical standards along with obligations for AI producers, deployers, and users, thereby fostering innovation while upholding safety and ethical standards [9]. The EAIB is crucial for standardization and preventing regulatory fragmentation within the EU, ensuring rules are equally applicable in all member states.

3.2 Delegated Acts and Technical Guidelines

Delegated acts empower the European Commission with the agility to dynamically calibrate the regulation in response to evolving technological paradigms, obviating the need for cumbersome formal amendments. These acts meticulously delineate granular technical prerequisites for high-risk AI systems, ensuring precision in compliance amid rapid innovation, while prescribing sophisticated protocols for risk evaluation and rigorous testing regimes to fortify systemic resilience against emergent threats. Concomitantly, they refine the compendium of proscribed practices and incorporate nascent AI archetypes, perpetually aligning regulatory contours with technological frontiers. Technical guidelines serve producers and users as a practical guide, explaining how to meet regulatory requirements.

3.3 Role of EU Standardization Bodies

The Commission collaborates with CEN (European Committee for Standardization) and CENELEC (European Committee for Electrotechnical Standardization) to develop technical standards for AI systems. Standardization seamlessly encompasses testing and risk assessment methodologies, alongside meticulously crafted procedures for verifying algorithmic transparency, while also furnishing comprehensive guidelines for human oversight and the pivotal human factor in critical decisions. This integrated framework ensures that technical rigor dovetails with ethical imperatives, fostering robust compliance across AI deployments. By harmonizing these elements, standardization not only mitigates risks but also elevates the reliability of AI systems in high-stakes environments [15]. The aim is to ensure alignment of technical and legal requirements, reduce regulatory uncertainty, and facilitate cross-border application of AI systems.

4 Comparative Approaches in AI Regulation

While the EU AI Act establishes a binding framework based on the principle of proportional risk, other jurisdictions develop diverse approaches reflecting their legal, political, and economic priorities. Analysis of these approaches enables better understanding of the global context and potential challenges in harmonizing rules.

4.1 United States Fragmented AI Regulatory Approach

In the US, AI regulation remains fragmented and sector-oriented to date. No single law regulates all AI systems; instead, various regulations apply in areas such as finance, healthcare, transportation, and national security [16].

The US AI regulatory model adopts a distinctly laissez-faire, decentralized approach through key instruments that prioritize voluntary guidance and sector-specific rules over comprehensive federal mandates: the Federal Trade Commission (FTC) guidance on AI,[1] which outlines principles for fair, ethical algorithm

[1] See more at: available at https://www.ftc.gov/ai.

use by prohibiting deceptive practices, bias, and discrimination under existing consumer protection laws; the National AI Initiative Act (2020),[2] focused primarily on advancing research, development, and federal AI coordination without direct usage regulations; and sector-specific regulations like HIPAA for healthcare data privacy in AI applications,[3] collectively promoting innovation through antitrust, privacy enforcement, and targeted oversight rather than centralized control [17].

The US emphasizes an innovation-first approach with minimal state intervention in early technology development stages. The risk is that such policy may lead to inadequate protection of user rights, particularly in areas of discrimination and data privacy.

4.2 Chinese Distinctive Regulatory Model

The Chinese AI regulatory model contrasts sharply with the more laissez-faire American approach, mirroring Europe's comprehensiveness through stringent, multi-faceted controls while incorporating significant political oversight to prioritize state security and social stability. Key components include the Regulations on Algorithmic Recommendation Services (2022),[4] which require platforms to prevent discrimination, user manipulation, and enable opt-outs from personalized recommendations; the Personal Information Protection Law (PIPL 2021),[5] imposing strict rules on data processing, consent, and cross-border transfers akin to GDPR; and directives on AI security and oversight in critical areas like public services, mandating algorithmic transparency, risk assessments, and government approvals for high-stakes systems to balance innovation with centralized governance.

China focuses on controlling AI's societal impact, often through centralized oversight mechanisms [18]. This model prioritizes national security and political stability, while individual rights protection takes a backseat.

4.3 International Organizations in AI Governance

The most prominent act among international organizations in the AI sphere is the Council of Europe's Framework Convention on Artificial Intelligence, fully aligned with the EU AI Act. Opened for signature in September 2024, this treaty is the first legally binding international instrument regulating AI's effects on human rights, democracy, and the rule of law. It requires states to assess AI risks, ensure transparency in AI systems, and protect fundamental rights like privacy

---

[2] See more at: https://www.congress.gov/bill/116th-congress/senate-bill/1558.

[3] See more at: https://www.hhs.gov/hipaa/index.html.

[4] See more at: https://www.lawinfochina.com/display.aspx?id=37324&lib=law.

[5] See more at: https://www.chinalawtranslate.com/personal-information-protection-law/.

and non-discrimination, while allowing flexibility for innovation and national security.[6] Building upon this foundation, the UNESCO Recommendation on the Ethics of AI [19] provides a comprehensive global framework that prioritizes respect for human rights and social justice.[7] These documents serve as a soft regulatory framework, often in the form of recommendations and guidelines, but lack binding legal force [19].

Criterion	EU AI act	USA	China	International organizations
Legal status	Binding law	Guidelines and sector-specific regulations	Binding national regulations + political oversight	Recommendations and principles
Approach	Risk-based/ proportional	Innovation first/ minimal regulation	Societal impact control	Ethical/principles
Focus	Rights protection + innovation	Innovation and market	Stability and security	Human rights and ethics
Oversight bodies	National AI regulator + EAIB	FTC/ sector-specific bodies	State oversight and regulators	No binding bodies

The European approach, though stricter, provides clear legal predictability, while the US and Chinese models demonstrate advantages and weaknesses in innovation and societal contexts [11].

The application of artificial intelligence in social and economic processes raises a range of legal and ethical issues, particularly when algorithms make decisions directly affecting human rights, privacy, and social equality. The EU AI Act and similar regulations seek to establish legal and ethical boundaries that balance innovation and citizen protection.

5 Privacy and Data Protection

AI systems often require large volumes of personal data for training and optimization. According to Article 10 of the EU AI Act, data must be relevant, representative, and non-discriminatory [9]. However, the practical application of privacy principles encounters formidable technical and legal challenges. Anonymization and pseudonymization techniques, though foundational, often prove insufficient, as advanced algorithms can ingeniously reconstruct identities from seemingly innocuous data combinations. Concurrently, the exercise of fundamental rights, such as access and deletion under GDPR Article 17, must harmonize with the imperatives

[6] See more at: https://rm.coe.int/1680afae3c.

[7] See more at: https://www.unesco.org/en/articles/recommendation-ethics-artificial-intelligence.

of automated systems, necessitating sophisticated technical procedures to balance individual protections with operational efficacy.

In Montenegro, where data protection legislation is in the process of aligning with the GDPR, these issues further complicate the implementation of high-risk AI systems [20].

The EU AI Act's risk classification system is directly linked to a clear allocation of responsibilities among AI providers, deployers, and users, yet it raises pressing practical dilemmas that test the boundaries of accountability in real-world applications. Key challenges include determining who bears liability for harm caused by autonomous systems, whether the programmer who designs the initial algorithms, the producer who integrates and certifies the system, or the user who deploys it operationally, and how to legally address decisions from self-learning models that autonomously adapt over time, potentially evolving beyond their original parameters and introducing unforeseen risks. These questions are particularly important in sectors such as healthcare, justice, and finance, where AI system decisions can directly impact human life or property [7].

One of the most significant challenges in AI deployment is algorithmic discrimination and bias, which perpetuates or amplifies social inequalities by embedding historical prejudices into automated decision-making processes. This issue manifests clearly in employment systems that favor certain genders or age groups, such as resume-screening tools trained on skewed historical data that systematically disadvantage women or older candidates; credit algorithms that reject applications from minority groups by using proxies like zip codes correlated with race or socioeconomic status rather than genuine creditworthiness; and facial recognition technologies exhibiting lower accuracy for darker-skinned individuals [21].

The EU AI Act requires producers to develop systems for evaluating and removing bias, including data testing, independent review, and decision transparency. The fundamental principle of the EU AI Act is human-in-the-loop: human oversight must be present for high-risk systems to prevent harmful or unethical decisions [9]. This entails mandatory intervention by qualified persons at critical moments, clearly defined procedures for overriding or correcting AI decisions, and regular monitoring and evaluation of system performance in the real world [22].

6 Challenges of Transposition in Montenegro and the Region

While Montenegro and the Western Balkans actively pursue EU accession and digital alignment, applying AI regulation encounters significant hurdles that demand targeted reforms. The application of AI regulation in Montenegro and the Western Balkans faces the following obstacles: (a) Lack of an AI regulatory framework, existing laws are fragmented and do not cover all risks, (b) Limited technical capacities, institutions lack sufficient experts for algorithm evaluation, (c) Ethnic and social dimension, AI systems can inadvertently amplify existing social inequalities, requiring special attention when defining local standards [11].

A regional approach, through cooperation among Western Balkan states including experience exchange and joint training, can significantly facilitate the implementation of international standards and EU AI Act principles.

The EU accession process for Western Balkan countries imposes the eventual obligation to fully apply the EU AI Act upon membership. In accordance with these strategic aims, Montenegro has already begun this alignment by updating its data protection and digital transformation laws, creating a solid foundation for incorporating AI Act requirements like risk assessments and transparency obligations; simultaneously, regional projects and specialized training programs can build essential regulatory capacities among authorities and stakeholders, enabling effective oversight of AI deployment. Ultimately, adopting high-risk system standards and algorithmic transparency measures not only enhances citizen trust through demonstrable accountability and fairness but also attracts foreign investments in the burgeoning AI sector, positioning the region as a reliable hub for ethical innovation within Europe's broader digital ecosystem [11].

Applying EU AI Act principles in Montenegro and the broader Western Balkans region represents a complex task due to the legal, institutional, and technical framework that is not yet fully developed. Key challenges and recommendations stand out here. First and foremost are legal challenges, which include:

(a) Legislative fragmentation—Existing laws on information technologies, data protection, and e-government partially cover AI issues but do not provide a comprehensive framework for automated decision responsibility, high-risk system risk assessment, and legal protection of users and citizens [20].
(b) Non-alignment with GDPR and EU standards—Application of European privacy and data protection principles is partial, complicating the introduction of high-risk systems without formal legislative alignment.
(c) Definition of responsibility—Lack of clear rules on liability for AI system decisions can limit the use of advanced technologies in critical sectors such as healthcare, finance, and public services.

Furthermore, there are institutional challenges, such as the lack of regulatory bodies. Montenegro and regional countries lack specialized institutions for AI system oversight, complicating the application of high-risk standards. Issues of limited capacities and expertise are also present. Education of regulators, technical experts, and auditors for algorithm model assessment is needed. Regional coordination is also necessary, as experience exchange and procedure standardization can facilitate transposition of EU rules.

Special attention must be directed to technical challenges, which involve risk assessment and testing, implementing tests for high-risk systems, requiring advanced technical infrastructure and tools for data and algorithm evaluation.

Transparency and algorithm auditing are also of great importance. Systems like generative AI or self-learning models require continuous review, which can be challenging in resource-limited environments.

In this regard, some recommendations for implementing this model would be: (a) Legal harmonization- adopting a comprehensive AI law or regulation aligned with the EU AI Act, including definitions and classification of AI systems, obligations of producers and users, oversight and sanctioning mechanisms; (b) Capacities and education- establishing specialized bodies and training regulators, IT experts, and public officials; (c) Regional cooperation- forming a network for exchanging experiences, standards, and best practices in the region to accelerate harmonization and implementation of EU principles; (d) Pilot projects and testing- applying pilot projects in key sectors (healthcare, education, public administration) to assess the effectiveness of AI systems and regulatory procedures; (e) Integration of ethical standards- adopting mechanisms for human oversight and decision transparency to protect fundamental rights and prevent discrimination.

Implementation of the EU AI Act in Montenegro and the Western Balkans transcends mere regulatory compliance, evolving into a strategic process that strengthens citizen trust in digital technologies by demonstrating commitment to ethical, transparent AI governance; simultaneously attracts investments and fosters innovation through alignment with proven European standards that reassure global stakeholders; and establishes a competitive, secure AI systems market capable of thriving within the EU digital single market, positioning the region as a reliable hub for high-quality, rights-respecting technological advancement [11]. A regional framework including legal, institutional, and technical harmonization represents a key prerequisite for the long-term successful integration of artificial intelligence into public and private sectors.

7 Conclusion

Regulation of artificial intelligence represents one of the most important legal-technical and ethical issues of contemporary society. The dynamic development of algorithmic and autonomous systems leads to complex challenges regarding privacy, accountability, discrimination, and human control.

Analysis of the EU AI Act shows that the proportional risk approach is key to balancing innovation and protection of fundamental rights. Prohibited and high-risk systems are subject to strict control, while minimal-risk AI systems enable free application, fostering innovative development. In implementation, the combination of national supervisory bodies, the European AI Board (EAIB), and technical standards creates a robust framework capable of keeping pace with rapidly changing technologies.

The comparative overview reveals diverse approaches by global actors:

- The US favors innovation and minimal regulation, increasing technological dynamism but leaving risks to citizens' rights.
- China prioritizes control of societal impact and political stability over individual rights.

- International organizations provide guidelines and ethical standards but lack binding force.

For Montenegro and the Western Balkans region, the main challenges are:

- Legislative fragmentation,
- Limited technical and regulatory capacities,
- Need for regional harmonization and expert education.

Recommendations include legal harmonization with the EU AI Act, development of national supervisory bodies, education, and pilot projects in key sectors, as well as integration of ethical principles into AI system design and application. Artificial intelligence is not merely a technological phenomenon but also a legal-societal challenge. Application of regulatory standards, particularly those defined by the EU AI Act, represents an opportunity to create a safe, transparent, and ethically responsible digital environment. In the Western Balkans region, strategic implementation of these principles can enhance citizen trust, attract investments, and integrate the region into the European digital ecosystem.

References

1. Davenport, T.H., Kalakota, R.: The potential for artificial intelligence in healthcare. Future Health J. **6**(2) (2019)
2. European Commission.: Proposal for a Regulation of the European Parliament and of the Council Laying Down Harmonised Rules on Artificial Intelligence (Artificial Intelligence Act), COM/2021/206 final
3. Hildebrandt, M.: Law for computer scientists: an introductory overview. Legal Stud. (2020)
4. Regulation (EU) 2016/679 of the European Parliament and of the Council (GDPR), Official Journal of the European Union (2016)
5. Laus, J., Ruschemeier, H.: Automation Bias in the AI act on the legal implications of attempting to De-Bias human oversight of AI (2025)
6. Gasiola, G.G.: Rebuilding the pyramid: the AI act's risk-based approach using a binary decision diagram. Comput. Law Secur. Rev. (2025)
7. Wischmeyer, T., Rademacher, T., (eds.): Regulating Artificial Intelligence. Springer (2020)
8. European Commission.: Artificial Intelligence Act—Questions and Answers, Brussels (2024)
9. Regulation (EU) 2024/1689 of the European Parliament and of the Council of 13 June 2024 laying down harmonised rules on artificial intelligence and amending Regulations (EC) No 300/2008, (EU) No 167/2013, (EU) No 168/2013, (EU) 2018/858, (EU) 2018/1139 and (EU) 2019/2144 and Directives 2014/90/EU, (EU) 2016/797 and (EU) 2020/1828 (Artificial Intelligence Act), OJ L 2024/1689, 12.7.2024
10. European Commission.: Explanatory Memorandum to the Artificial Intelligence Act (2024)
11. Renda, A.: The European approach to artificial intelligence: regulatory innovation and global leadership. CEPS Policy Brief (2023)
12. Hacker, P.: AI Regulation in the European Union: the AI Act and Beyond. Oxford University Press (2024)
13. Foomany, F., Product Architect, Security Research, Understanding EU AI Act Risk Categories (2024)
14. European Parliament.: Timeline for implementation of the AI act. Legislative Brief (2024)
15. CEN-CENELEC.: Standards for AI systems—Technical specifications (2023)

16. Wilson, H.: AI Governance in the United States. Brookings Institution (2022)
17. Davtyan, T., The U.S. approach to AI regulation: Federal fragmentation and industry self-governance. J. Law Technol. Innov (Examines federal initiatives, agency rules, and challenges of decentralized enforcement) (2025)
18. Chen, L.: Artificial Intelligence Regulation in China: legal and Political Dimensions. Tsinghua University Press (2023)
19. UNESCO.: Recommendation on the Ethics of Artificial Intelligence (2021)
20. Ministry of Public Administration of Montenegro.: Digital Transformation Strategy 2022–2026
21. Raji, I.D., Buolamwini, J.: Actionable Auditing: investigating AI for bias. In: FAT/ML Conference (2019)
22. Van Bekkum, M.: Using sensitive data to de-bias AI systems: Article 10(5) of the EU AI Act. Comput. Law Secur. Rev. (2025)

New Threats to Corporate Security and Critical Infrastructure Caused by the Development of Modern Technologies and How to Counter Them

Zoran Marjanović and Marija Mićović

Abstract This scientific research examines the impact of modern conflicts - and the new technologies that have emerged with them on the emergence of new threats to corporate security and critical infrastructure. A great number of conflicts around the world largely driven by the ongoing activities of major powers and, to some extent, smaller states and the rapid development of new technologies in weapons systems, military equipment, and various means of countering adversaries have given rise to new forms of threats, such as hypersonic missiles, unmanned aerial vehicles, and counter-UAV systems. Corporations and large technical systems are not, and never will be, immune to threats emerging from conflicts between states or great powers, which inevitably spill over into and shape their business and operational environment. The development of new systems and equipment, enabled by the application of advanced technologies in conflicts where their real-world use and effectiveness can truly be observed creates an obligation for large technical systems to adapt their protection measures. Such adaptation involves changing how threats are countered, as well as introducing new methods and tools to protect business operations, assets, personnel, and staff within corporate security services environment.

Keywords Critical infrastructure · Private security · Corporate security · Development of new technologies · Forms of threats · Countermeasures

Z. Marjanović
University of Defence, Ministry of Defence of the Republic of Serbia, Belgrade, Republic of Serbia

M. Mićović (✉)
University of Criminal Investigation and Police Studies, Belgrade, Republic of Serbia
e-mail: marija.blagojevic@kpu.edu.rs

Z. Keković et al. (eds.), *Critical Infrastructure: Risk and Crisis Management in the New Reality*, Advanced Sciences and Technologies for Security Applications,
https://doi.org/10.1007/978-3-032-20948-1_17

1 Introduction

Rapid changes in international relations, the increase in security challenges, risks and threats, as well as the lack of a security culture, are causing the complexity of the global security picture, while critical infrastructure is taking on new dimensions and increasing importance at the national and international levels. The rapid development and improvement of information technologies are changing the security environment. As critical infrastructure has become an important segment of national security, the protection of critical infrastructure has begun to develop and today it represents one of the main priorities of every state. The traditional approach to security views military threats as the greatest dangers to societies and states, while the modern approach finds security challenges and risks in the information sphere.

As security is the top priority for the company, every organization should have a corporate security team who can work better align to risk management activities to the security risk profiles of the customers. Periodically a survey can be conducted and using survey data, the security team in the organization will examine the services provided, the corresponding potential data exposure, and the associated contract liability to determine the level of oversight that company will provide to individual customers [1].

When we talk about critical infrastructure The worldwide categorization of critical infrastructure is based on extensive research. The authors analyzed official documents, including national security strategies, laws and ministry websites [2].

Newly formed forms of threats to corporations are the result of the massive use of new means and weapons in conflicts and their positive results in conflicts. The types of forms of threats to corporations are not only related to the massive use but also to the intended manner of use and the unknowns to which the use of new means can lead. Corporate security [3] is not a property of corporations understood as large business companies. It is a property of a much broader scope of units, so it is justified to understand corporations rather as legal organizations whether a single organization or a set of organizations that are a whole in themselves.

2 Methodology

Through scientific description, contemporary forms of corporate security threats have been identified, as well as possible ways of counteracting them. The data sources in this research are normative documents, scientific papers containing previous experiences in corporate security matters, previous knowledge of experts published in their works, as well as works of famous theorists. This work, due to a comprehensive review of theoretical knowledge on the phenomenology and etiology of corporate security, offers a synthesis of knowledge on the emergence and development of corporate security, its characteristics, place and role in the national security system and its contribution to national security, thereby contributing to the increase of the scientific fund in security sciences.

3 Security in Private Organizations

According to Nikač, "under the term private security stands the provision of security services carried out by privately owned companies or the services themselves, at the request of private individuals or privately owned organizations" [4]. Other authors who examine this segment of security define private security in a similar manner. In the second half of the nineteenth century, the first private security and detective companies/agencies were founded in France and the United States. In France, this was *Le bureau des renseignements universels,* founded by *Eugene-François Vidocq* in 1832 in Paris. In the United States, the Pinkerton's National Detective Agency was established in 1850 in Chicago, Illinois, by former government special agent *Allan Pinkerton.* The primary criterion for classifying private security is the area in which these actors operate. Thus, Nikač states that, depending on the services a corporation or firm provides, as well as its internal organization, we can distinguish the following subjects of private security: Private Military Companies (PMC), Private Security Companies (PSC), and Private Military-Security Companies (PMSC), which represent particularly interesting corporate entities [4]. The specificity of these private security actors lies in the fact that their engagement is predominantly commercial, although other motives cannot be excluded (e.g., covert engagement by a state, political elites, or individuals). Profit is generally the main driving motive behind the existence of such corporations, which carry out military, security-related, information and telecommunications, and various other tasks. Most often, they appear as "mediators" in controversial events in the world. The composition of such companies is usually made up of former members of elite special military and police units or former members of the security services.

An example of a severe violation of regulations is the conduct of members of a U.S. private military company during the war in Iraq, when many people were killed (dozens of civilian casualties) and wounded. This refers to the engagement of the American private military company Blackwater during the Iraq War, which subsequently changed its name several times, likely to avoid the negative connotations associated with the incident. Another notable example of a private military company involved in more recent conflicts is the Russian private military company Wagner Group. According to media reports (e.g., BBC News, bbc.com), the company has been linked to certain unlawful activities committed by its members in Ukraine in 2022, in the Central African Republic, and in Libya in 2020 (although we were not able to confirm these allegations).

When speaking about the Republic of Serbia, the term refers to private agencies providing physical and technical protection of people and property, as well as private detective agencies [4].

In his review report of the textbook Corporate Security, states that it requires great courage to study—and especially to write and publish scientifically—anything related to the field of corporate security. This observation by further motivated the authors to dedicate themselves to this area from a scientific and research-oriented perspective. Stajić argues that in a modern business corporation, economic security is achieved through: "affirming the principles of legality in management and operations; continuous attention to compliance with established standards and operational procedures of the company; ongoing collection of business information for the purpose of decision-making; integrating corporate security measures into everyday organizational, supervisory, contractual, and planning activities; monitoring and providing necessary support to employees in ensuring the security of business processes and each workplace; suppressing corruption and other forms of crime; continuously preventing and obstructing negative security processes and occurrences; ongoing care for corporate assets; organized measures for the protection of business secrets; security training and education of employees; fostering a security culture and business ethics; and constant, direct, and open cooperation with the corporate security service" [5]. Each of these elements is extremely important for the proper functioning of life and work within a corporation, and together they form a whole in which the absence or inadequate application of any element leads to the endangerment of corporate security.

Đukić approaches the concept of corporate security through a spatial dimension, which may appear at the national, regional, or global level. The connection between security and the economy is delicate, as well as their mutual dependence. Over the past several decades, corporate security has become increasingly dependent on non-state security actors, civil society, and corporations as sources of power. When observed in relation to the broader environment, corporate security depends on international security, while within a state it also depends on national security, to which it inherently belongs. In the practice of modern states, state–public and non-state–private partnerships are becoming increasingly present. With the growing commercialization and specialization of security services, the question arises as to how much the role of the state has changed. Since these are private organizations, it is logical that demand for and use of private security services has increased, and it is reasonable to assume that this trend may someday challenge the current role and position of the state, as well as the proportion between public and private presence in the field of security. Over the past few decades, there has been an expansion in the need for and use of private security and defense services. In the case of great powers, this segment of security organizations has developed alongside economic entities that represent prominent global business organizations, widely known as corporations. According to contemporary understanding, corporate security seeks to unite the security functions of both the state and non-state sectors with corporate responsibility. This modern concept implies the social responsibility of corporations to support state structures, nonprofit organizations, groups, and individuals in addressing needs arising from healthcare, education, environmental protection, and similar areas [6].

Contemporary interpretations of corporate security in the literature increasingly view a corporation less as a social entity and more as an economic subject. In order to fully understand the concept of corporate security, it must be seen as a set of elements within the structure of interests that reflect security as its overall condition. This definition suggests that corporate security encompasses the protection of all elements within the structure of interests of a corporation in relation to all challenges, risks, and threats [7]. That the term corporation is an Anglo-American concept used to denote all organizations with recognized legal capacity, regardless of whether they belong to the public or private, profit or non-profit sector. He concludes that the term legal entity is the most appropriate designation when speaking within domestic legal terminology [8]. After analyzing the concepts of corporation and security, we can concludes that "corporate security refers to the property of a legal entity, manifested as a state, process, and/or instrument of protecting persons, assets, and business operations (objects of protection) from all sources, actors, and forms of endangerment, achieved through the human, material, and organizational capacities of that entity. The primary goal of corporate security is the economic security of business operations by protecting the essential values of the legal entity (objects of protection), which should result in the improvement and development of the legal entity, safer and better working conditions, a better market position, and an increase in its profit" [6].

Certain conceptual terms, when translated from other languages, may pose difficulties due to their specificity within linguistic or cultural contexts. One such term is corporate security. We can conclusion that the fundamental function of corporate security lies in protecting a company's property, employees, and business operations. This protection enables a private organization to achieve its planned goals. The development of new technologies (primarily information technologies), along with significant political and economic changes, has greatly influenced the transformation of activities carried out within corporate security. These changes can be described as a modernized function of corporate security, which includes: preventing damage to the company, protecting persons, property, and business operations, and ensuring business continuity [9].

Similarly, Trivan and colleagues state that the scope of corporate security functions is substantively defined through "personnel security, information security, preparedness for contingencies and business continuity planning, physical security, economic security, internal control, and investigations" [10]. Each of these functions includes the following subfunctions: Personnel security (security vetting, internal order, employee behavior, and workplace violence prevention), Information security (ICT security, protection of trade secrets, classified information, personal data, intellectual and industrial property), Preparedness for contingencies and business continuity planning (protection and rescue, fire protection, defense preparations, occupational health and safety, environmental protection, business continuity, crisis management, critical infrastructure protection), Physical security (physical protection, technical security), Economic security (loss prevention, business intelligence activities), and Internal control and investigations (internal controls and investigative activities.

Experts agree that corporate security, or rather its functions, can and should be the subject of a national legislative framework, especially in the area of critical infrastructure protection.

The primary task of corporate security is to ensure, in an efficient manner, that a company or corporation becomes as resilient as much as possible to threats. This allows the corporation management to more easily focus on the key objectives that the corporation needs to achieve. Considering the ownership structure and the way corporate security is organized, it is important to clearly distinguish and emphasize that corporate security is a part of private security, and therefore a part of national security [10]. At the end of this part of the research, we may conclude that, in the literature of the Republic of Serbia, the concept of corporate security has almost as many meanings as there are authors who write about it [7].

4 New Forms of Threats to Private Organizations Caused by Changes in Modern Conflicts

In the previous period, a wide variety of challenges, risks, and threats (forms of endangerment) that influence the immediate corporate environment have been identified. When observing a corporation, its forms of endangerment can be grouped into three categories: natural (arising from natural forces, most commonly natural disasters such as earthquakes, floods, fires, extreme temperatures, etc.), business-related (financial risks and the overall economic–political situation in the market), and threats arising from operations (conditioned by the organization's activities and affecting the personnel, assets, and operations of the corporation) [11]. All risks that affect the safety of people, assets, and business operations, in a broader sense, represent security risks, regardless of whether they are naturally or socially conditioned. When the source of risk is the human factor, we speak of intentional and unintentional risks [12].

On 25 May 2025, Klabin reported that the American private military company (PMC UG Solutions) from North Carolina had hired 96 former U.S. special operations personnel to organize the management of a checkpoint in Gaza. Allegedly, this was the only arrangement acceptable to the conflicting parties during the ceasefire between Israel and Hamas. Many states in the vicinity of Gaza appeared unwilling to deploy their own security personnel to this conflict-affected region, so the American private military company accepted the task, as did the parties to the conflict. The area in question is one of the most unstable conflict zones in the world. Members of this company are expected to supervise access to humanitarian aid and be capable of defending themselves. Until now, it has not been known that a company under this name (UG Solutions) has been deployed in conflict zones. The company was founded in 2023 [13]. Its ability to adapt to complex situations and risks in such an unstable region remains uncertain—unless this is simply a case of rebranding and/

or absorbing part or most of the personnel of another company previously engaged in similar activities.

The sheer number and type of challenges, risks, and threats point to the complexity of the security function within a corporation, which must ensure safe and secure operations (its resilience) in a secure environment, thus creating conditions for high-quality business functioning [10]. Misunderstanding or lacking knowledge in the field of corporate security can very easily lead to threats to a company's operations. When discussing forms of endangerment affecting private companies, enterprises, and corporations, we must note that the ways conflicts are conducted globally are changing at an incredible speed. This includes the emergence of new technologies in the field of IT (cyber) technologies as well as in weaponry and military equipment, primarily among major powers. Since corporations must operate both in peacetime and during armed conflicts, we now observe new forms of threats to corporate security, which in turn complicate the ability to defend the corporation and/or mitigate consequences. These rapid changes are extremely difficult for state institutions specialized in these matters to follow, as the inertia and sluggishness of large systems require significant time to build resilience. These new forms of endangerment, along with the inability of small and medium-sized states to adapt to such rapid developments that threaten state, organizational, and corporate security, inevitably manifest negatively in corporate business operations. It is important to emphasize that these forms of corporate endangerment also include the (mis)use of artificial intelligence aimed at harming corporate operations. Under the influence of such diverse types of threats to corporate security, all entities within a state citizen as well as organizations are affected.

When discussing terrorism, we refer to a threat whose influence on private organizations, companies, and corporation's manifests through terrorist acts (killings, abductions, kidnappings, planting explosive devices, etc.) that may deprive corporations of the ability to operate; this represents the traditional form of terrorism's impact on the business activities of private organizations. However, newer forms of endangerment targeting private organizations include cyberterrorism, corruption, organized crime, and transnational organized crime [6].

Private organizations, viewed globally, possess different types of assets that may be easily exposed to threats that can lead to various forms of endangerment. Certain forms of endangerment are characteristic of specific time periods (e.g., transition), while others are specific to certain geographical regions (e.g., the Far East). Sources of endangerment may be natural (the influence of nature, which cannot typically be prevented, though consequences can sometimes be mitigated), artificial (everything created by human activity), and social (social actors who aim to take lives or inflict injuries, cause material damage, disclose business secrets, etc.) [14]. We may conclude that theft is not the activity that produces the greatest damage to a corporation; rather, for an action to constitute a threat, there are far more dangerous forms such as the disclosure of confidential business data or other sensitive corporate information, or the neglect of obligations toward the corporation.

5 The Way of Countering New Forms of Threats to Corporation

The job of corporate security services has evolved together with changes in societies and states, which have introduced new challenges, risks, and threats, resulting in shifts in operations while their core responsibilities have remained unchanged. When discussing the responsibilities of a corporate security service, we primarily refer to the protection of the corporation's assets, personnel, and business operations, with the aim of ensuring favorable security conditions for achieving the corporation's goals—something that remains unchanged to this day. However, changes influenced by socio-political and economic processes have significantly altered the tasks performed within the corporate security function. Globalization, for instance, has reshaped the structure and speed of corporate activities, pushing companies to operate in increasingly risky markets [15].

The foundation of corporate security operations is the corporate security risk assessment, which evaluates the vulnerability of the corporation (personnel, assets, operations). Based on the identified challenges, risks, and threats, appropriate corporate protection measures are proposed. It is essential that these initial identifications be accurate; otherwise, inadequate protective measures may be proposed from the start, rendering them ineffective for protecting the corporation or mitigating consequences [10]. The pinnacle of protecting the national interests of a state lies in preventive action against threats, and in a similar manner, corporations ensure adequate business functioning by implementing preventive protective measures [16].

It must be emphasized that the fate of a corporation, company, or critical infrastructure operator is determined by its top management. How a company or corporation will respond to modern challenges, risks, and threats primarily depends on the quality of its security. When discussing corporate security, we principally refer to the protection of assets, personnel, and business operations—upon which business success most directly depends. A corporation's success is directly tied to the degree of security endangerment: unfavorable security influences reduce the performance of business organizations, companies, and corporations. These influences may originate from the external environment or from within the organization itself. External influences range from adverse societal trends to natural disasters, which may exert greater or lesser effects on business operations and business success [6]. Internal influences, however, are primarily linked to human behavior. Actions that deviate from established organizational or national norms (theft, disclosure of business secrets, industrial espionage, non-compliance with internal procedures, etc.) threaten confidential business information and may seriously endanger the operations of a private organization, company, or corporation.

The emergence of new technologies, means, and equipment used in modern armed conflicts has created new demands and needs in corporate security, requiring the training of personnel to operate and defend against highly sophisticated systems and devices. These include highly specialized technologies used by major military powers such as supersonic missiles, air-defense systems, drones and unmanned aerial vehicles, firearms and various counter-drone systems, aircraft, computers, and ICT systems which, in turn, necessitate the use of certain technologies or defensive systems within corporations. For this reason, corporations must train expert personnel capable of responding to these threats. Conflicts in a region do not always discriminate between objects and personnel, meaning that corporate security must anticipate and prepare for this form of endangerment.

Your observations and research are interesting that today form the basis for corporate security services when identifying new threats faced by companies and corporations. He highlights the emergence of the English term business intelligence, which first appeared in 1989 and refers to the process of collecting data within the business environment. This process business intelligence represents a legal and public business information gathering activity aimed at preserving established business positions and ensuring business success. It does not refer to business espionage, which is an illegal and prohibited activity. Unlike business intelligence, business espionage involves the use of unlawful or unethical methods with the goal of obtaining sensitive, protected, and confidential information. Information essential to the life and functioning of a company provides crucial support to management in making informed decisions aligned with corporate policies and the business environment [17]. For this reason, it is impossible to conduct any successful business process without comprehensive control and management of business data and information relevant to operations. The most important data used in business intelligence represents a strategic resource for corporate management—without them, a company would be "blind" in its operations. Corporate security encompasses the entire security structure of a company or corporation, with the goal of ensuring business success. Thus, Jusufranić concludes that business intelligence is an integral part of corporate security. Business achievements are simultaneously exposed to numerous threats in the business sphere, requiring corporate leaders to anticipate future events and implement measures to restrain, mitigate, or neutralize them. The primary tool for achieving this goal is business intelligence [17]. Establishing institutionalized corporate security within a company or corporation requires many years of active work and substantial material support. Investments are significant, and results are not immediately visible, which may pose challenges in forming such a service [17]. The need to prioritize the use of new technologies in corporate security particularly for acquiring essential business data—follows directly from the above. After analyzing such data and making managerial decisions, corporations achieve business objectives. In short, business intelligence encompasses the collection, storage, analysis, and application of business information to implement measures that improve corporate performance and achieve strategic goals.

It is also essential not to overlook the greatest threat to the business of a company or corporation: industrial espionage. This activity functions in a covert mode (usually invisible), encompassing exactly what Zemljanov depicts as the structure of the intelligence process (Fig. 1). Zemljanov conditionally classifies corporate espionage into human (agent-based) and technical. Human espionage obtains data through human agents, while technical espionage relies primarily on technological means. Its existence is inherently secretive, and countering such an adversary must likewise be

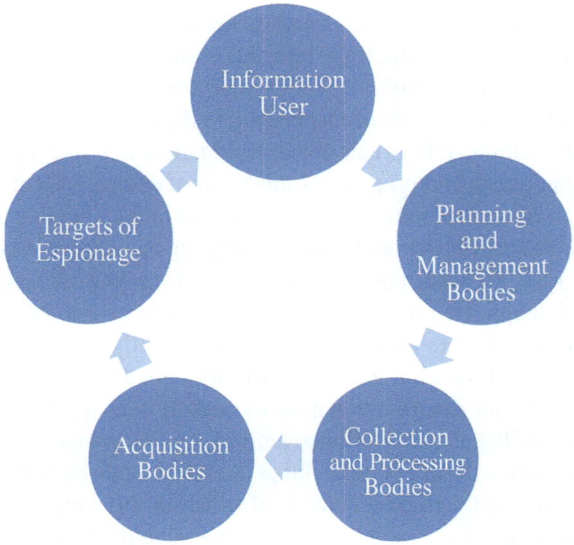

Fig. 1 Structure of the intelligence process [18]

carried out covertly. This represents one of the most damaging threats to corporate operations, and the only effective response to espionage is the implementation of counterintelligence measures.

The corporate security function must:

Expanding its operations beyond the boundaries of the traditional security portfolio to areas including business continuity, business reputation, risk management, corporate responsibility, which creates additional pressure on management emphasizing the importance of network and change management skills [19].

6 Conclusion

Corporate security, as a term adopted in the literature in our region, is related to certain actions of security professionals to protect the operations of not only large corporations (present all over the world) but also in all private organizations that have a sufficient need to protect certain individuals, property, and operations of that organization from identified challenges, risks, and threats. Corporate security can be performed by professional organizations designed to provide these and similar security services, or they can already be part of the structure of a private organization or corporation. When it comes to external security service providers, the access to work for members of the corporate security service is more difficult, but it is a much more objective approach to working with identified threats to the security of the corporation. Unlike external, when we talk about internal security service providers, we have a much easier variant of access to the corporation, but also the correctness of the actions of the head of the corporate security service can be easily questioned, where his proposals, if they conflict with the proposals of the organization's top management, can easily lead to the presentation of an incorrect picture of the threat to the corporation that has been identified, and therefore of what needs to be done to neutralize it or at least minimize its consequences.

The importance of corporate security in the national security system is particularly important when considering organizations whose functioning provides essential products and services that are of fundamental importance for the survival and long-term development of the state and society. These are systems, as well as all facilities and locations that are designated and identified as critical infrastructure. Corporate security, through the protection of critical infrastructure, directly contributes to maintaining national security by ensuring continuity in the provision of essential products and services, increasing resilience, protecting economic assets and ensuring a coordinated response to threats.

The development of new technologies, means and equipment used in conflicts today have led to extreme changes in the issues of defense and corporate security. The engagement of American private military companies as well as other forms of private security in the Gaza Strip to secure "points" represents the official arrival of American "private" forces in this territory, and may be one of the reasons for potential preparations for physical approach to Iran's nuclear facilities in order to achieve the desired needs in the coming period. Modern corporate security activities include measures and actions to detect and prevent: bribery, corruption, industrial espionage, environmental incidents, abuse of power, money laundering, mobbing, as well as work on protecting information systems, crisis management activities, risk management, development of business continuity plans in the event of major incidents and other related areas [20].

In the former Yugoslav countries, there are generally no separate organizational units dealing with "corporate security" (this term does not exist in the regulations of this area, but only in theory), but in larger private organizations and companies, certain corporate security functions are performed by engaging internal resources (engaging mainly former members of the police, military, security services, and other specialized persons) and/or by concluding contracts with specialized organizations for a specific activity.

It is important to point out that the corporate security activity is predominantly preventive in nature and that the efforts of the company, i.e. its representatives, are aimed precisely at trying to base the focus of the corporate security activity precisely on preventive work.

References

1. Mukherjee, S.: Overview of the importance of corporate security in business. Int. J. Importance Corp. Secur. Bus. **8**(4) (2019). https://doi.org/10.15680/IJIRSET.2019.0804002
2. Weber, V., Pericàs Riera, M., Laumann, E.: Mapping the world's critical infrastructure sectors. DGAP Policy Brief 35. German Council on Foreign Relations (2023)
3. Vujaklija, M.: Leksikon stranih jezika i izraza. Prosveta, Beograd (1996/1997)

4. Nikač, Ž., Korajlić, N., Ahić, J. i Bećirović, M.: Pravna regulativa privatne bezbednosti na prostorima nekadašnje SFRJ, sa osvrtom na poslednje promene u Srbiji. Časopis za kriminalistiku, kriminologiju i sigurnosne studije 27–53 (2013)
5. Stajić, L.:. Recenzija udžbenika-Korporativna bezbednost, autora: Goran J. Mandić i Petar Stanojević, Fakultet bezbednosti, Beograd (2019)
6. Đukić, S.: Korporativna bezbednost i menadžment vlasničkog kapitala. Vojno delo, 3/2017, Beograd (2017)
7. Marković, I.S.: Filozofija korporativne bezbednosti. CIVITAS broj 7 MMXIV 9–22 (2014)
8. Mandić, G.: Korporativna bezbednost-poslovna funkcija pravnog lica. Nauke bezbednosti: vrste i oblici, Fakultet bezbednosti, Beograd, str. 167–182 (2020)
9. Trbojević, M.: Korporativna bezbednost na prostoru Zapadnog Balkana. Kultura polisa, god. XVII (2020), br. 41, str. 391–402 (2020)
10. Trivan, D., Jakišić, A. Pavlović, D., Radović, L., Jovanović, V., Srdić, S.: Priručnik za menadžere korporativne bezbednosti. Srpska asocijacija menadžera korporativne bezbednosti, Centar za multidisciplinarne studije, Lampas, Beograd (2024)
11. Milošević, O., Minić, S i Jakić, B.: Korporativna bezbednost organizacije i rizici i pretnje po poslovanje kompanija. Naučna konferencija, decembar 2016. dostupno 30.05.2025. godine na Internet sajtu (2016). www.researchgate.net/publication/340547682
12. Mićović, M.: Specifičnosti kritične infrastrukture u Republici Srbiji, monografija, Kriminalističko-policijska akademija, Beograd (2020)
13. Klabin N.: Private military companies find a role in Gaza's security void. Trusted Mideast News Wednesday 05/28/2025 a доступан 30.05.2025 (2025). https://themedialine.org/top-stories/private-military-companies-find-a-role-in-gazas-security-void/
14. Đukić, S.: Menadžment korporativne bezbednosti i kriminalitet kao oblik i izvor ugrožavanja bezbednosti korporacija. Vojno delo 5/2018, Beograd (2018)
15. Nešković, S.: Paradigm of modern corporate security and methodology implementation of crisis management in the Western Balkans. In: XIV Međunarodna konferencija, Corporate Security In BiH And The Western Balkan Countries From Economic, Legal And Communication Aspect, 16–17. December 2016, Travnik, BiH, p. 35–45 (2016)
16. Marjanović, Z. i Mićović, M.: Američki koncept odvraćanja u posthladnoratovskom periodu, časopis Bezbednost, broj 1/2025, LXVII, Ministarstvo unutrašnjih poslova Republike Srbije, Beograd, str. 87–96 (2025)
17. Jusufranić, I.: Corporate security in BiH and in the Western balkans countries from economic, legal, traffic and communication aspect. In: XIV International Conference, Corporate Security In BiH and the Western Balkan Countries From Economic, Legal and Communication Aspect, 16–17. December 2016, Travnik, BiH, p. 16–35 (2016)
18. Zemljanov, M.V.: Kontraobaveštajni rad-praktični vodič, Fakultet bezbednosti i Institut za evropske studije, Beograd (2022)
19. Briggs, R., Edwards, C. The Business of Resilience Corporate Security for the 21st Century, London (2006)
20. Pejanović, L., Stojanović, S., i Komarčević, M.: Corporate and system coordinated integration of security services of state and private sector. In: Čaleta, D. (ed.) Contemporary Concept of Corporate Security, Beopress, Belgrade (2018)

The manufacturer's authorised representative in the EU is Springer
Nature Customer Service Centre GmbH, Europaplatz 3, 69115 Heidelberg,
Germany. If you have any concerns regarding our products, please
contact ProductSafety@springernature.com

Printed and bound by CPI Group (UK) Ltd, Croydon, CR0 4YY
16/09/2026
02217815-0005